COMPUTER LOGIC DESIGN

COMPUTER LOGIC DESIGN

M. MORRIS MANO

Professor of Engineering
California State University, Los Angeles

PRENTICE-HALL, INC., Englewood Cliffs, New Jersey

Library of Congress Cataloging in Publication Data

MANO, M. MORRIS
Computer logic design

(Prentice-Hall series in automatic computation)
Includes bibliographies.
1. Switching theory. 2. Electronic digital computers–Circuits. I. Title.
TK7868.S9M275 621.3819'535 72-668
ISBN 0-13-165472-1

10 9 8 7 6

Prentice-Hall International, Inc., London
Prentice-Hall of Australia, Pty, Ltd., Sydney
Prentice-Hall of Canada, Ltd., Toronto
Prentice-Hall of India Private Limited, New Delhi
Prentice-Hall of Japan, Inc., Tokyo

to my wife

CONTENTS

PREFACE

A digital system is a system that processes discrete elements of information. The best known example of a digital system is the general purpose digital computer. *Logic design* is concerned with the interconnections between digital components and modules and is a term used to denote the design of digital systems.

The purpose of this book is to present the basic concepts used in the design and analysis of digital systems and to introduce the principles of digital computer organization. The viewpoint of the book is primarily tutorial. It provides various methods and techniques from which the reader hopefully will develop a design philosophy applicable to any digital system problem.

Electronic circuits used in digital systems are invariably manufactured in integrated circuit form. Many integrated circuit packages available commercially contain a large amount of interconnected components within a single silicon chip and provide a specific and complete digital function. It is very important that the logic designer be familiar with the various digital functions commonly encountered in integrated circuit packages. For this reason, many such circuits are introduced throughout the book and their logical function fully explained.

Logic design and switching circuit theory are two interrelated branches of study and the two names are interchangeable when simple digital systems are considered. Both logic design and switching circuit theory are concerned with the analysis and synthesis of combinational and sequential circuits. Complex digital systems, however, have two different widely used interpretations. From the switching circuit theory point of view, a digital system is described by a *state system* which is a direct extention of the definition given to a sequential circuit. This point of view asserts that, at any instant of time, a digital system is capable of being in one of many possible abstract states. The behavior of the system is then specified by a transition function that determines the next state from the current state and the inputs. A large system such as a digital computer may be represented, in principle, as a state system, but the number of states is far too large to make this representation practical. When dealing with the logic design of complex digital systems, it is convenient to formulate and define the system by means of *register transfers*. The components in this representation are computer registers and modules together with a set of functional transfers between registers. With this representation, the digital system is decomposed into register subunits and specified by the operations executed in each register. Chapter 1 to 7 of this book contain a practical approach to the subject which is sometimes classified under the heading of basic switching circuits. What makes this book different from a switching circuit theory book is the subject matter covered in Chapters 8 through 12 where the register transfer concept is developed and used to describe and design digital systems.

Chapter 1 introduces various binary systems such as binary numbers, binary codes, binary storage and binary logic. Chapter 2 covers Boolean algebra and Chapter 3 presents the map and tabulation methods for simplifying Boolean functions. The first three chapters serve as background information and contain basic material needed for the rest of the book.

Chapter 4 presents design and analysis procedures for combinational circuits. Various functions that are available in integrated circuit packages such as adders, comparators, decoders and multiplexers are introduced as design or analysis examples. Chapter 5 starts with a discussion of integrated circuits and their impact on logic design. The chapter continues with a presentation of techniques for NOR and NAND logic implementation and of various other gate implementations.

Chapter 6 on sequential logic starts by presenting the properties of various types of flip-flops such as RS, JK, T and D. The tools for analysing clocked sequential circuits; the state diagram, state table and state equations, are then presented. The logic diagrams of some common clocked sequential circuits such as binary and decimal counters, and registers are shown and their function explained. The examples given in this analysis chapter are chosen from those that have specific functions and, because of their widespread use, are commonly available in integrated circuit packages.

Chapter 7 introduces procedures for the design of clocked sequential circuits. The topics of state reduction and state assignment are mentioned briefly without bringing any manipulative methods. The emphasis in this chapter is on the methods by which the Boolean functions to the inputs of flip-flops are derived from the state table and simplified by means of maps. Numerous examples are given including the design of counters.

Chapter 8 introduces a symbolic notation for register transfer operations. The memory unit is explained in terms of transfers among a collection of storage registers and the types of data represented in registers is discussed. The first half of Chapter 9 introduces methods for binary addition. The design of a general purpose accumulator register is then undertaken, starting with a specified set of elementary operations and culminating in a logic diagram. A discussion of various accumulator operations is then presented to emphasize the concept of a general purpose register and its place in large-scale integration design.

Chapter 10 on computer organization presents a description of the most common type of digital system, i.e., the stored program digital computer. The implementation of a simple general purpose computer is explained in terms of register transfer operations. The rest of the chapter discusses many important concepts found in general purpose computers.

A simple digital computer is designed in Chapter 11 starting from a specified set of machine instructions and hardware registers and culminating in a list of Boolean functions that specify the interconnections between the various digital components. The system description is formalized by register transfer symbology and a procedure is mechanized for translating the register transfer relations into Boolean functions that represent the combinational networks among the registers. Chapter 12 contains examples that demonstrate the design of control logic and the development of algorithms for the implementation of digital functions.

In summary, the book covers the following main topics. Chapter 1 gives an introduction to binary systems. Chapters 2-5 are concerned with combinational circuits. Chapters 6-7 deal with synchronous clocked sequential circuits. Chapters 8-9 introduce the concept of registers, their function and design. Chapters 10-11 cover the organization and design of general purpose digital computers and Chapter 12 is a summary of the various design techniques introduced throughout the book. Answers to most of the problems appear in the back of the book to provide an aid for the student and to help the independent reader. A solutions manual is available for the instructor from the publisher.

The book is suitable for a two-term course in computer logic design introducing the methods and techniques of analysis and design of digital systems. It can be used in an Electrical Engineering or in a Computer Science curriculum. The book is also suitable for self study by engineers or computer scientists who need to acquire a knowledge of logic design.

The book can also be used in a one-term course in a variety of ways. (1) As a first course in logic design or switching circuits covering Chapters 1 through 8. (2) As a course in computer organization and design with pre-requisite of a course in basic switching circuit theory by covering Chapters 8 through 12 with a review of Chapters 6 and 7. (3) As a course in computer organization in a Computer Science Department by covering Chapters 1 through 10 and omitting some of the design sections in Chapters 4, 5, 7 and 9. In fact, this book covers about 80% of the material of the computer organization course number I3 recommended by the report "ACM Curriculum 68" published in the March 1968 issue of the *Communications of the ACM.*

I wish to express my thanks to Dean Eugene Kopp for his encouragement. Thanks also to Mrs. Ernestine Wellenstein and Mrs. Pat Anderson for typing parts of the manuscript. My greatest thanks go to my wife Sandra for editing the entire manuscript, for typing most of it and for her encouragement and support of the entire project.

M. MORRIS MANO

1 BINARY SYSTEMS

1-1 DIGITAL COMPUTERS AND DIGITAL SYSTEMS

Digital computers have made possible many scientific, industrial, and commercial advances that would have been unattainable otherwise. Our space program would have been impossible without real time, continuous computer monitoring, and many business enterprises function efficiently only with the aid of automatic data processing. Computers are used in scientific calculations, commercial and business data processing, air traffic control, space guidance, the educational field, and many other areas. The most striking property of a digital computer is its generality. It can follow a sequence of instructions, called a *program*, that operates on given data. The user can specify and change programs and/or data according to the specific need. As a result of this flexibility, general purpose digital computers can perform a wide variety of information processing tasks.

The general purpose digital computer is the best known example of a digital system. Other examples include: telephone switching exchanges, digital voltmeters, frequency counters, calculating machines, and teletype machines. Characteristic of a digital system is its manipulation of *discrete elements* of information. Such discrete elements may be electric impulses, the decimal digits, the letters of an alphabet, arithmetic operations, punctuation marks, or any other set of meaningful symbols. The juxtaposition of discrete elements of information represents a quantity of information. For example, the letters *d*, *o*, and *g* form the word *dog*. The digits 237 form a number. Thus, a sequence of discrete elements forms a language; that is, a discipline that conveys information. Early digital computers were used mostly for numerical computations. In this case the discrete elements used are the digits. From this application, the term *digital computer* has emerged. A more appropriate name for a digital computer would be a "discrete information processing system."

Discrete elements of information are represented in a digital system by physical quantities called *signals.* Electrical signals such as voltages and currents are the most common. The signals in all present-day electronic digital systems have only two discrete values and are said to be *binary.* The digital system designer is restricted to the use of binary signals because of the lower reliability of many-valued electronic circuits. In other words, a circuit with ten states, using one discrete voltage value for each state, can be designed, but it would possess a very low reliability of operation. In contrast, a transistor circuit that is either on or off has two possible signal values and can be constructed to be extremely reliable. Because of this physical restriction of components, and because human logic tends to be binary, digital systems that are constrained to take discrete values are further constrained to take binary values.

Discrete quantities of information emerge either from the nature of the process or may be purposely quantized from a continuous process. For example, a payroll schedule is an inherently discrete process that contains employee names, social security numbers, weekly salaries, income taxes, etc. An employee's paycheck is processed using discrete data values such as letters of the alphabet (names), digits (salary), and special symbols such as $. On the other hand, a research scientist may observe a continuous process but record only specific quantities in tabular form. The scientist is thus quantizing his continuous data. Each number in his table is a discrete element of information.

Many physical systems can be described mathematically by differential equations whose solutions as a function of time give the complete mathematical behavior of the process. An *analog computer* performs a direct *simulation* of a physical system. Each section of the computer is the analog of some particular portion of the process under study. The variables in the analog computer are represented by continuous signals, usually electric voltages that vary with time. The signal variables are considered analogous to those of the process and behave in the same manner. Thus measurements of the analog voltage can be substituted for variables of the process. The term *analog signal* is sometimes substituted for *continuous signal* because "analog computer" has come to mean a computer that manipulates continuous variables.

To simulate a physical process in a digital computer, the quantities must be quantized. When the variables of the process are presented by real time continuous signals, the latter are quantized by an analog to digital conversion device. A physical system whose behavior is described by mathematical equations is simulated in a digital computer by means of numerical methods. When the problem to be processed is inherently discrete, as in commercial applications, the digital computer manipulates the variables in their natural form.

Both analog and digital computers have their advantages. Analog computers are used when problems require fast solutions with limited accuracy or when large numbers of repetitive calculations are required with variations of parameters. A digital computer is used when the data are in discrete form; when high accuracy, logical decision, and control capabilities are required; and when a general purpose machine is advantageous. A *hybrid computer* is an interconnection of both an analog and a digital computer. This combination possesses the advantages of both computers and is very useful for simulation studies of physical systems.

A block diagram of a small digital computer is shown in Fig. 1-1. The memory unit stores programs as well as input, output, and intermediate data. The arithmetic unit performs the required processing tasks on data obtained from the memory unit. The control unit supervises the flow of information between the various units. The input and output devices accept programs and data (either prepared by the user or generated by the computer) and transfer them to and from the memory unit. The input and output devices are special purpose digital systems driven by an electromechanical mechanism and controlled by electronic digital circuits. Since they are special purpose machines, they will not be covered in this book. The operational characteristics of a memory unit are explained in Sec. 8-3. The organization of a small digital computer, viewed especially from the

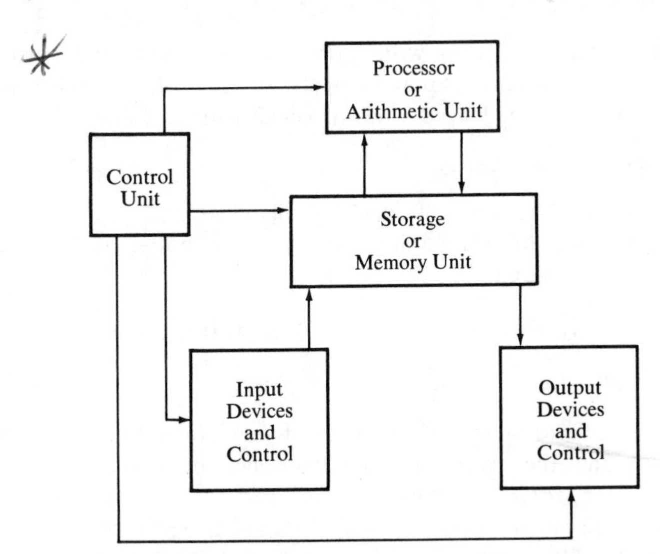

Figure 1-1 Block diagram of a small digital computer

operation of the processor and control units, is presented in some detail in Ch. 10. The processor, memory, and control units of a large digital computer are essentially the same as those shown in Fig. 1-1. On the other hand, a large computer has a wide range of input and output equipment and uses special processors to control the information flow between the memory unit and external devices. Such a computer is introduced and discussed in more detail in Sec. 10-3.

It has already been mentioned that a digital computer manipulates discrete elements of information and that these elements are represented in the binary form. Operands used for calculations may be expressed in the binary number system. Other discrete elements, including the decimal digits, are represented in binary codes. Data processing is carried out by means of binary logic elements using binary signals. Quantities are stored in binary storage elements. The purpose of this chapter is to introduce the various binary concepts as a frame of reference for further detailed study in the succeeding chapters.

1-2 BINARY NUMBERS

A decimal number such as 7392 represents a quantity equal to 7 thousands plus 3 hundreds, plus 9 tens, plus 2 units. The thousands, hundreds, etc., are powers of 10 implied by the position of the coefficients. To be more exact, 7392 should be written as

$$7 \times 10^3 + 3 \times 10^2 + 9 \times 10^1 + 2 \times 10^0$$

However, the convention is to write only the coefficients and from their position deduce the necessary powers of 10. In general, a number with a decimal point is represented by a series of coefficients as follows:

$$a_5 a_4 a_3 a_2 a_1 a_0 \, . \, a_{-1} a_{-2} a_{-3}$$

The a_j coefficients are one of the 10 digits (0, 1, 2, . . . , 9), and the subscript value j gives the place value and hence, the power of 10 by which the coefficient must be multiplied:

$$10^5 a_5 + 10^4 a_4 + 10^3 a_3 + 10^2 a_2 + 10^1 a_1 + 10^0 a_0 + 10^{-1} a_{-1} + 10^{-2} a_{-2} + 10^{-3} a_{-3}$$

The decimal number system is said to be of *base*, or *radix*, 10 because it uses 10 digits and the coefficients are multiplied by powers of 10. The *binary* system is a different number system. The coefficients of the binary number system have two possible values: 0 and 1. Each coefficient a_j is multiplied by 2^j. For example, the decimal equivalent of the binary number 11010.11 is 26.75, as shown from the multiplication of the coefficients by powers of 2:

$$1 \times 2^4 + 1 \times 2^3 + 0 \times 2^2 + 1 \times 2^1 + 0 \times 2^0 + 1 \times 2^{-1} + 1 \times 2^{-2} = 26.75$$

In general, a number expressed in base r system has coefficients multiplied by powers of r:

$$a_n \cdot r^n + a_{n-1} \cdot r^{n-1} + \ldots + a_2 \cdot r^2 + a_1 \cdot r + a_0 + a_{-1} \cdot r^{-1} + a_{-2} \cdot r^{-2} + \ldots + a_{-m} \cdot r^{-m}$$

The coefficients a_j range in value from 0 to $r - 1$. To distinguish between numbers of different bases, we enclose the coefficients in parentheses and write a subscript equal to the base used (except sometimes for decimal numbers where the content makes it obvious that it is decimal). An example of a base 5 number is:

$$(4021.2)_5 = 4 \times 5^3 + 0 \times 5^2 + 2 \times 5^1 + 1 \times 5^0 + 2 \times 5^{-1} = (511.4)_{10}$$

Note that coefficient values for base 5 can be only 0, 1, 2, 3, and 4.

It is customary to borrow the needed r digits for the coefficients from the decimal system when the base of the number is less than 10. The letters of the alphabet are used to supplement the 10 decimal digits when the base of the number is greater than 10. For example, in the *hexadecimal* (base 16) number system, the first 10 digits are borrowed from the decimal system. The letters *A, B, C, D, E,* and *F* are used for digits 10, 11, 12, 13, 14, and 15, respectively. An example of a hexadecimal number is:

$$(B65F)_{16} = 11 \times 16^3 + 6 \times 16^2 + 5 \times 16 + 15 = (46687)_{10}$$

The first 16 numbers in the decimal, binary, octal, and hexadecimal systems are listed in Table 1-1.

Table 1-1 Numbers With Different Bases

Decimal (base 10)	*Binary (base 2)*	*Octal (base 8)*	*Hexadecimal (base 16)*
00	0000	00	0
01	0001	01	1
02	0010	02	2
03	0011	03	3
04	0100	04	4
05	0101	05	5
06	0110	06	6
07	0111	07	7
08	1000	10	8
09	1001	11	9
10	1010	12	A
11	1011	13	B
12	1100	14	C
13	1101	15	D
14	1110	16	E
15	1111	17	F

Arithmetic operations with numbers in base r follow the same rules as for decimal numbers. When other than the familiar base 10 is used, one must be careful to use only the r allowable digits. Examples of addition, subtraction, and multiplication of two binary numbers are shown below:

augend:	101101	minuend:	101101	multiplicand:	1011
addend:	100111	subtrahend:	100111	multiplier:	101
sum:	1010100	difference:	000110		1011
					0000
					1011
				product:	110111

The sum of two binary numbers is calculated by the same rules as in decimals, except that the digits of the sum in any significant position can be only 0 or 1. Any "carry" obtained in a given significant position is used by the pair of digits one significant position higher. The subtraction is slightly more complicated. The rules are still the same as in decimal except that the "borrow" in a given significant position adds 2 to a minuend digit. (A borrow in the decimal system adds 10 to a minuend digit.) Multiplication is very simple. The multiplier digits are always 1 or 0. Therefore, the partial products are equal either to the multiplicand or to 0.

1-3 NUMBER BASE CONVERSIONS

A binary number can be converted to decimal by forming the sum of the powers of 2 of those coefficients whose value is 1. For example:

$$(1010.011)_2 = 2^3 + 2^1 + 2^{-2} + 2^{-3} = (10.375)_{10}$$

The binary number has four 1's and the decimal equivalent is found from the sum of four powers of 2. Similarly, a number expressed in base r can be converted to its decimal equivalent by multiplying each coefficient with the corresponding power of r and adding. The following is an example of octal to decimal conversion:

$$(630.4)_8 = 6 \times 8^2 + 3 \times 8 + 4 \times 8^{-1} = (408.5)_{10}$$

The conversion from decimal to binary or to any other base r system is more convenient if the number is separated into an *integer part* and a *fraction part* and the conversion of each part done separately. The conversion of an *integer* from decimal to binary is best explained by example.

EXAMPLE 1-1. Convert decimal 41 to binary. First, 41 is divided by 2 to give an integer quotient of 20 and a remainder of 1/2. The quotient is again divided by 2 to give a new quotient and remainder. This process is continued until the integer quotient becomes 0. The *coefficients* of the desired binary number are obtained from the *remainders* as follows:

		integer quotient		*remainder*	*coefficient*
$\frac{41}{2}$	=	20	+	$\frac{1}{2}$	$a_0 = 1$
$\frac{20}{2}$	=	10	+	0	$a_1 = 0$
$\frac{10}{2}$	=	5	+	0	$a_2 = 0$
$\frac{5}{2}$	=	2	+	$\frac{1}{2}$	$a_3 = 1$
$\frac{2}{2}$	=	1	+	0	$a_4 = 0$
$\frac{1}{2}$	=	0	+	$\frac{1}{2}$	$a_5 = 1$

ANSWER: $(41)_{10} = (a_5a_4a_3a_2a_1a_0)_2 = (101001)_2$

The arithmetic process can be manipulated more conveniently as follows:

integer	*remainder*
41	
20	1
10	0
5	0
2	1
1	0
0	1

101001 = answer

The conversion from decimal integers to any base r system is similar to the above example except that division is done by r instead of 2.

EXAMPLE 1-2. Convert decimal 153 to octal. The required base r is 8. First, 153 is divided by 8 to give an integer quotient of 19 and a remainder of 1. Then 19 is divided by 8 to give an integer quotient of 2 and a remainder of 3. Finally, 2 is divided by 8 to give a quotient of 0 and a remainder of 2. This process can be conveniently manipulated as follows:

153	
19	1
2	3
0	2

$= (231)_8$

The conversion of a decimal *fraction* to binary is accomplished by a method similar to that used for integers. However, multiplication is used instead of division, and integers are accumulated instead of remainders. Again the method is best explained by example.

EXAMPLE 1-3. Convert $(0.6875)_{10}$ to binary. First, 0.6875 is multiplied by 2 to give an integer and a fraction. The new fraction is multiplied by 2 to give a new integer and a new fraction. This process is continued until the fraction becomes 0 or until the number of digits have sufficient accuracy. The coefficients of the binary number are obtained from the integers as follows:

	integer		*fraction*	*coefficient*
0.6875 × 2 =	1	+	0.3750	$a_{-1} = 1$
0.3750 × 2 =	0	+	0.7500	$a_{-2} = 0$
0.7500 × 2 =	1	+	0.5000	$a_{-3} = 1$
0.5000 × 2 =	1	+	0.0000	$a_{-4} = 1$

ANSWER: $(0.6875)_{10} = (0.a_{-1}a_{-2}a_{-3}a_{-4})_2 = (0.1011)_2$

In converting a decimal fraction to a number expressed in base r, the procedure is similar. Multiplication is by r instead of 2, and the coefficients found from the integers may range in value from 0 to $r - 1$ instead of 0 and 1.

EXAMPLE 1-4. Convert $(0.513)_{10}$ to octal.

$0.513 \times 8 = 4.104$

$0.104 \times 8 = 0.832$

$0.832 \times 8 = 6.656$

$0.656 \times 8 = 5.248$

$0.248 \times 8 = 1.984$

$0.984 \times 8 = 7.872$

The answer, to seven significant figures, is obtained from the integer part of the products:

$$(0.513)_{10} = (0.406517\ldots)_8$$

The conversion of decimal numbers with both integer and fraction parts is done by converting the integer and fraction separately and then combining the two answers together. Using the results of Exs. 1-1 and 1-3 we obtain:

$$(41.6875)_{10} = (101001.1011)_2$$

From Exs. 1-2 and 1-4, we have:

$$(153.513)_{10} = (231.406517)_8$$

1-4 OCTAL AND HEXADECIMAL NUMBERS

The conversion from and to binary, octal, and hexadecimal plays an important part in digital computers. Since $2^3 = 8$ and $2^4 = 16$, each octal digit corresponds to three binary digits and each hexadecimal digit corresponds to four binary digits. The conversion from binary to octal is easily accomplished by partitioning the binary number into groups of three digits, each starting from the binary point and proceeding to the left and to the right. The corresponding octal digit is then assigned to each group. The following example illustrates the procedure:

$$(\underbrace{10}_{2}\,\underbrace{110}_{6}\,\underbrace{001}_{1}\,\underbrace{101}_{5}\,\underbrace{011}_{3}\;\cdot\;\underbrace{111}_{7}\,\underbrace{100}_{4}\,\underbrace{000}_{0}\,\underbrace{110}_{6})_2 = (26153.7406)_8$$

Conversion from binary to hexadecimal is similar except that the binary number is divided into groups of four digits:

$$(\underbrace{10}_{2}\,\underbrace{1100}_{C}\,\underbrace{0110}_{6}\,\underbrace{1011}_{B}\;\cdot\;\underbrace{1111}_{F}\,\underbrace{0010}_{2})_2 = (2C6B.F2)_{16}$$

The corresponding hexadecimal (or octal) digit for each group of binary digits is easily remembered after studying the values listed in Table 1-1.

Conversion from octal or hexadecimal to binary is done by a procedure reverse to the above. Each octal digit is converted to its three-digit binary equivalent. Similarly, each hexadecimal digit is converted to its four-digit binary equivalent. This is illustrated in the following examples:

$$(673.124)_8 = (\underbrace{110}_{6}\,\underbrace{111}_{7}\,\underbrace{011}_{3}\;\cdot\;\underbrace{001}_{1}\,\underbrace{010}_{2}\,\underbrace{100}_{4})_2$$

$$(306.D)_{16} = (\underbrace{0011}_{3}\,\underbrace{0000}_{0}\,\underbrace{0110}_{6}\;\cdot\;\underbrace{1101}_{D})_2$$

Binary numbers are difficult to work with because they require three or four times as many digits as their decimal equivalent. For example, the binary number 111111111111 is equivalent to decimal 4095. However, digital computers use binary numbers and it is sometimes necessary for the human operator or user to communicate directly with the machine by means of binary numbers. One scheme that retains the binary system in the computer but reduces the number of digits the human must consider utilizes the relationship between the binary number system and the octal or hexadecimal systems. By this method, the human thinks in terms of octal or hexadecimal numbers and performs the required conversion by inspection when direct communication with the machine is necessary. Thus the binary number 111111111111 has 12 digits and is expressed in octal as 7777 (four digits) or in hexadecimal as *FFF* (three digits). During communication between people (about binary numbers in the computer), the octal or hexadecimal representation is more desirable because it can be expressed more compactly with 1/3 or 1/4 the number of digits required for the equivalent binary number. When the human communicates with the machine (through console switches or indicator lights or by means of programs written in *machine language*), the conversion from octal or hexadecimal to binary and vice versa is done by inspection by the human user.

1-5 COMPLEMENTS

Complements are used in digital computers for simplifying the subtraction operation and for logical manipulations. There are two types of complements for each base r system: (a) the r's complement and (b) the $(r - 1)$'s complement. When the value of the base is substituted, the two types receive the names: 2's and 1's complement for binary numbers, or 10's and 9's complement for decimal numbers.

The r's Complement

Given a positive number N in base r with an integer part of n digits. The r's complement of N is defined to be: $r^n - N$ for $N \neq 0$, and 0 for $N = 0$. The following numerical example will help clarify the definition.

The 10's complement of $(52520)_{10}$ is $10^5 - 52520 = 47480$.
The number of digits in the number is $n = 5$.
The 10's complement of $(0.3267)_{10}$ is $1 - 0.3267 = 0.6733$.
No integer part, so $10^n = 10^0 = 1$.
The 10's complement of $(25.639)_{10}$ is $10^2 - 25.639 = 74.361$.
The 2's complement of $(101100)_2$ is $(2^6)_{10} - (101100)_2 = (1000000 - 101100)_2 = 010100$.
The 2's complement of $(0.0110)_2$ is $(1 - 0.0110)_2 = 0.1010$.

From the definition and the examples it is clear that the 10's complement of a decimal number can be formed by leaving all least significant zeros unchanged, subtracting the first nonzero least significant digit from 10, and then subtracting all other higher significant digits from 9. The 2's complement can be formed by leaving all least significant zeros and the first nonzero digit unchanged, and then replacing 1's by 0's and 0's by 1's in all other higher significant digits. A third simpler method for obtaining the r's complement is given after the definition of the $(r - 1)$'s complement.

The r's complement of a number exists for any base r (r greater than but not equal to 1) and may be obtained from the definition given above. The examples listed here use numbers with $r = 10$ (decimal) and $r = 2$ (binary) because these are the two bases of most interest to us. The name of the complement is related to the base of the number used. For example, the $(r - 1)$'s complement of a number in base 11 is named the 10's complement, since $r - 1 = 10$ for $r = 11$.

The $(r - 1)$'s Complement

Given a positive number N in base r with an integer part of n digits and a fraction part of m digits. The $(r - 1)$'s complement of N is defined to be: $r^n - r^{-m} - N$. Some numerical examples follow:

The 9's complement of $(52520)_{10}$ is $(10^5 - 1 - 52520) = 99999 - 52520 = 47479$.
No fraction part, so $10^{-m} = 10^0 = 1$.
The 9's complement of $(0.3267)_{10}$ is $(1 - 10^{-4} - 0.3267) = 0.9999 - 0.3267 = 0.6732$.
No integer part, so $10^n = 10^0 = 1$.
The 9's complement of $(25.639)_{10}$ is $(10^2 - 10^{-3} - 25.639) = 99.999 - 25.639 = 74.360$.
The 1's complement of $(101100)_2$ is $(2^6 - 1) - (101100) = (111111 - 101100)_2 = 010011$.
The 1's complement of $(0.0110)_2$ is $(1 - 2^{-4})_{10} - (0.0110)_2 = (0.1111 - 0.0110)_2 = 0.1001$.

From the examples, we see that the 9's complement of a decimal number is formed simply by subtracting every digit from 9. The 1's complement of a binary number is even simpler to form: the 1's are changed into 0's and the 0's into 1's. Since the $(r - 1)$'s complement is very easily obtained, it is sometimes convenient to use it when the r's complement is desired. From the definitions, and from a comparison of the results obtained in the examples, it follows that the r's complement can be obtained from the $(r - 1)$'s complement after the addition of r^{-m} to the least significant digit. For example, the 2's complement of 10110100 is obtained from the 1's complement 01001011 by adding 1 to give 01001100.

It is worth mentioning that the complement of the complement restores the number to its original value. The r's complement of N is $r^n - N$ and the complement of $(r^n - N)$ is $r^n - (r^n - N) = N$; and similarly for the 1's complement.

Subtraction with r's Complement

The direct method of subtraction taught in elementary schools uses the borrow concept. By this method, we borrow a 1 from a higher significant position when the minuend digit is smaller than the corresponding subtrahend digit. This seems to be easiest when people perform subtraction with paper and pencil. When subtraction is implemented by means of digital components, this method is found to be less efficient than the method that uses complements and addition as stated below.

The subtraction of two positive numbers $(M - N)$, both of base r, may be done as follows:

(1) Add the minuend M to the r's complement of the subtrahend N.

(2) Inspect the result obtained in step (1) for an end carry:
  (a) If an end carry occurs, discard it.
  (b) If an end carry does not occur, take the r's complement of the number obtained in step (1) and place a negative sign in front.

The following examples illustrate the procedure:

EXAMPLE 1-5. Using 10's complement, subtract 72532 − 3250.

$$\begin{array}{rl} M = & 72532 \\ N = & 03250 \\ \text{10's complement of } N = & 96750 \end{array} \qquad \begin{array}{rr} & 72532 \\ + & 96750 \\ \hline \text{end carry} \rightarrow 1 & 69282 \end{array}$$

ANSWER: 69282

EXAMPLE 1-6. Subtract: $(3250 - 72532)_{10}$.

$$\begin{array}{rl} M = & 03250 \\ N = & 72532 \\ \text{10's complement of } N = & 27468 \end{array} \qquad \begin{array}{rr} & 03250 \\ + & 27468 \\ \hline \text{no carry} & 30718 \end{array}$$

ANSWER : −69282 = − (10's complement of 30718)

EXAMPLE 1-7. Use 2's complement to perform $M - N$ with the given binary numbers.

(a)

$$\begin{array}{rl} M = & 1010100 \\ N = & 1000100 \\ \text{2's complement of } N = & 0111100 \end{array} \qquad \begin{array}{rr} & 1010100 \\ + & 0111100 \\ \hline \text{end carry} \rightarrow 1 & 0010000 \end{array}$$

ANSWER: 10000

(b) M = 1000100 $\quad$ N = 1010100

2's complement of N = 0101100

$$\begin{array}{r} 1000100 \\ +\ 0101100 \\ \hline \text{no carry} \quad 1110000 \end{array}$$

ANSWER : -10000 = - (2's complement of 1110000)

The proof of the procedure is: The addition of M to the r's complement of N gives $(M + r^n - N)$. For numbers having an integer part of n digits, r^n is equal to a 1 in the $(n + 1)$th position (what has been called the "end carry"). Since both M and N are assumed to be positive, then:

(a) $(M + r^n - N) \geqslant r^n$ if $M \geqslant N$, or

(b) $(M + r^n - N) < r^n$ if $M < N$

In case (a) the answer is positive and equal to $M - N$, which is directly obtained by discarding the end carry r^n. In case (b) the answer is negative and equal to $-(N - M)$. This case is detected from the absence of an end carry. The answer is obtained by taking a second complement and adding a negative sign: $-[r^n - (M + r^n - N)] = -(N - M)$.

Subtraction with $(r - 1)$'s Complement

The procedure for subtraction with the $(r - 1)$'s complement is exactly the same as the one used with the r's complement except for one variation, called "end around carry," as shown below. The subtraction of $M - N$, both positive numbers in base r, may be calculated in the following manner:

(1) Add the minuend M to the $(r - 1)$'s complement of the subtrahend N.

(2) Inspect the result obtained in step (1) for an end carry.

  (a) If an end carry occurs, add 1 to the least significant digit (end around carry).

  (b) If an end carry does not occur, take the $(r - 1)$'s complement of the number obtained in step (1) and place a negative sign in front.

The proof of this procedure is very similar to the one given for the r's complement case and is left as an exercise. The following examples illustrate the procedure.

EXAMPLE 1-8. Repeat Exs. 1-5 and 1-6 using 9's complements.

(a) M = 72532 $\quad$ N = 03250

9's complement of N = 96749

$$\begin{array}{r} 72532 \\ +\ 96749 \\ \hline \text{end around carry} \quad 1 \ \ 69281 \\ +\ \quad 1 \\ \hline 69282 \end{array}$$

ANSWER: 69282

(b) $M = 03250$ $\quad$ 03250
$N = 72532$ $\quad$ + 27467
no carry $\quad$ 30717

9's complement of N = 27467

ANSWER: $-69282 = -$ (9's complement of 30717)

EXAMPLE 1-9. Repeat Ex. 1-7, using 1's complement.

(a) $M = 1010100$ $\quad$ 1010100
$N = 1000100$ $\quad$ + 0111011
end around carry 1 0001111
+ 1
0010000

1's complement of N = 0111011

ANSWER: 10000

(b) $M = 1000100$ $\quad$ 1000100
$N = 1010100$ $\quad$ + 0101011
no carry $\quad$ 1101111

1's complement of N = 0101011

ANSWER: $-10000 = -$ (1's complement of 1101111)

Comparison Between 1's and 2's Complements

A comparison between 1's and 2's complements reveals the advantages and disadvantages of each. The 1's complement has the advantage of being easier to implement by digital components since the only thing that must be done is to change 0's into 1's and 1's into 0's. The implementation of the 2's complement may be obtained in two ways: 1) by adding 1 to the least significant digit of the 1's complement, or 2) by leaving all leading 0's in the least significant positions and the first 1 unchanged, and only then changing all 1's into 0's and all 0's into 1's. During subtraction of two numbers by complements, the 2's complement is advantageous in that only one arithmetic addition operation is required. The 1's complement requires two arithmetic additions when an end around carry occurs. The 1's complement has the additional disadvantage of possessing two arithmetic zeros: one with all 0's and one with all 1's. To illustrate this fact, consider the subtraction of the two equal binary numbers $1100 - 1100 = 0$

Using 1's complement

$$\begin{array}{r} 1100 \\ + \underline{0011} \\ + 1111 \end{array}$$

complement again to obtain: −0000

Using 2's complement

$$\begin{array}{r} 1100 \\ + \underline{0100} \\ + 0000 \end{array}$$

While the 2's complement has only one arithmetic zero, the 1's complement zero can be positive or negative, which may complicate matters.

Complements, very useful for arithmetic manipulations in digital computers, are discussed more in Chs. 8 and 9. However, the 1's complement is also useful in logical manipulations (as will be shown later), since the change of 1's to 0's and vice versa is equivalent to a logical inversion operation. The 2's complement is used only in conjunction with arithmetic applications. Consequently, it is convenient to adopt the following convention: when the word *complement*, without mention of the type, is used in conjunction with a nonarithmetic application, the type is assumed to be the 1's complement.

1-6 BINARY CODES

Electronic digital systems use signals that have two distinct values and circuit elements that have two stable states. There is a direct analogy among binary signals, binary circuit elements, and binary digits. A binary number of n digits, for example, may be represented by n binary circuit elements, each having an output signal equivalent to a 0 or a 1. Digital systems represent and manipulate not only binary numbers, but also many other discrete elements of information. Any discrete element of information distinct among a group of quantities can be represented by a binary code. For example, *red* is one distinct color of the spectrum. The letter A is one distinct letter of the alphabet.

A *bit*, by definition, is a binary digit. When used in conjunction with a binary code, it is better to think of it as denoting a binary quantity equal to 0 or 1. To represent a group of 2^n distinct elements in a binary code requires a minimum of n bits. This is because it is possible to arrange n

bits in 2^n distinct ways. For example, a group of four distinct quantities can be represented by a two-bit code with each quantity assigned one of the following bit combinations: 00, 01, 10, 11. A group of eight elements requires a three-bit code with each element assigned to one and only one of the following: 000, 001, 010, 011, 100, 101, 110, 111. The examples show that the distinct bit combinations of an n-bit code can be found by counting in binary from 0 to $(2^n - 1)$. Some bit combinations are unassigned when the number of elements of the group to be coded is not a multiple of power of two. The 10 decimal digits 0, 1, 2, . . . , 9 are an example of such a group. A binary code that distinguishes among 10 elements must contain at least four bits; three bits can distinguish a maximum of eight elements. Four bits can form 16 distinct combinations, but since only 10 digits are coded, the remaining six combinations are unassigned and not used.

Although the *minimum* number of bits required to code 2^n distinct quantities is n, there is no *maximum* number of bits that may be used for a binary code. For example, the 10 decimal digits can be coded with 10 bits, and each decimal digit assigned a bit combination of nine 0's and a 1. In this particular binary code, the digit 6 is to be assigned the bit combination 0001000000.

Decimal Codes

Binary codes for decimal digits require a minimum of four bits. Numerous different codes can be obtained by arranging four or more bits in 10 distinct possible combinations. A few possibilities are shown in Table 1-2.

Table 1-2 Binary Codes for the Decimal Digits

Decimal digit	*(BCD) 8421*	*Excess-3*	*84-2-1*	*2421*	*(Biquinary) 5043210*
0	0000	0011	0000	0000	0100001
1	0001	0100	0111	0001	0100010
2	0010	0101	0110	0010	0100100
3	0011	0110	0101	0011	0101000
4	0100	0111	0100	0100	0110000
5	0101	1000	1011	1011	1000001
6	0110	1001	1010	1100	1000010
7	0111	1010	1001	1101	1000100
8	1000	1011	1000	1110	1001000
9	1001	1100	1111	1111	1010000

The BCD (binary coded decimal) is a straight assignment of the binary equivalent. It is possible to assign weights to the binary bits according to their position. The weights in the BCD code are 8, 4, 2, 1. The bit assignment 0110, for example, can be interpreted by the weights to represent the decimal digit 6 because: $0 \times 8 + 1 \times 4 + 1 \times 2 + 0 \times 1 = 6$. It is also possible to assign negative weights to a decimal code, as shown by the 8,4,–2,–1 code. In this case, the bit combination 0110 is interpreted as the decimal digit 2, as obtained from: $0 \times 8 + 1 \times 4 + 1 \times (-2) + 0 \times (-1) = 2$. Two other weighted codes shown in the table are the 2421 and the 5043210. A decimal code that has been used in some old computers is the excess-3 code. This is an unweighted code; its code assignment is obtained from the corresponding value of BCD after the addition of 3.

Numbers are represented in digital computers either in binary or in decimal through a binary code. When a user specifies his data, he likes to give it in decimal form. The input decimal numbers are stored internally in the computer by means of a decimal code. Each decimal digit requires at least four binary storage elements. The decimal numbers are converted to binary when arithmetic operations are done internally with numbers represented in binary. It is also possible to perform the arithmetic operations directly in decimal with all numbers left in a coded form throughout. For example, the decimal number 395, when converted to binary, is equal to 110001011 and consists of nine binary digits. The same number, when represented internally in the BCD code, occupies four bits for each decimal digit for a total of 12 bits: 001110010101. The first four bits represent a 3; the next four a 9; and the last four a 5.

It is very important to understand the difference between *conversion* of a decimal number to binary and the binary *coding* of a decimal number. In each case, the final result is a series of bits. The bits obtained from conversion are binary digits. Bits obtained from coding are combinations of 1's and 0's arranged according to the rules of the code used. Therefore, it is extremely important to realize that a series of 1's and 0's in a digital system may sometimes represent a binary number, and at other times represent some other discrete quantity of information as specified by a given binary code. The BCD code, for example, has been chosen to be both a code and a direct binary conversion as long as the decimal numbers are integers from 0 to 9. For numbers greater than 9, the conversion and the coding are completely different. This concept is so important that it is worth repeating with another example. The binary conversion of decimal 13 is 1101; the coding of decimal 13 with BCD is 00010011.

From the five binary codes listed in Table 1-2, the BCD seems the most natural to use and is indeed the one most commonly encountered. The other four-bit codes listed have one characteristic in common not found in

BCD. The excess-3; 2,4,2,1; and 8,4,−2,−1 are self complementary codes. That is, the 9's complement of the decimal number is easily obtained by changing 1's to 0's and 0's to 1's. For example, the decimal 395 is represented in the 2,4,2,1 code by 001111111011. Its 9's complement 604 is represented by 110000000100, which is easily obtained from the replacement of 1's by 0's and 0's by 1's. This property is useful when arithmetic operations are internally done with decimal numbers (in a binary code) and subtraction is calculated by means of 9's complement.

The biquinary code shown in Table 1-2 is an example of a seven-bit code with error detection properties. Each decimal digit consists of five 0's and two 1's placed in the corresponding weighted columns. The error detection property of this code may be understood when one realizes that digital systems represent binary 1 by one distinct signal and binary 0 by a second distinct signal. During transmission of signals from one location to another, an error may occur. One or more bits may change value. A circuit in the receiving side can detect the presence of more (or less) than two 1's and, if the received combination of bits does not agree with the allowable combination, an error is detected.

Error Detection Codes

Binary information, be it pulse modulated signals or digital computer input or output, may be transmitted through some form of communication medium such as wires or radio waves. Any external noise introduced into a physical communication medium changes bit values from 0 to 1 or vice versa. An error detection code can be used to detect errors during transmission. The detected error cannot be corrected but its presence is indicated. The usual procedure is to observe the frequency of errors. If errors occur only once in awhile at random without a pronounced effect on the overall information transmitted, then either nothing is done or the particular erroneous message is transmitted again. If the error occurs so often as to distort the meaning of the received information, the system is checked for malfunction.

A *parity* bit is an extra bit included with a message to make the total number of 1's either odd or even. A message of four bits and a parity bit P is shown in Table 1-3. In (a) the bit P is chosen in such a way as to make the sum of all 1's odd (in all five bits). In (b) the P bit is chosen to make the sum of all 1's even. During transfer of information from one location to another, the parity bit is handled as follows: In the sending end, the message (in this case the first four bits) is applied to a "parity generation" network where the required P bit is generated. The message, including the parity bit, is transferred to its destination. In the receiving end, all the incoming bits (in this case five) are applied to a "parity check" network to check the proper parity adopted. An error is detected if the checked parity

Table 1-3 Parity Bit Generation

(a) Message	*P (odd)*	*(b) Message*	*P (even)*
0000	1	0000	0
0001	0	0001	1
0010	0	0010	1
0011	1	0011	0
0100	0	0100	1
0101	1	0101	0
0110	1	0110	0
0111	0	0111	1
1000	0	1000	1
1001	1	1001	0
1010	1	1010	0
1011	0	1011	1
1100	1	1100	0
1101	0	1101	1
1110	0	1110	1
1111	1	1111	0

does not correspond to the adopted one. The parity method detects the presence of one, three, or any odd combination of errors. An even combination of errors is undetectable.

The Reflected Code

Digital systems can be designed to process data in discrete form only. Many physical systems supply continuous output data. This data must be converted into digital or discrete form before it is applied to a digital system. Continuous or analog information is converted into digital form by means of an analog to digital converter. It is sometimes convenient to use the reflected code as shown in Table 1-4 to represent the digital data converted from the analog data. The advantage of the reflected code over pure binary numbers is that a number in the reflected code changes by only one bit as it proceeds from one number to the next. A typical application of the reflected code occurs when the analog data is represented by a continuous change of a shaft position. The shaft is partitioned into segments. Each segment is assigned a number. If adjacent segments are made to correspond to adjacent reflected code numbers, ambiguity is reduced when detection is sensed in the line that separates any two segments. The reflected code shown in Table 1-4 is only one of many possible such codes. To obtain a different reflected code, one can start with any bit combination and proceed to obtain the next bit combination by changing only one bit from 0 to 1 or 1 to 0 in any desired random fashion, as long as two numbers do not have identical code assignments.

Table 1-4 4-Bit Reflected Code

Reflected code	*Decimal equivalent*
0000	0
0001	1
0011	2
0010	3
0110	4
0111	5
0101	6
0100	7
1100	8
1101	9
1111	10
1110	11
1010	12
1011	13
1001	14
1000	15

Alphanumeric Codes

Many applications of digital computers require the handling of data that consists not only of numbers, but also of letters. For instance, an insurance company with millions of policy holders may use a digital computer to process its files. To represent the policy holder's name in binary form, it is necessary to have a binary code for the alphabet. In addition, the same binary code must represent decimal numbers and some other special characters. An alphanumeric (sometimes abbreviated *alphameric*) code is a binary code of a group of elements consisting of the 10 decimal digits, the 26 letters of the alphabet, and a certain number of special symbols such as $. The total number of elements in an alphanumeric group is greater than 36. Therefore, it must be coded with a minimum of six bits ($2^6 = 64$, but $2^5 = 32$ is insufficient).

One possible arrangement of a six-bit alphanumeric code is shown in Table 1-5 under the name of internal code. With a few variations, it is used in many computers to represent alphanumeric characters internally. The need to represent more than 64 characters (the lower case letters and special control characters for the transmission of digital information) gave rise to seven- and eight-bit alphanumeric codes. One such code is known as ASCII (American Standard Code for Information Interchange); another is known as EBCDIC (Extended BCD Interchange Code). The ASCII code listed in Table 1-5 consists of seven bits but is, for all practical purposes, an eight-bit code because an eighth bit is invariably added for parity. When discrete information is transferred through punched cards, the alphanumeric

Table 1-5 Alphanumeric Character Codes

Character	*6-bit internal code*	*7-bit ASCII code*	*8-bit EBCDIC code*	*12-bit card code*
A	010 001	100 0001	1100 0001	12,1
B	010 010	100 0010	1100 0010	12,2
C	010 011	100 0011	1100 0011	12,3
D	010 100	100 0100	1100 0100	12,4
E	010 101	100 0101	1100 0101	12,5
F	010 110	100 0110	1100 0110	12,6
G	010 111	100 0111	1100 0111	12,7
H	011 000	100 1000	1100 1000	12,8
I	011 001	100 1001	1100 1001	12,9
J	100 001	100 1010	1101 0001	11,1
K	100 010	100 1011	1101 0010	11,2
L	100 011	100 1100	1101 0011	11,3
M	100 100	100 1101	1101 0100	11,4
N	100 101	100 1110	1101 0101	11,5
O	100 110	100 1111	1101 0110	11,6
P	100 111	101 0000	1101 0111	11,7
Q	101 000	101 0001	1101 1000	11,8
R	101 001	101 0010	1101 1001	11,9
S	110 010	101 0011	1110 0010	0,2
T	110 011	101 0100	1110 0011	0,3
U	110 100	101 0101	1110 0100	0,4
V	110 101	101 0110	1110 0101	0,5
W	110 110	101 0111	1110 0110	0,6
X	110 111	101 1000	1110 0111	0,7
Y	111 000	101 1001	1110 1000	0,8
Z	111 001	101 1010	1110 1001	0,9
0	000 000	011 0000	1111 0000	0
1	000 001	011 0001	1111 0001	1
2	000 010	011 0010	1111 0010	2
3	000 011	011 0011	1111 0011	3
4	000 100	011 0100	1111 0100	4
5	000 101	011 0101	1111 0101	5
6	000 110	011 0110	1111 0110	6
7	000 111	011 0111	1111 0111	7
8	001 000	011 1000	1111 1000	8
9	001 001	011 1001	1111 1001	9
blank	110 000	010 0000	0100 0000	no punch
.	011 011	010 1110	0100 1011	12,8,3
(	111 100	010 1000	0100 1101	12,8,5
+	010 000	010 1011	0100 1110	12,8,6
$	101 011	010 0100	0101 1011	11,8,3
*	101 100	010 1010	0101 1100	11,8,4
)	011 100	010 1001	0101 1101	11,8,5
–	100 000	010 1101	0110 0000	11
/	110 001	010 1111	0110 0001	0,1
,	111 011	010 1100	0110 1011	0,8,3
=	001 011	011 1101	0111 1110	8,6

characters use a 12-bit binary code. A punched card consists of 80 columns and 12 rows. Each column represents an alphanumeric character by punching holes in the appropriate rows. A hole is sensed as a 1 and the absence of a hole is sensed as a 0. The 12 rows are marked, starting from the top, as the 12, 11, 0, 1, 2, . . ., 9 punch. The first three are called the *zone* punch and the last 9 are called the *numeric* punch. The 12-bit card code shown in Table 1-5 lists the rows where a hole is punched (giving the 1's). The remaining unlisted rows are assumed to be 0's. The 12-bit card code is inefficient with respect to the number of bits used. Most computers translate the input code into an internal six-bit code. As an example, the internal code representation of the name "John Doe" is:

100001	100110	011000	100101	110000	010100	100110	010101
J	O	H	N	blank	D	O	E

1-7 BINARY STORAGE AND REGISTERS

The discrete elements of information in a digital computer must have a physical existence in some information storage medium. Furthermore, when discrete elements of information are represented in binary form, the information storage medium must contain binary storage elements for storing individual bits. A *binary cell* is a device that possesses two stable states and is capable of storing one bit of information. The input to the cell receives excitation signals that set it to one of the two states. The output of the cell is a physical quantity that distinguishes between the two states. The information stored in a cell is a 1 when it is in one stable state and a 0 when in the other stable state. Examples of binary cells are electronic flip-flop circuits, ferrite cores used in memories, and positions punched with a hole or not punched in a card.

Registers

A *register* is a group of binary cells. Since a cell stores one bit of information, it follows that a register with n cells can store any discrete quantity of information that contains n bits. The *state* of a register is an n-tuple number of 1's and 0's with each bit designating the state of one cell in the register. The *content* of a register is a function of the interpretation given to the information stored in it. Consider for example, the following 16-cell register:

1	1	0	0	0	0	1	1	1	1	0	0	1	0	0	1
1	2	3	4	5	6	7	8	9	10	11	12	13	14	15	16

Physically, one may think of the register as composed of 16 binary cells, with each cell storing either a 1 or a 0. Suppose that the bit configuration stored in the register is as shown. The state of the register is the 16-tuple number 1100001111001001. Clearly, a register with n cells can be in one of 2^n possible states. Now, if one assumes that the content of the register represents a binary integer, then obviously the register can store any binary number from 0 to $2^{16} - 1$. For the particular example shown, the content of the register is the binary equivalent of the decimal number 50121. If it is assumed that the register stores alphanumeric characters of an eight-bit code, the content of the register is any two meaningful characters (unassigned bit combinations do not represent meaningful information). In the EBCDIC code, the above example represents the two characters C (left eight bits) and I (right eight bits). On the other hand, if one interprets the content of the register to be four decimal digits represented by a four-bit code, the content of the register is a four-digit decimal number. In the excess-3 code, the above example is the decimal number 9096. The content of the register is meaningless in BCD since the bit combination 1100 is not assigned to any decimal digit. From this example, it is clear that a register can store one or more discrete elements of information and that the same bit configuration may be interpreted differently for different types of elements of information. It is important that the user store meaningful information in registers and that he program the computer to process this information according to the *type* of information stored.

Register Transfer

A digital computer is characterized by its registers. The memory unit (Fig. 1-1) is merely a collection of thousands of registers for storing digital information. The processor unit is composed of various registers that store operands upon which operations are performed. The control unit uses registers to keep track of various computer sequences, and every input or output device must have at least one register to store the information transferred to or from the device. An *inter-register transfer* operation, a basic operation in digital systems, consists of a transfer of the information stored in one register into another. Figure 1-2 illustrates the transfer of information among registers and demonstrates pictorially the transfer of binary information from a teletype keyboard into a register in the memory unit. The input teletype unit is assumed to have a keyboard, a control circuit, and an input register. Each time a key is struck, the control enters into the input register an equivalent eight-bit alphanumeric character code. We shall assume that the code used is the ASCII code with an odd parity eighth bit. The information from the input register is transferred into the eight least significant cells of a processor register. After every transfer, the

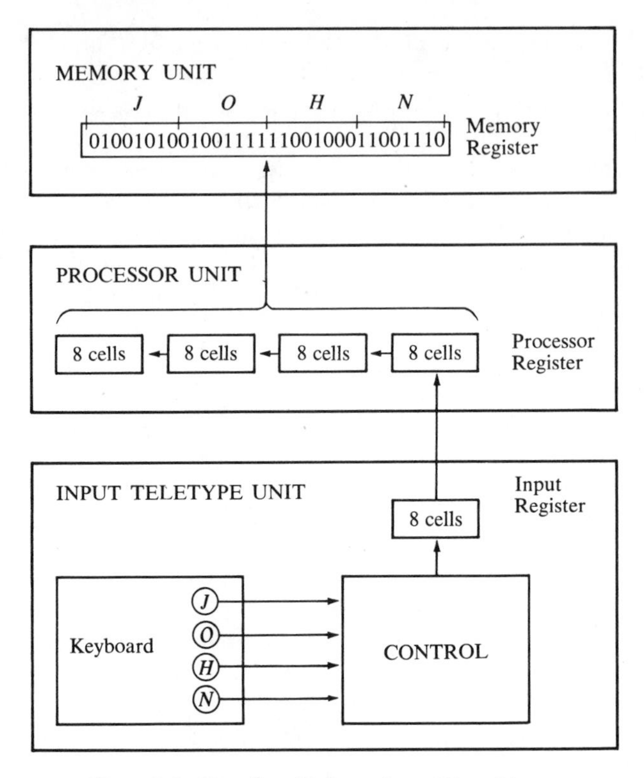

Figure 1-2 Transfer of information with registers

input register is cleared to enable the control to insert a new eight-bit code when the keyboard is struck again. Each eight-bit character transferred to the processor register is preceeded by a shift of the previous character to the next eight-cells on its left. When a transfer of four characters is completed, the processor register is full; and its contents are transferred into a memory register. The content stored in the memory register shown in Fig. 1-2 came from the transfer of the characters JOHN after striking the four appropriate keys.

To process discrete quantities of information in binary form, a computer must be provided with: (a) devices that hold the data to be processed and (b) circuit elements that manipulate individual bits of information. The device most commonly used for holding data is a register. Manipulation of binary variables is done by means of digital logic circuits. Figure 1-3 illustrates the process of adding two 10-bit binary numbers. The memory unit, which normally consists of thousands of registers, is shown in the diagram with only three of its registers. The part of the processor unit shown consists of three registers, R1, R2, and R3, together with digital logic circuits that manipulate the bits of R1 and R2 and transfer into R3 a binary number equal to their arithmetic sum. Memory registers store infor-

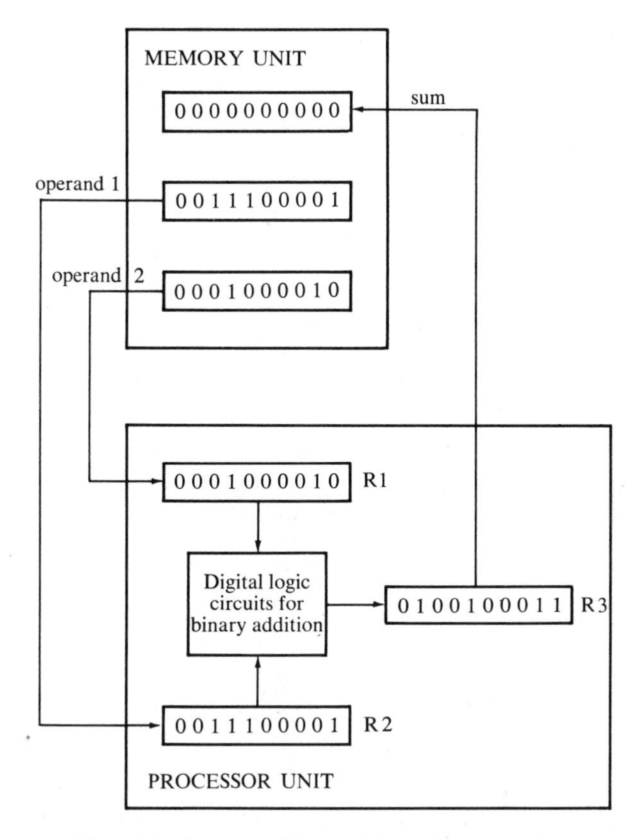

Figure 1-3 Example of binary information processing

mation and are incapable of processing the two operands. However, the information stored in memory can be transferred to processor registers. Results obtained in processor registers can be transferred back into a memory register for storage until needed again. The diagram shows the contents of two operands transferred from two memory registers into R1 and R2. The digital logic circuits produce the sum, which is transferred to register R3. The contents of R3 can now be transferred back to one of the memory registers.

The last two examples have demonstrated the information flow capabilities of a digital system in a very simple manner. The registers of the system are the basic elements for storing and holding the binary information. The digital logic circuits process the information. Digital logic circuits and their manipulative capabilities are introduced in the next section. The subject of registers and register transfer operations is taken up again in Ch. 8.

1-8 BINARY LOGIC

Binary logic deals with variables that take on two discrete values and with operations that assume logical meaning. The two values the variables take may be called by different names (such as *true* and *false*, *yes* and *no*, etc.), but for our purpose it is convenient to think in terms of bits and assign the values of 1 and 0. Binary logic is used to describe, in a mathematical way, the manipulation and processing of binary information. It is particularly suited for the analysis and design of digital systems. For example, the digital logic circuits of Fig. 1-3 that perform the binary arithmetic are circuits whose behavior is most conveniently expressed by means of binary variables and logical operations. The binary logic to be introduced in this section is equivalent to an algebra called Boolean algebra. The formal presentation of a two-valued Boolean algebra is covered in more detail in Ch. 2. The purpose of this section is to introduce Boolean algebra in a heuristic manner and relate it to digital logic circuits and binary signals.

Definition of Binary Logic

Binary logic consists of binary variables and logical operations. The variables are designated by letters of the alphabet such as A, B, C, x, y, z, etc., with each variable having two and only two distinct possible values, 1 and 0. There are three basic logical operations, AND, OR, and NOT.

1. AND: This operation is represented by a dot or an absence of an operator. For example, $x \cdot y = z$ or $xy = z$ is read: "x AND y is equal to z." The logical operation AND is interpreted to mean that $z = 1$ if and only if $x = 1$ *and* $y = 1$; otherwise $z = 0$. (Remember that x, y, and z are binary variables and can be equal either to 1 or 0, and nothing else).

2. OR: This operation is represented by a plus sign. For example, $x + y = z$ is read: "x OR y is equal to z," meaning that $z = 1$ if $x = 1$ *or* if $y = 1$ *or* if both $x = 1$ and $y = 1$. If both $x = 0$ and $y = 0$, then $z = 0$.

3. NOT: This operation is represented by a prime (sometimes by a bar). For example, $x' = z$ (or $\bar{x} = z$) is read: "x not is equal to z," meaning that z is what x is not; that is, if $x = 1$; then $z = 0$; but if $x = 0$, then $z = 1$.

Binary logic resembles binary arithmetic and the operations AND and OR have some similarities to multiplication and addition, respectively. In fact, the symbols used for AND and OR are the same as those used for multiplication and addition. However, binary logic should not be confused with binary arithmetic. One should realize that an arithmetic variable designates a number that may consist of many digits. A logic variable is always either a 1 or a 0. For example, in binary arithmetic we have $1 + 1 = 10$ (read: "one plus one is equal to 2"), while in binary logic we have $1 + 1 = 1$ (read: "one OR one is equal to one").

For each combination of the value of x and y, there is a value of z specified by the definition of the logical operation. These definitions may be listed in a compact form using *truth tables*. A truth table is a table of all possible combinations of the variables showing the relation between the values that the variables may take and the result of the operation. For example, the truth tables for the operations AND and OR with variables x and y are obtained by listing all possible values that the variables may have when combined in pairs. The result of the operation for each combination is then listed in a separate row. The truth tables for AND, OR, and NOT are listed in Table 1-6. These tables clearly demonstrate the definitions of the operations.

Table 1-6 Truth Tables of Logical Operations

AND

x	y	$x \cdot y$
0	0	0
0	1	0
1	0	0
1	1	1

OR

x	y	$x + y$
0	0	0
0	1	1
1	0	1
1	1	1

NOT

x	x'
0	1
1	0

Switching Circuits and Binary Signals

The use of binary variables and the application of binary logic is demonstrated by the simple switching circuits of Fig. 1-4. Let the manual switches A and B represent two binary variables with values equal to 0 when the switch is open and 1 when the switch is closed. Similarly, let the lamp L represent a third binary variable equal to 1 when the light is on and 0 when off. For the switches in series, the light turns on if *A and B* are closed. For the switches in parallel, the light turns on if *A or B* is closed. It is obvious that the two circuits can be expressed by means of binary logic with the AND and OR operations, respectively:

$L = A \cdot B$ for the circuit of Fig. 1-4(a)

$L = A + B$ for the circuit of Fig. 1-4(b)

Electronic digital circuits are sometimes called *switching circuits* because they behave like a switch with the active element such as a transistor either

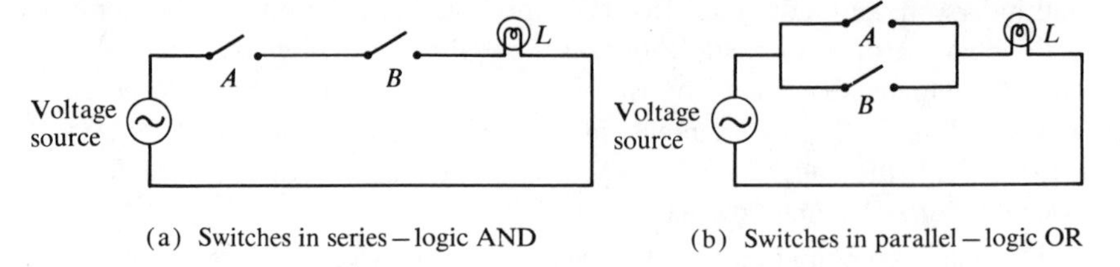

(a) Switches in series—logic AND (b) Switches in parallel—logic OR

Figure 1-4 Switching circuits that demonstrate binary logic

conducting (switch closed) or not conducting (switch open). Instead of changing the switch manually, an electronic switching circuit uses binary signals to control the conduction or nonconduction state of the active element. Electrical signals such as voltages or currents exist throughout a digital system in either one of two recognizable values (except during transition). Voltage operated circuits, for example, respond to two separate voltage levels which represent a binary variable equal to logic-1 or logic-0. For example, a particular digital system may define logic-1 as a signal with a nominal value of 3 volts, and logic-0 as a signal with a nominal value of 0 volt. As shown in Fig. 1-5, each voltage level has an acceptable deviation from the nominal. The intermediate region between the allowed regions is crossed only during state transitions. The input terminals of digital circuits accept binary signals within the allowable tolerances and respond at the output terminal with binary signals that fall within the specified tolerances.

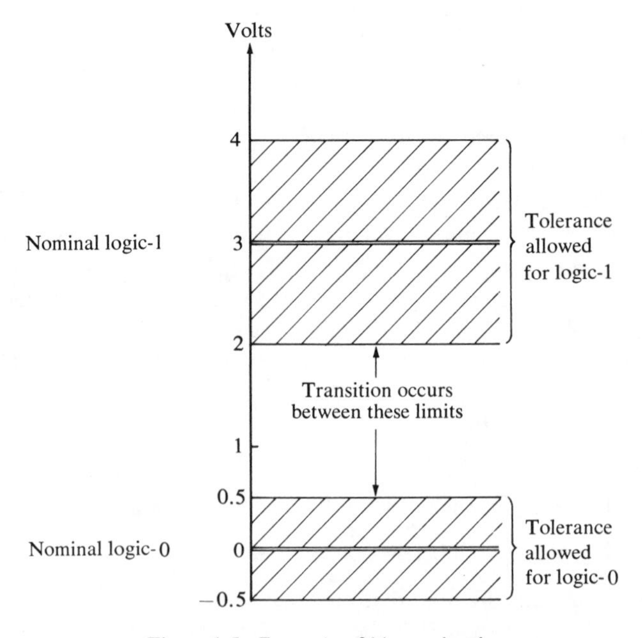

Figure 1-5 Example of binary signals

Logic Gates

Electronic digital circuits are also called *logic circuits* because, with the proper input, they establish logical manipulation paths. Any desired information for computing or control can be operated upon by passing binary signals through various combinations of logic circuits, each signal representing a variable and carrying one bit of information. Logic circuits that perform the logical operations of AND, OR, and NOT are shown with their symbols in Fig. 1-6. These circuits, called *gates* are blocks of hardware that produce a logic-1 or logic-0 output signal if input logic requirements are satisfied. Note that four different names have been used for the same type of circuits: digital circuits, switching circuits, logic circuits, and gates. All four names are widely used, but we shall refer to the circuits as AND, OR, and NOT gates. The NOT gate is sometimes called an *inverter circuit* since it inverts a binary signal.

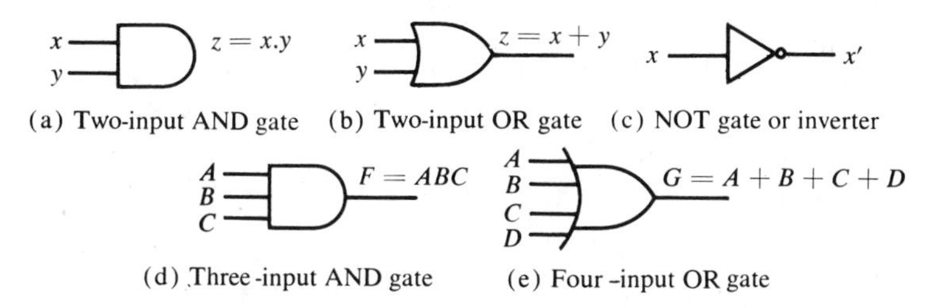

Figure 1-6 Symbols for digital logic circuits

The input signals x and y in the two-input gates of Fig. 1-6 may exist in one of four possible states: 00, 10, 11, or 01. These input signals are shown in Fig. 1-7, together with the output signals for the AND and OR gates. The timing diagrams of Fig. 1-7 illustrate the response of each circuit to each of the four possible input binary combinations. The reason for the name inverter for the NOT gate is apparent from a comparison of the signal x (input of inverter) and that of x' (output of inverter).

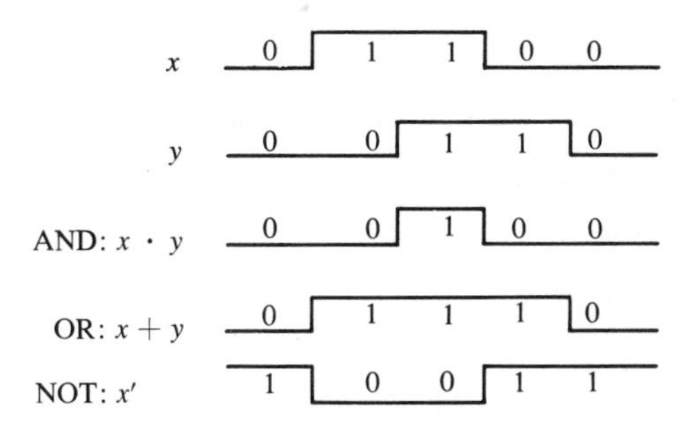

Figure 1-7 Input-output signals for gates (a), (b), and (c) of Figure 1-6

AND and OR gates may have more than two inputs. An AND gate with three inputs and an OR gate with four inputs are shown in Fig. 1-6. The three-input AND gate responds with a logic-1 output if all three input signals are logic-1. The output produces a logic-0 signal if any input is logic-0. The four-input OR gate responds with a logic-1 when any input is a logic-1. Its output becomes logic-0 if all input signals are logic-0.

The mathematical system of binary logic is better known as Boolean, or switching, algebra. This algebra is conveniently used to describe the operation of complex networks of digital circuits. Designers of digital systems use Boolean algebra to transform circuit diagrams to algebraic expressions and vice versa. Chapters 2 and 3 are devoted to the study of Boolean algebra, its properties and manipulative capabilities. We shall return to the subject of logic gates in Ch. 4 and show how Boolean algebra may be used to express mathematically the interconnections among networks of gates.

PROBLEMS

1-1. Write the first 20 decimal digits in base 3.

1-2. Add and multiply the following numbers in the given base without converting to decimal.

(a) $(1230)_4$ and $(23)_4$

(b) $(135.4)_6$ and $(43.2)_6$

(c) $(367)_8$ and $(715)_8$

(d) $(296)_{12}$ and $(57)_{12}$

1-3. Convert the decimal number 250.5 to base 3, base 4, base 7, base 8, and base 16.

1-4. Convert the following decimal numbers to binary: 12.0625, 10^4, 673.23, and 1998.

1-5. Convert the following binary numbers to decimal: 10.10001, 101110.0101, 1110101.110, 1101101.111.

1-6. Convert the following numbers from the given base to the required bases:

(a) decimal 225.225 to binary, octal, and hexadecimal

(b) binary 11010111.110 to decimal, octal, and hexadecimal

(c) octal 623.77 to decimal, binary, and hexadecimal

(d) hexadecimal $2AC5.D$ to decimal, octal, and binary

1-7. Convert the following numbers to decimal.

(a) $(1001001.011)_2$

(b) $(12121)_3$

(c) $(1032.2)_4$

(d) $(4310)_5$

(e) $(0.342)_6$

(f) $(50)_7$

(g) $(8.3)_9$

(h) $(198)_{12}$

1-8. Obtain the 1's and 2's complement of the following binary numbers: 1010101, 0111000, 0000001, 10000, 00000.

1-9. Obtain the 9's and 10's complement of the following decimal numbers: 13579, 09900, 90090, 10000, 00000.

1-10. Find the 10's complement of $(935)_{11}$.

1-11. Perform the subtraction with the following decimal numbers using, (1) 10's complement and (2) 9's complement. Check the answer by straight subtraction.

(a) 5250 − 321 (c) 753 − 864

(b) 3570 − 2100 (d) 20 − 1000

1-12. Perform the subtraction with the following binary numbers using (1) 2's complement and (2) 1's complement. Check the answer by straight subtraction.

(a) 11010 − 1101 (c) 10010 − 10011

(b) 11010 − 10000 (d) 100 − 110000

1-13. Prove the procedure stated in Sec. 1-5 for the subtraction of two numbers with $(r - 1)$'s complement.

1-14. For the weighted codes (a) 3,3,2,1 and (b) 4,4,3,-2 for the decimal digits, determine all possible tables so that the 9's complement of each decimal digit is obtained by changing 1's to 0's and 0's to 1's.

1-15. Represent the decimal number 8620 (a) in BCD, (b) in excess-3 code, (c) in 2,4,2,1 code, and (d) as a binary number.

1-16. A binary code uses 10 bits to represent each of the 10 decimal digits. Each digit is assigned a code of nine 0's and a 1. The code for digit 6, for example, is 0001000000. Determine the binary code for the remaining decimal digits.

1-17. Obtain the weighted binary code for the base 12 digits using weights of 5421.

1-18. Determine the odd parity bit generated when the message is the 10 decimal digits in the 8,4,-2,-1 code.

1-19. Determine two other combinations for a reflected code other than the one shown in Table 1-4.

1-20. Obtain a binary code to represent all base 6 digits so that the 5's complement is obtained by replacing 1's by 0's and 0's by 1's in the bits of the code.

1-21. Assign a binary code in some orderly manner to the 52 playing cards. Use the minimum number of bits.

1-22. Write your first name, middle initial, and last name in an eight-bit code made up of the seven ASCII bits of Table 1-5 and an even parity bit in the most significant position. Include blanks between names and a period after the middle initial.

1-23. Show the bit configuration of a 24-cell register when its content represents (a) the number $(295)_{10}$ in binary, (b) the decimal number 295 in BCD, (c) the characters $XY5$ in EBCDIC.

1-24. The state of a 12-cell register is 010110010111. What is its content if it represents (a) three decimal digits in BCD, (b) three decimal digits in excess-3 code, (c) three decimal digits in 2,4,2,1 code, (d) two characters in the internal code of Table 1-5?

1-25. Show the contents of all registers in Fig. 1-3 if the two binary numbers added have the decimal equivalent of 257 and 1050.

1-26. Express the following switching circuit in binary logic notation.

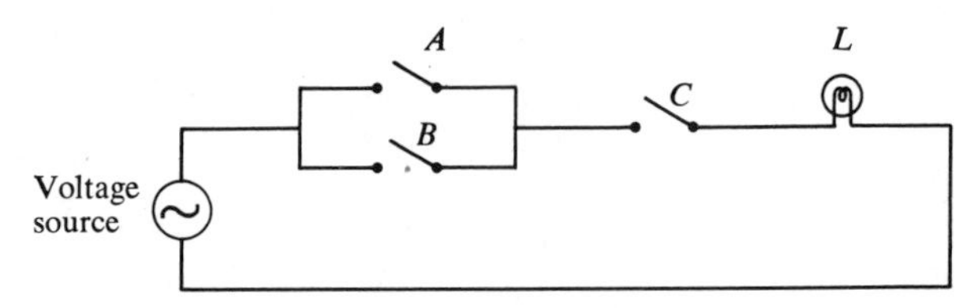

1-27. Show the signals (by means of a diagram similar to Fig. 1-7) of the outputs F and G in Fig. 1-6. Use arbitrary binary signals for the inputs A, B, C, and D.

2 BOOLEAN ALGEBRA

2-1 BASIC DEFINITIONS

Boolean algebra like any other deductive mathematical system may be defined with a set of elements, a set of operators, and a number of unproved axioms or postulates. A *set* of elements is any collection of objects having a common property. If S is a set, and x and y are certain objects, then $x \in S$ denotes that x is a member of the set S, while $y \notin S$ denotes that y is not an element of S. A set with a denumerable number of elements is specified by curly brackets: $A = \{1, 2, 3, 4\}$; i.e., the elements of set A are the numbers 1, 2, 3, and 4. A *binary operator* defined on a set S of elements is a rule that assigns to each pair of elements from S a unique element from S. As an example, consider the relation $a * b = c$, we say that $*$ is a binary operator if it specifies a rule for finding c from the pair (a, b) and also if $a, b, c \in S$. However, $*$ is not a binary operator if $a, b \in S$, while the rule finds $c \notin S$.

The postulates of a mathematical system form the basic assumptions from which it is possible to deduce the rules, theorems, and properties of the system. The most common postulates used to formulate various algebraic structures are:

1. *Closure.* A set S is closed with respect to a binary operator if for every pair of elements of S, the binary operator specifies a rule for obtaining a unique element of S. For example, the set of natural numbers $N = \{1, 2, 3, 4, \ldots\}$ is closed with respect to the binary operator plus (+) by the rules of arithmetic addition, since for any $a, b \in N$ we obtain a unique $c \in N$ by the operation $a + b = c$. The set of

natural numbers is not closed with respect to the binary operator minus (–) by the rules of arithmetic subtraction because $2 - 3 = -1$ and $2, 3 \in N$, while $(-1) \notin N$.

2. *Associative law.* A binary operator $*$ on a set S is said to be associative whenever:

$$(x * y) * z = x * (y * z) \quad \text{for all } x, y, z \in S$$

3. *Commutative law.* A binary operator $*$ on a set S is said to be commutative whenever:

$$x * y = y * x \quad \text{for all } x, y \in S$$

4. *Identity element.* A set S is said to have an identity element with respect to a binary operation $*$ on S if there exists an element $e \in S$ with the property:

$$e * x = x * e = x \quad \text{for every } x \in S$$

Example. The element 0 is an identity element with respect to operation + on the set of integers $I = \{\ldots -3, -2, -1, 0, 1, 2, 3, \ldots\}$ since

$$x + 0 = 0 + x = x \quad \text{for any } x \in I$$

The set of natural numbers N has no identity element since 0 is excluded from the set.

5. *Inverse.* A set S having the identity element e with respect to a binary operator $*$ is said to have an inverse whenever, for every $x \in$ S, there exists an element $y \in S$ such that

$$x * y = e$$

Example. In the set of integers I with $e = 0$, the inverse of an element (a) is $(-a)$ since $a + (-a) = 0$.

6. *Distributive law.* If $*$ and $\cdot$ are two binary operators on a set S, $*$ is said to be distributive over $\cdot$ whenever

$$x * (y \cdot z) = (x * y) \cdot (x * z)$$

An example of an algebraic structure is a *field*. A field is a set of elements, together with two binary operators, each having properties 1 to 5 and both operators combined to give property 6. The set of real numbers together with the binary operators + and $\cdot$ form the field of real numbers. The field of real numbers is the basis for arithmetic and ordinary algebra. The operators and postulates have the following meanings:

The binary operator + defines addition.
The additive identity is 0.

The additive inverse defines subtraction.
The binary operator $\cdot$ defines multiplication.
The multiplicative identity is 1.
The multiplicative inverse of $a = 1/a$ defines division; i.e., $a \cdot 1/a = 1$.
The only distributive law applicable is that of $\cdot$ over +:

$$a \cdot (b + c) = (a \cdot b) + (a \cdot c)$$

2-2 AXIOMATIC DEFINITION OF BOOLEAN ALGEBRA

George Boole (1) in 1854 introduced a systematic treatment of logic and developed for this purpose an algebraic system now called *Boolean algebra.* C. E. Shannon (2) in 1938 introduced a two-valued Boolean algebra called *switching algebra*, in which he demonstrated that the properties of bistable electrical switching circuits can be represented by this algebra. For the formal definition of Boolean algebra, we shall employ the postulates formulated by E. V. Huntington (3) in 1904. These postulates or axioms are not unique for defining Boolean algebra. Other sets of postulates have been used.* Boolean algebra is an algebraic structure defined on a set of elements B together with two binary operators + and $\cdot$ provided the following (Huntington) postulates are satisfied:

1(a) Closure with respect to the operator +.

(b) Closure with respect to the operator $\cdot$.

2(a) An identity element with respect to +, designated by 0: $x + 0 = 0 + x = x$.

(b) An identity element with respect to $\cdot$, designated by 1: $x \cdot 1 = 1 \cdot x = x$.

3(a) Commutative with respect to +: $x + y = y + x$.

(b) Commutative with respect to $\cdot$: $x \cdot y = y \cdot x$.

4(a) $\cdot$ is distributive over +: $x \cdot (y + z) = (x \cdot y) + (x \cdot z)$.

(b) + is distributive over $\cdot$: $x + (y \cdot z) = (x + y) \cdot (x + z)$.

5 For every element $x \in B$, there exists an element $x' \in B$ (called the complement of x) such that

$$\text{(a)}\quad x + x' = 1 \text{ and (b)}\quad x \cdot x' = 0$$

6 There exists at least two elements $x, y \in B$ such that $x \neq y$.
Comparing Boolean algebra with arithmetic and ordinary algebra (the field of real numbers) we note the following differences:

*See for example Birkoff and Bartee (4), Chap. 5.

(a) Huntington postulates do not include the associative law. However, this law holds for Boolean algebra and can be derived (for both operators) from the other postulates.

(b) The distributive law of + over $\cdot$; i.e., $x + (y \cdot z) = (x + y) \cdot (x + z)$, is valid for Boolean algebra, but not for ordinary algebra.

(c) Boolean algebra does not have additive or multiplicative inverses; therefore, there are no subtraction or division operations.

(d) Postulate 5 defines an operator called complement which is not available in ordinary algebra.

(e) Ordinary algebra deals with the real numbers, which constitute an infinite set of elements. Boolean algebra deals with the as yet undefined set of elements *B*, but in the two-valued Boolean algebra defined below (and of interest in our subsequent use of this algebra), *B* is defined to be a set with only two elements, 0 and 1.

Boolean algebra resembles ordinary algebra in some respects. The choice of symbols + and $\cdot$ is intentional to facilitate Boolean algebraic manipulations by persons already familiar with ordinary algebra. Although one can use some of his knowledge from ordinary algebra to deal with Boolean algebra, the beginner must be careful not to substitute the rules of ordinary algebra where they are not applicable.

It is important to distinguish between the elements of the set of an algebraic structure and the variables of an algebraic system. For example, the elements of the field of real numbers are numbers, while variables such as *a, b, c*, etc., used in ordinary algebra, are symbols that stand for real numbers. Similarly in Boolean algebra, one defines the elements of the set *B* and variables such as *x, y, z* are merely symbols that represent the elements. At this point, it is important to realize that in order to have a Boolean algebra, one must show:

(a) the elements of the set *B*,
(b) the rules of operation for the two binary operators, and
(c) that the set of elements *B* together with the two operators satisfy the six Huntington postulates.

One can formulate many Boolean algebras, depending on the choice of elements of *B* and the rules of operation.* In our subsequent work, we shall deal only with a two-valued Boolean algebra; i.e., one with only two elements. Two-valued Boolean algebra has applications in set theory (the algebra of classes) and in propositional logic. Our interest here is with the application of Boolean algebra to gate-type circuits.

*See for example Hohn (6), Whitesitt (7), or Birkhoff and Bartee (4).

Two-Valued Boolean Algebra

A two-valued Boolean algebra is defined on a set of two elements, $B = \{0, 1\}$, with rules for the two binary operators + and $\cdot$ as shown in the following operator tables (the rule for the complement operator is for verification of postulate 5):

x y	$x \cdot y$
0 0	0
0 1	0
1 0	0
1 1	1

x y	$x + y$
0 0	0
0 1	1
1 0	1
1 1	1

x	x'
0	1
1	0

These rules are exactly the same as the AND, OR, and NOT operations, respectively, defined in Table 1-6. We must now show that the Huntington postulates are valid for the set $B = \{0, 1\}$ and the two binary operators defined above.

1. *Closure* is obvious from the tables since the result of each operation is either 1 or 0 and $1, 0 \in B$.

2. From the tables we see that

   (a) $0 + 0 = 0 \qquad 0 + 1 = 1 + 0 = 1$
   (b) $1 \cdot 1 = 1 \qquad 1 \cdot 0 = 0 \cdot 1 = 0$

   which establishes the two *identity elements* 0 for + and 1 for $\cdot$ as defined by postulate 2.

3. The *commutative* laws are obvious from the symmetry of the binary operator tables.

4. (a) The distributive law $x \cdot (y + z) = (x \cdot y) + (x \cdot z)$ can be shown to hold true from the operator tables by forming a truth table of all possible values of x, y, and z. For each combination we derive $x \cdot (y + z)$ and show that the value is the same as $(x \cdot y) + (x \cdot z)$.

x y z	$y + z$	$x \cdot (y + z)$	$x \cdot y$	$x \cdot z$	$(x \cdot y) + (x \cdot z)$
0 0 0	0	0	0	0	0
0 0 1	1	0	0	0	0
0 1 0	1	0	0	0	0
0 1 1	1	0	0	0	0
1 0 0	0	0	0	0	0
1 0 1	1	1	0	1	1
1 1 0	1	1	1	0	1
1 1 1	1	1	1	1	1

(b) The *distributive* law of + over · can be shown to hold true by means of a truth table similar to the one above.

5. From the complement table it is easily shown that:

   (a) $x + x' = 1$, since $0 + 0' = 0 + 1 = 1$ and $1 + 1' = 1 + 0 = 1$

   (b) $x \cdot x' = 0$, since $0 \cdot 0' = 0 \cdot 1 = 0$ and $1 \cdot 1' = 1 \cdot 0 = 0$ which verifies postulate 5.

6. Postulate 6 is satisfied because the two-valued Boolean algebra has two distinct elements 1 and 0 with $1 \neq 0$.

We have just established a two-valued Boolean algebra having a set of two elements, 1 and 0, two binary operators with operation rules equivalent to the AND and OR operations, and a complement operator equivalent to the NOT operator. Thus, Boolean algebra has been defined in a formal mathematical manner and has been shown to be equivalent to the binary logic presented heuristically in Sec. 1-8. The heuristic presentation is helpful in understanding the application of Boolean algebra to gate-type circuits. The formal presentation is necessary for developing the theorems and properties of the algebraic system. The two-valued Boolean algebra defined in this section is also called "switching algebra" by engineers. In order to emphasize the similarities between two-valued Boolean algebra and other binary systems, this algebra was called "binary logic" in Sec. 1-8. From here on, we shall drop the adjective "two-valued" from Boolean algebra in subsequent discussion.

2-3 BASIC THEOREMS AND PROPERTIES OF BOOLEAN ALGEBRA

Duality

The Huntington postulates have been listed in pairs and designated by part (a) and part (b). One part may be obtained from the other if the binary operators and the identity elements are interchanged. This important property of Boolean algebra is called the *duality principle.* It states that every algebraic expression deducible from the postulates of Boolean algebra remains valid if the operators and identity elements are interchanged. In a two-valued Boolean algebra, the identity elements and the elements of the set B are the same; 1 and 0. The duality principle has many applications. If the *dual* of an algebraic expression is desired, we simply interchange OR and AND operators and replace 1's by 0's and 0's by 1's.

Basic Theorems

Table 2-1 lists six theorems of Boolean algebra and four of its postulates. The notation is simplified by omitting the · whenever this does not lead to

Table 2-1 Postulates and Theorems of Boolean Algebra

KNOW

Post. 2	(a) $x + 0 = x$	(b) $x \cdot 1 = x$
Post. 5	(a) $x + x' = 1$	(b) $x \cdot x' = 0$
Theorem 1	(a) $x + x = x$	(b) $x \cdot x = x$
Theorem 2	(a) $x + 1 = 1$	(b) $x \cdot 0 = 0$
Theorem 3 involution:	$(x')' = x$	
Post. 3 commutative:	(a) $x + y = y + x$	(b) $xy = yx$
Theorem 4 associative:	(a) $x + (y + z) = (x + y) + z$	(b) $x(yz) = (xy)z$
Post. 4 distributive:	(a) $x(y + z) = xy + xz$	(b) $x + yz = (x + y)(x + z)$
Theorem 5 DeMorgan:	(a) $(x + y)' = x'y'$	(b) $(xy)' = x' + y'$
Theorem 6 absorption:	(a) $x + xy = x$	(b) $x(x + y) = x$

confusion. The theorems and postulates listed are the most basic relations of Boolean algebra. The reader is advised to become familiar with them as soon as possible. The theorems, like the postulates, are listed in pairs; each relation is the dual of the one paired with it. The postulates are basic axioms of the algebraic structure and need no proof. The theorems must be proven from the postulates. The proofs of the theorems with one variable are presented below. At the right-hand side of the page is listed the number of the postulate which justifies each step of the proof.

THEOREM 1(a): $x + x = x$

	by postulate:
$x + x = (x + x) \cdot 1$	2(b)
$= (x + x)(x + x')$	5(a)
$= x + xx'$	4(b)
$= x + 0$	5(b)
$= x$	2(a)

THEOREM 1(b): $x \cdot x = x$

	by postulate:
$x \cdot x = xx + 0$	2(a)
$= xx + xx'$	5(b)
$= x(x + x')$	4(a)
$= x \cdot 1$	5(a)
$= x$	2(b)

Note that theorem 1(b) is the dual of theorem 1(a) and that each step of the proof in part (b) is the dual of part (a). Any dual theorem can be similarly derived from the proof of its corresponding pair.

THEOREM 2(a): $x + 1 = 1$

	by postulate:
$x + 1 = 1 \cdot (x + 1)$	2(b)
$= (x + x')(x + 1)$	5(a)
$= x + x' \cdot 1$	4(b)
$= x + x'$	2(b)
$= 1$	5(a)

THEOREM 2(b): $x \cdot 0 = 0$ by duality.

THEOREM 3: $(x')' = x$

From postulate 5, we have $x + x' = 1$ and $x \cdot x' = 0$, which defines the complement of x. The complement of x' is x and is also $(x')'$. Therefore, since the complement is unique, we have that $(x')' = x$.

The theorems involving two or three variables may be proven algebraically from the postulates and the theorems which have already been proven. Take for example the absorption theorem.

THEOREM 6(a): $x + xy = x$

$x + xy = x \cdot 1 + xy$	by postulate:	2(b)
$= x(1 + y)$		4(a)
$= x(y + 1)$		3(a)
$= x \cdot 1$	by theorem:	2(a)
$= x$	by postulate:	2(b)

THEOREM 6(b): $x(x + y) = x$ by duality.

The theorems of Boolean algebra can be shown to hold true by means of truth tables. In truth tables, both sides of the relation are checked to yield identical results for all possible combinations of variables involved. The following truth table verifies the first absorption theorem.

x	y	xy	$x + xy$
0	0	0	0
0	1	0	0
1	0	0	1
1	1	1	1

(The x and $x + xy$ columns are marked =.)

The algebraic proofs of the associative law and De Morgan's theorem are long and will not be shown here. However, their validity is easily shown with truth tables. For example, the truth table for the first De Morgan's theorem $(x + y)' = x'y'$ is shown below.

x	y	$x + y$	$(x + y)'$	x'	y'	$x'y'$
0	0	0	1	1	1	1
0	1	1	0	1	0	0
1	0	1	0	0	1	0
1	1	1	0	0	0	0

Operator Precedence

The operator precedence for evaluating Boolean expressions is: (1) parentheses, (2) NOT, (3) AND, (4) OR. In other words, the expression inside the parentheses must be evaluated before all other operations. The next operation that holds precedence is the complement, then follows the AND and finally the OR. As an example, consider the truth table for De Morgan's theorem. The left side of the expression is $(x + y)'$. Therefore, the expression inside the parentheses is evaluated first and the result then complemented. The right side of the expression is $x'y'$. Therefore, the complement of x and the complement of y are both evaluated first and the result is then ANDed. Note that in ordinary arithmetic the same precedence holds (except for the complement) when multiplication and addition are replaced by AND and OR, respectively.

Venn Diagram

A helpful illustration that may be used to visualize the relationship among the variables of a Boolean expression is the *Venn diagram.* This diagram consists of a rectangle such as shown in Fig. 2-1, inside of which are drawn overlapping circles, one for each variable. Each circle is labeled by a variable. We designate all points inside a circle as belonging to the named variable and all points outside a circle as not belonging to the variable. Take for example the circle labeled x. If we are inside the circle, we say that $x = 1$; when outside, we say $x = 0$. Now, with two overlapping circles, there are four distinct areas inside the rectangle: The area not belonging to either x or y $(x'y')$, the area inside circle y but outside x $(x'y)$, the area inside circle x but outside y (xy'), and the area inside both circles (xy).

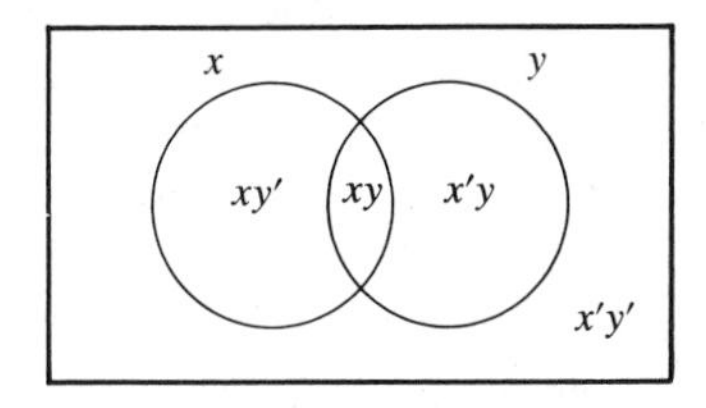

Figure 2-1 Venn diagram for two variables

Venn diagrams may be used to illustrate the postulates of Boolean algebra or to show the validity of theorems. Figure 2-2, for example, illustrates that the area belonging to xy is inside the circle x, and therefore, $x + xy = x$. Figure 2-3 illustrates the distributive law $x(y + z) = xy + xz$. In this diagram we have three overlapping circles, one for each of the

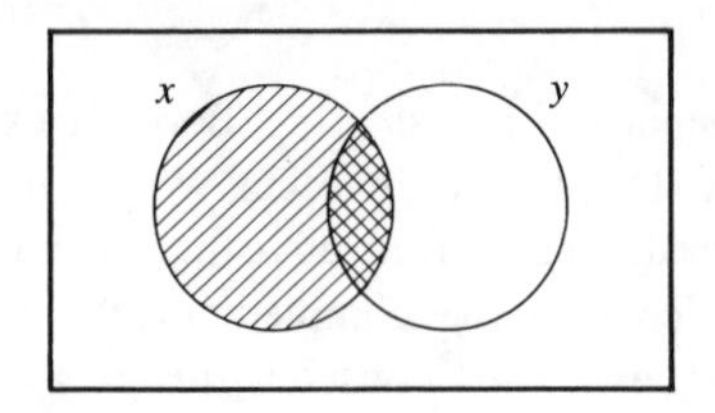

Figure 2-2 Venn diagram illustration $x = xy + x$

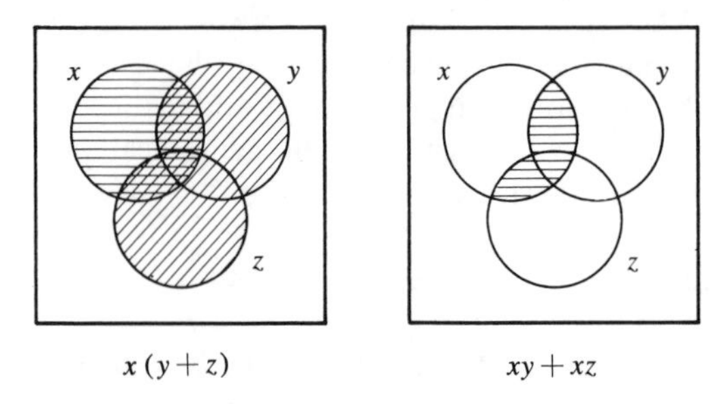

Figure 2-3 Venn diagram illustration of the distributive law

variables x, y, and z. It is possible to distinguish eight distinct areas in a three-variable Venn diagram. For this particular example, the distributive law is demonstrated by noting that the area intersecting the circle x with the area enclosing y or z is the same area belonging to xy or xz.

2-4 BOOLEAN FUNCTIONS

A binary variable can take the value of 0 or 1. A Boolean function is an expression formed with binary variables, the two binary operators OR and AND, the unary operator NOT, parentheses, and equal sign. For a given value of the variables, the function can be either 0 or 1. Consider for example the Boolean function:

$$F_1 = xyz'$$

The function F_1 is equal to 1 if $x = 1$ *and* $y = 1$ *and* $z' = 1$, otherwise $F_1 = 0$. The above is an example of a Boolean function represented as an algebraic expression. A Boolean function may also be represented in a truth table. To represent a function in a truth table, we need a list of the 2^n combinations of 1's and 0's of the n binary variables, and a column showing the combinations for which the function is equal to 1 or 0. As shown in Table 2-2, there are eight possible distinct combinations for assigning bits to three variables. The column labeled F_1 is either a 0 or a 1 for each of these combinations. The table shows that the function F_1 is equal

Table 2-2 Truth Tables for $F_1 = xyz'$, $F_2 = x + y'z$, $F_3 = x'y'z + x'yz + xy'$, and $F_4 = xy' + x'z$

x	y	z	F_1	F_2	F_3	F_4
0	0	0	0	0	0	0
0	0	1	0	1	1	1
0	1	0	0	0	0	0
0	1	1	0	0	1	1
1	0	0	0	1	1	1
1	0	1	0	1	1	1
1	1	0	1	1	0	0
1	1	1	0	1	0	0

to 1 only when $x = 1$, $y = 1$, and $z = 0$. It is equal to 0 otherwise. (Note that the statement $z' = 1$ is equivalent to saying that $z = 0$.) Consider now the function:

$$F_2 = x + y'z$$

$F_2 = 1$ if $x = 1$ or if $y = 0$, while $z = 1$. In Table 2-2, $x = 1$ in the last four rows, and $yz = 01$ in rows 001 and 101. The latter combination applies also for $x = 1$. Therefore, there are five combinations that make $F_2 = 1$. As a third example, the function

$$F_3 = x'y'z + x'yz + xy'$$

is shown in Table 2-2 with four 1's and four 0's. F_4 is the same as F_3 and is considered below.

Any Boolean function can be represented in a truth table. The number of rows in the table is 2^n, where n is the number of binary variables in the function. The 1's and 0's combinations for each row is easily obtained from the binary numbers by counting from 0 to 2^n-1. For each row of the table, there is a value for the function equal to either 1 or 0. The question now arises, is an algebraic expression of a given Boolean function unique? In other words, is it possible to find two algebraic expressions that specify the same function? The answer to this question is yes. As a matter of fact, the manipulation of Boolean algebra is applied mostly to the problem of finding simpler expressions for the same function. Consider for example the function

$$F_4 = xy' + x'z$$

From Table 2-2, we find that F_4 is the same as F_3, since both have identical 1's and 0's for each combination of values of the three binary variables. In general, two functions of n binary variables are said to be equal if they have the same value for all possible 2^n combinations of the n variables.

A Boolean function may be transformed from an algebraic expression into a logic diagram composed of AND, OR, and NOT gates. The implementation of the four functions introduced in the previous discussion is shown in Fig. 2-4. The logic diagram includes an inverter circuit for every variable present in its complement form. (The inverter is unnecessary if the complement of the variable is available.) There is an AND gate for each term in the expression, and an OR gate is used to combine two or more terms. From the diagrams it is obvious that the implementation of F_4 requires less gates and less inputs than F_3. Since F_4 and F_3 are equal Boolean functions, it is more economical to implement the F_4 form than the F_3 form. To find simpler circuits, one must know how to manipulate Boolean functions to obtain equal and simpler expressions. What constitutes the best form of a Boolean function depends on the particular application. In this section, consideration is given to the criterion of equipment minimization.

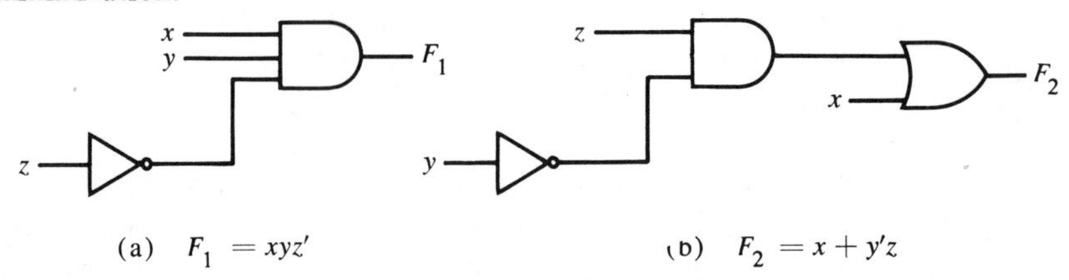

(a) $F_1 = xyz'$

(b) $F_2 = x + y'z$

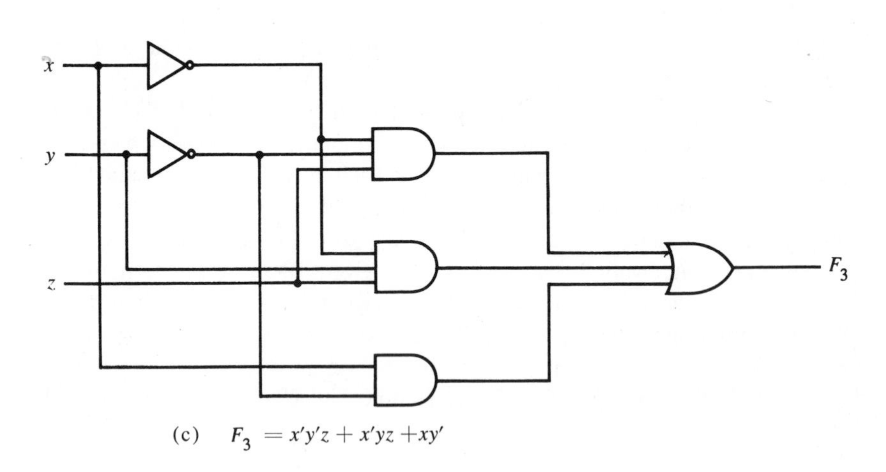

(c) $F_3 = x'y'z + x'yz + xy'$

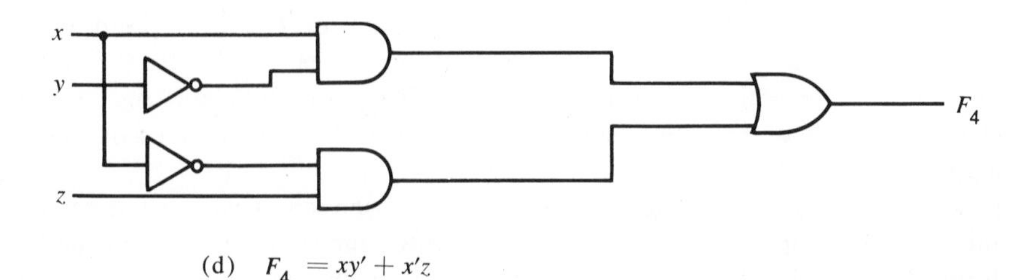

(d) $F_4 = xy' + x'z$

Figure 2-4 Implementation of Boolean functions with gates

Algebraic Manipulation

A *literal* is a primed or unprimed variable. When a Boolean function is implemented with logic gates, each literal in the function designates an input to a gate, and each term is implemented with a gate. The minimization of the number of literals and the number of terms will result in a circuit with less equipment. It is not always possible to minimize both simultaneously; usually, further criteria must be available. At the moment, we shall narrow the minimization criterion to literal minimization. We shall discuss other criteria in Ch. 5. The number of literals in a Boolean function can be minimized by algebraic manipulations. Unfortunately, there are no specific rules to follow that will guarantee the final answer. The only method available is a cut-and-try procedure employing the postulates, the basic theorems, and any other manipulation method which becomes familiar with use. The following examples illustrate this procedure.

EXAMPLE 2-1. Simplify the following Boolean functions to a minimum number of literals.

1. $x + x'y = (x + x')(x + y) = 1 \cdot (x + y) = x + y$
2. $x(x' + y) = xx' + xy = 0 + xy = xy$
3. $x'y'z + x'yz + xy' = x'z(y' + y) + xy' = x'z + xy'$
4. $xy + x'z + yz = xy + x'z + yz(x + x')$
   $= xy + x'z + xyz + x'yz$
   $= xy(1 + z) + x'z(1 + y)$
   $= xy + x'z$
5. $(x + y)(x' + z)(y + z) = (x + y)(x' + z)$ by duality from 4.

Functions 1 and 2 are the dual of each other and use dual expressions in corresponding steps. Function 3 shows the equality of the functions F_3 and F_4 discussed previously. The fourth illustrates the fact that an increase in the number of literals sometimes leads to a final simpler expression. Function 5 is not minimized directly but can be derived from the dual of the steps used to derive 4.

Complement of a Function

The complement of a function F is F' and is obtained from an interchange of 0's for 1's and 1's for 0's in the truth table. The complement of a function may be derived algebraically through De Morgan's theorem. This pair of theorems is listed in Table 2-1 for two variables. De Morgan's theorems can be extended to three or more variables. The three-variable form of the first De Morgan's theorem is derived below. The postulates and theorems are those listed in Table 2-1.

$$(A + B + C)' = (A + X)' \quad \text{Let } B + C = X$$
$$= A'X' \quad \text{by theorem 5(a) (De Morgan)}$$
$$= A' \cdot (B + C)' \quad \text{substitute } B + C = X$$
$$= A' \cdot (B'C') \quad \text{by theorem 5(a) (De Morgan)}$$
$$= A'B'C' \quad \text{by theorem 4(b) (associative)}$$

De Morgan's theorems for any number of variables resemble in form the two-variable case and can be derived by successive substitutions similar to the method used in the above derivation. These theorems can be generalized as follows:

$$(A + B + C + D + \cdots + F)' = A'B'C'D' \ldots F'$$

$$(ABCD \cdots F)' = A' + B' + C' + D' + \ldots + F'$$

The generalized form of De Morgan's theorem states that the complement of a function is obtained by interchanging AND and OR operators and complementing each literal.

EXAMPLE 2-2. Find the complement of the functions $F_1 = x'yz' + x'y'z$ and $F_2 = x(y'z' + yz)$.

Applying De Morgan's theorem as many times as necessary, the complements are obtained as follows:

$$F_1' = (x'yz' + x'y'z)' = (x'yz')' \cdot (x'y'z)' = (x + y' + z)(x + y + z')$$

$$F_2' = [x(y'z' + yz)]' = x' + (y'z' + yz)' = x' + (y'z')' \cdot (yz)'$$
$$= x' + (y + z)(y' + z')$$

A simpler procedure for deriving the complement of a function is to take the dual of the function and complement each literal. This method follows from the generalized De Morgan's theorem. Remember that the dual of a function is obtained from the interchange of AND and OR operators and 1's and 0's.

EXAMPLE 2-3. Find the complement of the functions F_1 and F_2 of Ex. 2-2 by taking their dual and complementing each literal.

1. $F_1 = x'yz' + x'y'z$
   The dual of F_1 is: $(x' + y + z')(x' + y' + z)$
   Complement each literal: $(x + y' + z)(x + y + z') = F_1'$

2. $F_2 = x(y'z' + yz)$
   The dual of F_2 is: $x + (y' + z')(y + z)$
   Complement each literal: $x' + (y + z)(y' + z') = F_2'$

2-5 CANONICAL FORMS

Minterms and Maxterms

A binary variable may appear either in its normal form (x) or in its complement form (x'). Now consider two binary variables x and y combined with an AND operation. Since each variable may appear in either form, there are four possible combinations: $x'y'$, $x'y$, xy', and xy. Each of these four AND terms represents one of the distinct areas in the Venn diagram of Fig. 2-1 and is called a *minterm*, or a *standard product*. In a similar manner, n variables can be combined to form 2^n minterms. The 2^n different minterms may be determined by a method similar to the one shown in Table 2-3 for three variables. The binary numbers from 0 to $2^n - 1$ are listed under the n variables. Each minterm is obtained from an AND term of the n variables, with each variable being primed if the corresponding bit of the binary number is a 0 and unprimed if a 1. A symbol for

Table 2-3 Minterms and Maxterms for Three Binary Variables

	Minterms		*Maxterms*	
$x\ y\ z$	*Term*	*Designation*	*Term*	*Designation*
0 0 0	$x'y'z'$	m_0	$x + y + z$	M_0
0 0 1	$x'y'z$	m_1	$x + y + z'$	M_1
0 1 0	$x'yz'$	m_2	$x + y' + z$	M_2
0 1 1	$x'yz$	m_3	$x + y' + z'$	M_3
1 0 0	$xy'z'$	m_4	$x' + y + z$	M_4
1 0 1	$xy'z$	m_5	$x' + y + z'$	M_5
1 1 0	xyz'	m_6	$x' + y' + z$	M_6
1 1 1	xyz	m_7	$x' + y' + z'$	M_7

each minterm is also shown in the table and is of the form m_j, where j denotes the decimal equivalent of the binary number of the minterm designated.

In a similar fashion, n variables forming an OR term, with each variable being primed or unprimed, provide 2^n possible combinations, called *maxterms*, or *standard sums*. The eight maxterms for three variables, together with their symbolic designation, are listed in Table 2-3. Any 2^n maxterms for n variables may be determined similarly. Each maxterm is obtained from an OR term of the n variables, with each variable being unprimed if the corresponding bit is a 0 and primed if a 1*. Note that each

*Some books define a maxterm as an OR term of the n variables, with each variable being unprimed if the bit is a 1 and primed if a 0. The definition adopted in this book is preferable as it leads to simpler conversions between maxterm and minterm type functions.

maxterm is the complement of its corresponding minterm and vice versa.

A Boolean function may be expressed algebraically from a given truth table by forming a minterm for each combination of the variables which produces a 1 in the function, and then taking the OR of all those terms. For example, the function f_1 in Table 2-4 is determined by expressing the combinations 001, 100, and 111 as $x'y'z$, $xy'z'$, and xyz, respectively. Since each one of these minterms results in $f_1 = 1$, we should have

$$f_1 = x'y'z + xy'z' + xyz = m_1 + m_4 + m_7$$

Similarly, it may be easily verified that

$$f_2 = x'yz + xy'z + xyz' + xyz = m_3 + m_5 + m_6 + m_7$$

These examples demonstrate an important property of Boolean algebra: any Boolean function can be expressed as a sum of minterms (by "sum" is meant the ORing of terms).

Now consider the complement of a Boolean function. It may be read from the truth table by forming a minterm for each combination that produces a 0 in the function and then ORing those terms. The complement of f_1 is read as:

$$f_1' = x'y'z' + x'yz' + x'yz + xy'z + xyz'$$

If we take the complement of f_1', we obtain the function f_1:

$$f_1 = (x + y + z)(x + y' + z)(x + y' + z')(x' + y + z')(x' + y' + z)$$
$$= M_0 \cdot M_2 \cdot M_3 \cdot M_5 \cdot M_6$$

Table 2-4 Functions of Three Variables

$x\ y\ z$	*Function* f_1	*Function* f_2
0 0 0	0	0
0 0 1	1	0
0 1 0	0	0
0 1 1	0	1
1 0 0	1	0
1 0 1	0	1
1 1 0	0	1
1 1 1	1	1

Similarly, it is possible to read the expression for f_2 from the table:

$$f_2 = (x + y + z)(x + y + z')(x + y' + z)(x' + y + z)$$
$$= M_0 M_1 M_2 M_4$$

These examples demonstrate a second important property of Boolean algebra: any Boolean function can be expressed as a product of maxterms (by "product" is meant the ANDing of terms). The procedure for obtaining the product of maxterms directly from the truth table is as follows. Form a maxterm for each combination of the variables which produces a 0 in the function, and then form the AND of all those maxterms. Boolean functions expressed as sum of minterms or product of maxterms are said to be in *canonical form*.

Sum of Minterms

It was previously stated that for n binary variables, one can obtain 2^n distinct minterms, and that any Boolean function can be expressed as a sum of minterms. The minterms whose sum defines the Boolean function are those that give the 1's of the function in a truth table. Since the function can be either 1 or 0 for each minterm, and since there are 2^n minterms, one can calculate the possible functions that can be formed with n variables to be 2^{2^n}. It is sometimes convenient to express the Boolean function in its sum of minterms form. If not in this form, it can be made so by first expanding the expression into a sum of AND terms. Each term is then inspected to see if it contains all the variables. If it misses one or more variables, it is ANDed with an expression like $x + x'$, where x is one of the missing variables. The following example clarifies this procedure.

EXAMPLE 2-4. Express the Boolean function $F = A + B'C$ in a sum of minterms. The function has three variables A, B, and C. The first term A is missing two variables, therefore:

$$A = A(B + B') = AB + AB'$$

This is still missing one variable:

$$A = AB(C + C') + AB'(C + C') = ABC + ABC' + AB'C + AB'C'$$

The second term $B'C$ is missing one variable:

$$B'C = B'C(A + A') = AB'C + A'B'C$$

Combining all terms we have:

$$F = A + B'C = ABC + ABC' + AB'C + AB'C' + AB'C + A'B'C$$

But $AB'C$ appears twice and according to theorem 1 ($x + x = x$), it is possible to remove one of them. Rearranging the minterms in ascending order, we finally obtain:

$$F = A'B'C + AB'C' + AB'C + ABC' + ABC = m_1 + m_4 + m_5 + m_6 + m_7$$

It is sometimes convenient to express the Boolean function, when in its sum of minterms, in the following short notation:

$$F(A, B, C) = \Sigma(1, 4, 5, 6, 7)$$

The summation symbol Σ stands for the ORing of terms; the numbers following it are the minterms of the function. The letters in parentheses with F form a list of the variables in the order taken when the minterm is converted to an AND term.

Product of Maxterms

Each of the 2^{2^n} functions of n binary variables can be also expressed as a product of maxterms. To express the Boolean function as a product of maxterms it must first be brought into a form of OR terms. This may be done by using the distributive law $x + yz = (x + y)(x + z)$. Then any missing variable x in each OR term is ORed with xx'. This procedure is clarified by the following example.

EXAMPLE 2-5. Express the Boolean function $F = xy + x'z$ in a product of maxterm form.

First convert the function into OR terms using the distributive law:

$$F = xy + x'z = (xy + x')(xy + z)$$
$$= (x + x')(y + x')(x + z)(y + z)$$
$$= (x' + y)(x + z)(y + z)$$

The function has three variables: x, y, and z. Each OR term is missing one variable, therefore:

$$x' + y = x' + y + zz' = (x' + y + z)(x' + y + z')$$
$$x + z = x + z + yy' = (x + y + z)(x + y' + z)$$
$$y + z = y + z + xx' = (x + y + z)(x' + y + z)$$

Combining all the terms and removing those that appear more than once, we finally obtain:

$$F = (x + y + z)(x + y' + z)(x' + y + z)(x' + y + z')$$
$$= M_0 M_2 M_4 M_5$$

A convenient way to express this function is as follows:

$$F(x, y, z) = \Pi(0, 2, 4, 5)$$

The product symbol denotes the ANDing of maxterms; the numbers are the maxterms of the function.

Conversion Between Canonical Forms

The complement of a function expressed as the sum of minterms equals the sum of minterms missing from the original function. This is because the original function is expressed by those minterms that make the function equal to 1, while its complement is a 1 for those minterms that the function is a 0. As an example, the function

$$F(A, B, C) = \Sigma(1, 4, 5, 6, 7)$$

has a complement that can be expressed as

$$F'(A, B, C) = \Sigma(0, 2, 3) = m_0 + m_2 + m_3$$

Now, if we take the complement of F' by De Morgan's theorem we obtain F back in a different form:

$$F = (m_0 + m_2 + m_3)' = m_0' \cdot m_2' \cdot m_3' = M_0 M_2 M_3 = \Pi(0, 2, 3)$$

The last conversion follows from the definition of minterms and maxterms as shown in Table 2-3. From the table, it is clear that the following relation holds true:

$$m_j' = M_j$$

That is, the maxterm with subscript j is a complement of the minterm with the same subscript j, and vice versa.

The last example has demonstrated the conversion between a function expressed in sum of minterms and its equivalent in product of maxterms. A similar argument will show that the conversion between the product of maxterms and the sum of minterms is similar. We now state a general conversion procedure. To convert from one canonical form to another, interchange the symbol Σ and Π and list those numbers missing from the original form. As another example, the function

$$F(x, y, z) = \Pi(0, 2, 4, 5)$$

is expressed in the product of maxterm form. Its conversion to sum of minterms is:

$$F(x, y, z) = \Sigma(1, 3, 6, 7)$$

Note that in order to find the missing terms, one must realize that the total number of minterms or maxterms is 2^n, where n is the number of binary variables in the function.

Standard Forms

The two canonical forms of Boolean algebra are basic forms that one obtains from reading a function from the truth table. These forms are very seldom the ones with the least number of literals. This is because each minterm or maxterm must contain, by definition, all the variables either primed or unprimed. A Boolean function expressed with terms containing any number of literals is said to be in a *standard form.* There are two standard forms: The sum of products (*disjunctive* form) and the product of sums (*conjunctive* form). An example of a function expressed in a sum of products is:

$$F_1 = xy + xyz' + y'$$

The product terms are AND terms of one, two, or any number of literals. The *sum* denotes the ORing of these terms. An example of a function expressed in a product of sums is:

$$F_2 = x\,(y' + z)\,(x' + y + z')$$

Again, the sum terms are OR terms of one or more variables and the *product* denotes the ANDing of these terms.

2-6 OTHER BINARY OPERATORS

When the binary operators AND and OR are placed between two variables, they form two Boolean functions, $x \cdot y$ and $x + y$, respectively. We have stated previously that there are 2^{2^n} functions for n variables. Therefore, these functions are only two out of a total of sixteen possible functions formed with two variables x and y. All 16 functions are listed in Table 2-5. In this table, each of the columns represents the truth table of one function F_j for $j = 0$ to 15. Some of the functions are shown with an operator symbol and are also known by a characteristic name as indicated below. The following list of 16 functions is expressed algebraically. When applicable, the function is also written with its special symbol and its characteristic name mentioned.

Table 2-5 Truth Tables for the Sixteen Functions of Two Binary Variables

x	y	F_0	F_1	F_2	F_3	F_4	F_5	F_6	F_7	F_8	F_9	F_{10}	F_{11}	F_{12}	F_{13}	F_{14}	F_{15}
0	0	0	0	0	0	0	0	0	0	1	1	1	1	1	1	1	1
0	1	0	0	0	0	1	1	1	1	0	0	0	0	1	1	1	1
1	0	0	0	1	1	0	0	1	1	0	0	1	1	0	0	1	1
1	1	0	1	0	1	0	1	0	1	0	1	0	1	0	1	0	1
Operator Symbol			$\cdot$	/		/		$\oplus$	+	$\downarrow$	$\odot$	′	$\subset$	′	$\supset$	$\uparrow$	

$F_0 = 0$		zero, or null function
$F_1 = x \cdot y$	AND	x and y
$F_2 = xy' = x/y$	INHIBITION,	x but not y
$F_3 = x$		the function is equal to x
$F_4 = x'y = y/x$	INHIBITION,	y but not x
$F_5 = y$		the function is equal to y
$F_6 = x'y + xy' = x \oplus y$	EXCLUSIVE-OR	x or y but not both
$F_7 = x + y$	OR	x or y
$F_8 = (x + y)' = x \downarrow y$	NOR	not-or
$F_9 = x'y' + xy = x \odot y$	EQUIVALENCE	x equals y
$F_{10} = y'$	COMPLEMENT	not y
$F_{11} = x + y' = x \subset y$	IMPLICATION	if y then x
$F_{12} = x'$	COMPLEMENT	not x
$F_{13} = x' + y = x \supset y$	IMPLICATION	if x then y
$F_{14} = (xy)' = x \uparrow y$	NAND	not-and
$F_{15} = 1$		one, or identity function

By checking Table 2-5 and the list of functions we see that the 16 functions of two variables define a total of eight different binary operators and one unary operator. The binary operators are: AND ($\cdot$), INHIBITION ($/$), EXCLUSIVE-OR ($\oplus$), OR ($+$), NOR ($\downarrow$), EQUIVALENCE ($\odot$), IMPLICATION ($\supset$), and NAND ($\uparrow$). The unary operator is the COMPLEMENT ($'$). The AND, OR, and COMPLEMENT operators have been used previously to define Boolean algebra. Four other functions have frequent use in computer logic. The EXCLUSIVE-OR is similar to OR but excludes the combination of both x and y. The EQUIVALENCE is a function that is a 1 when both variables are equal. Note that the EXCLUSIVE-OR and EQUIVALENCE functions are the complement of each other. The NOR function is the complement of OR, and the NAND is the complement of AND. Both these functions are of particular interest because of the frequent use of transistor circuits that implement these functions (see Ch. 5).

Boolean algebra has been defined in Sec. 2-2 to have two binary operators which we have called AND and OR and a unary operator NOT (complement). From the definitions, we have deduced a number of properties of these operators and now have defined other binary operators in terms of them. There is nothing unique about this procedure. We could have just as well started with the operator NOR ($\downarrow$), for example, and later defined AND, OR, and NOT in terms of it. There are, nevertheless, good reasons for introducing Boolean algebra in the way it has been introduced. The concepts of "and," "or," and "not" are familiar and are used by people to express everyday logical ideas. Moreover, the Huntington postulates reflect the dual nature of the algebra, emphasizing the symmetry of (+) and (·) with respect to each other.

PROBLEMS

2-1. Which of the six basic laws (closure, associative, commutative, identity, inverse, and distributive) are satisfied for the pair of binary operators listed below?

+	0	1	2
0	0	0	0
1	0	1	1
2	0	1	2

·	0	1	2
0	0	1	2
1	1	1	2
2	2	2	2

2-2. Show that the set of three elements $\{0, 1, 2\}$ and the two binary operators + and · as defined by the above table is not a Boolean algebra. State which of the Huntington postulates is not satisfied.

2-3. Demonstrate by means of truth tables the validity of the following theorems of Boolean algebra.

(a) The associative laws.

(b) De Morgan's theorems for three variables.

(c) The distributive law of + over ·.

2-4. Repeat Prob. 2-3 using Venn diagrams.

2-5. Simplify the following Boolean functions to a minimum number of literals.

(a) $xy + xy'$

(b) $(x + y)(x + y')$

(c) $xyz + x'y + xyz'$

(d) $zx + zx'y$

(e) $(A + B)' (A' + B')'$

(f) $y(wz' + wz) + xy$

2-6. Reduce the following Boolean expression to the required number of literals.

(a) $ABC + A'B'C + A'BC + ABC' + A'B'C'$ to five literals

(b) $BC + AC' + AB + BCD$ to four literals

(c) $[(CD)' + A]' + A + CD + AB$ to three literals

(d) $(A + C + D)(A + C + D')(A + C' + D)(A + B')$ to four literals

2-7. Find the complement of the following Boolean functions and reduce them to a minimum number of literals.

(a) $(BC' + A'D)(AB' + CD')$

(b) $B'D + A'BC' + ACD + A'BC$

(c) $[(AB)'A][(AB)'B]$

(d) $AB' + C'D' + A'CD' + DC'(AB + A'B') + DB(AC' + A'C)$

2-8. Given the functions below determine the functions $F_1 + G_1$ and $F_2 +$ G2 and reduce them to a minimum number of literals.

(a) $F_1 = D + ABC' + A'C$

$G_1 = D'(A' + B' + C)(A + C')$

(b) $F_2 = AB' + A'B' + C'D + A'CD'$

$G_2 = C + A'B'C' + AB'C' + BC'D'$

2-9. Obtain the truth table for $F_1 + G_1$ and $F_2 + G_2$ of Prob. 2-8.

2-10. Implement the simplified Boolean functions from Prob. 2-6 with logic gates.

2-11. Given the Boolean function:

$F = xy + x'y' + y'z$

(a) Implement it with AND, OR, and NOT gates.

(b) Implement it with *only* OR and NOT gates.

(c) Implement it with *only* AND and NOT gates.

2-12. Simplify the functions T_1 and T_2 to a minimum number of literals.

A	B	C	T_1	T_2
0	0	0	1	0
0	0	1	1	0
0	1	0	1	0
0	1	1	0	1
1	0	0	0	1
1	0	1	0	1
1	1	0	0	1
1	1	1	0	1

2-13. Express the following functions in a sum of minterms and a product of maxterms.

(a) $F(A, B, C, D) = D(A' + B) + B'D$

(b) $F(w, x, y, z) = y'z + wxy' + wxz' + w'x'z$

(c) $F(A, B, C, D) = (A + B' + C)(A + B')(A + C' + D')$
$(A' + B + C + D')(B + C' + D')$

(d) $F(A, B, C) = (A' + B)(B' + C)$

(e) $F(x, y, z) = 1$

(f) $F(x, y, z) = (xy + z)(y + xz)$

2-14. Convert the following to the other canonical form.

(a) $F(x, y, z) = \Sigma(1, 3, 7)$

(b) $F(A, B, C, D) = \Sigma(0, 2, 6, 11, 13, 14)$

(c) $F(x, y, z) = \Pi(0, 3, 6, 7)$

(d) $F(A, B, C, D) = \Pi(0, 1, 2, 3, 4, 6, 12)$

2-15. What is the difference between canonical form and standard form? Which form is preferable when implementing a Boolean function with gates? Which form is obtained when reading a function from a truth table?

2-16. The sum of all minterms of a Boolean function of n variables is 1.

(a) Prove the above statement for $n = 3$.

(b) Suggest a procedure for a general proof.

2-17. The product of all maxterms of a Boolean function of n variables is 0.

(a) Prove the above statement for $n = 3$.

(b) Suggest a procedure for a general proof. Can we use the duality principle after proving (b) of Prob. 2-16?

2-18. Show that the INHIBITION operation is not commutative.

2-19. Show that the NOR and NAND operations are not associative.

2-20. Show that the dual of the EXCLUSIVE-OR is equal to its complement.

REFERENCES

1. Boole, G., *An Investigation of the Laws of Thought.* New York: Dover Pub., 1954.

2. Shannon, C. E., "A Symbolic Analysis of Relay and Switching Circuits," *Trans. of the AIEE,* Vol. 57 (1938), 713-23.

3. Huntington, E. V., "Sets of Independent Postulates for the Algebra of Logic," *Trans. Am. Math. Soc.,* Vol. 5 (1904), 288-309.

4. Birkhoff, G., and T. C. Bartee, *Modern Applied Algebra.* New York: McGraw-Hill Book Co., 1970.

5. Birkhoff, G., and S. Maclane, *A Survey of Modern Algebra,* 3rd ed. New York: The Macmillan Co., 1965.

6. Hohn, F. E., *Applied Boolean Algebra,* 2nd ed. New York: The Macmillan Co., 1966.

7. Whitesitt, J. E., *Boolean Algebra and its Applications.* Reading, Mass.: Addison-Wesley Pub. Co., 1961.

3 SIMPLIFICATION OF BOOLEAN FUNCTIONS

3-1 THE MAP METHOD

The complexity of the digital logic gates that implement a Boolean function is directly related to the complexity of the algebraic expression from which the function is implemented. Although the truth-table representation of a function is unique, expressed algebraically, it can appear in many different forms. Boolean functions may be simplified by algebraic means as discussed in Sec. 2-4. However, this procedure of minimization is awkward because it lacks specific rules to predict each succeeding step in the manipulative process. The map method provides a simple straightforward procedure for minimizing Boolean functions. This method may be regarded either as a pictorial form of a truth table or as an extension of the Venn diagram. The map method, first proposed by Veitch (1) and slightly modified by Karnaugh (2), is also known as the "Veitch diagram" or the "Karnaugh map."

The map is a diagram made up of squares. Each square represents one minterm. Since any Boolean function can be expressed as a sum of minterms, it follows that a Boolean function is recognized graphically in the map from the area enclosed by those squares whose minterms are included in the function. In fact, the map presents a visual diagram of all possible ways a function may be expressed in a standard form. By recognizing various patterns, the user can derive alternate algebraic expressions for the same function, from which he can select the simplest one. We shall assume that the simplest algebraic expression is any one in a sum of products or product of sums that has a minimum number of literals. (This expression is not necessarily unique.)

3-2 TWO- AND THREE-VARIABLE MAPS

A two-variable map is shown in Fig. 3-1. There are four minterms for two variables, hence the map consists of four squares, one for each minterm. The map is redrawn in (b) to show the relation between the squares and the two variables. The 0's and 1's marked for each row and each column designate the value of variable x and y respectively. Notice that x appears primed in row 0 and unprimed in row 1. Similarly, y appears primed in column 0 and unprimed in column 1.

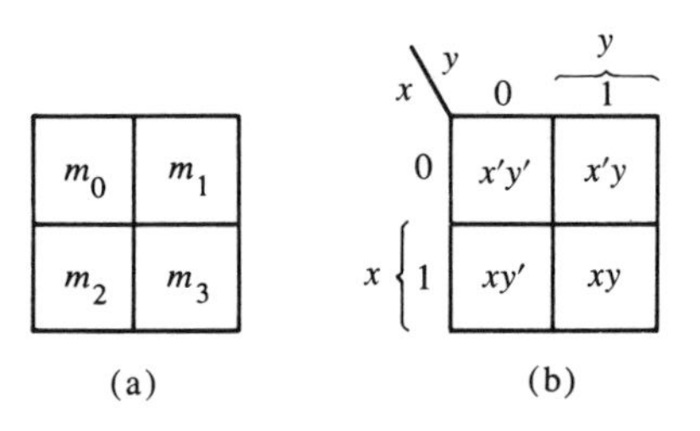

Figure 3-1 Two-variable map

If we mark the squares whose minterms belong to a given function, the two-variable map becomes another useful way for representing any one of the 16 Boolean functions of two variables. As an example, the function xy is shown in Fig. 3-2(a). Since xy is equal to m_3, a 1 is placed inside the square that belongs to m_3. Similarly, the function $x + y$ is represented in the map of Fig. 3-2(b) by three squares marked with 1's. These squares are found from the minterms of the function:

$$x + y = x'y + xy' + xy = m_1 + m_2 + m_3$$

The three squares could have also been determined from the intersection of variable x in the second row and variable y in the second column, which encloses the area belonging to x or y.

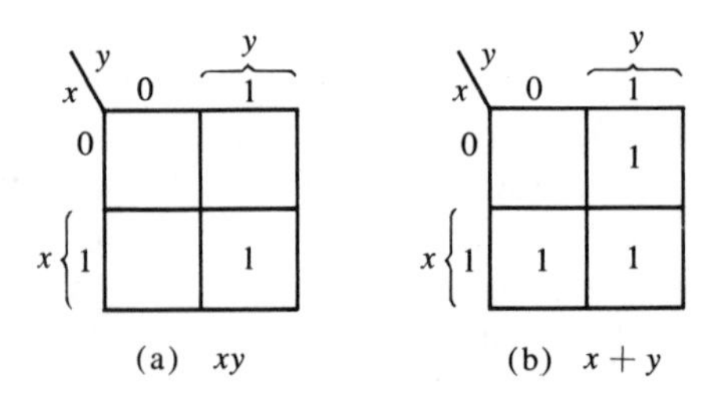

Figure 3-2 Representation of functions in the map

A three-variable map is shown in Fig. 3-3. There are eight minterms for three binary variables. Therefore, a map consists of eight squares. Note that the minterms are arranged, not in a binary sequence, but in a sequence similar to the reflected code listed in Table 1-4. The characteristic of this sequence is that only one bit changes from 1 to 0 or from 0 to 1 in the listing sequence. The map drawn in part (b) is marked with numbers in each row and each column to show the relation between the squares and the three variables. For example, the square assigned to m_5 corresponds to row 1 and column 01. When these two numbers are concatenated, they give the binary number 101, whose decimal equivalent is 5. Another way of looking at square $m_5 = xy'z$ is to consider it to be in the row marked x and the column belonging to $y'z$ (column 01). Note that there are four squares where each variable is equal to 1 and four where each is equal to 0. The variable appears unprimed in those four squares that it is equal to 1 and primed in those squares that it is equal to 0. For convenience, we write the variable with its letter symbol under the four squares where it is unprimed.

In order to understand the usefulness of the map for simplifying Boolean functions, we must recognize the basic property possessed by adjacent squares. Any two adjacent squares in the map differ by only one variable which is primed in one square and unprimed in the other. For example, m_5 and m_7 lie in two adjacent squares. Variable y is primed in m_5 and

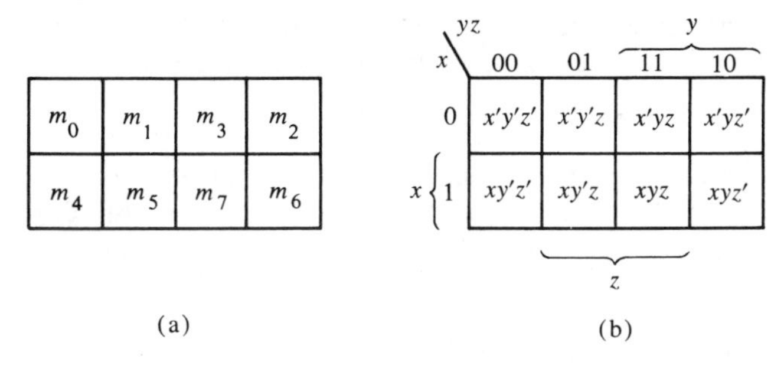

Figure 3-3 Three-variable map

unprimed in m_7, while the other two variables are the same in both squares. From the postulates of Boolean algebra, it follows that the sum of two minterms in adjacent squares can be simplified to a single AND term consisting of only two literals. To clarify this, consider the sum of two adjacent squares such as m_5 and m_7.

$$m_5 + m_7 = xy'z + xyz = xz\,(y' + y) = xz$$

Here the two squares differ by the variable y, which can be removed when the sum of the two minterms is formed. Thus any two minterms in adjacent squares that are ORed together will cause a removal of the different variable. The following example explains the procedure to be used when minimizing a Boolean function with a map.

EXAMPLE 3-1. Simplify the Boolean function

$$F = x'yz + x'yz' + xy'z' + xy'z$$

First, a 1 is marked in each square as needed to represent the function as shown in Fig. 3-4. This can be accomplished in two ways: either by converting each minterm to a binary number and then marking a 1 in the corresponding square, or by obtaining the coincidence of the variables in each term. For example, the term $x'yz$ has the corresponding binary number 011 and represents minterm m_3 in square 011. The second way of recognizing the square is by the coincidence of variables x', y, and z, which is found in the map by observing that x' belongs to the four squares in the first row, y belongs to the four squares in the two right columns, and z belongs to the four squares in the two middle columns. The area that belongs to all three literals is the single square in the first row and third column. In a similar manner, the other three squares belonging to the function F are marked by 1's in the map. The function is thus represented by an area containing four squares, each marked with a 1 as shown in Fig. 3-4. The next step is to subdivide the given area into adjacent squares. These are indicated in the map by two rectangles, each enclosing two 1's. The upper right rectangle represents the area enclosed by $x'y$; the lower left, the area enclosed by xy'. The sum of these two terms gives the answer

$$F = x'y + xy'$$

Next consider the two squares labeled m_0 and m_2 in Fig. 3-3(a) or $x'y'z'$ and $x'yz'$ in Fig. 3-3(b). These two minterms also differ by one variable y and their sum can be simplified to a two-literal expression:

$$x'y'z' + x'yz' = x'z'$$

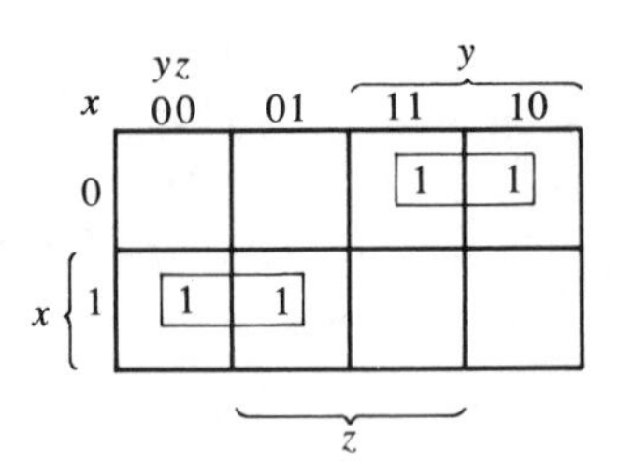

Figure 3-4 Map for Example 3-1; $x'yz + x'yz' + xy'z' + xy'z = x'y + xy'$

Consequently, we must modify the definition of adjacent squares to include this and other similar cases. This is done by considering the map as being drawn on a surface where the right and left edges touch each other to form adjacent squares.

EXAMPLE 3-2. Simplify the Boolean function

$$F = x'yz + xy'z' + xyz + xyz'$$

The map for this function is shown in Fig. 3-5. There are four squares marked with 1's, one for each minterm of the function. Two adjacent squares are combined in the third column to give a two-literal term yz. The

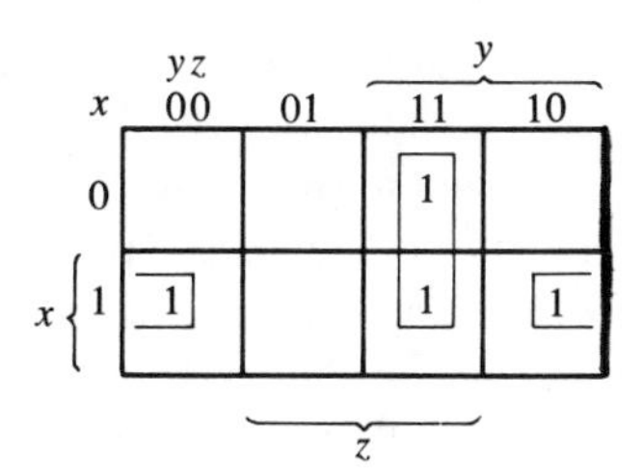

Figure 3-5 Map for Example 3-2; $x'yz + xy'z' + xyz + xyz' = yz + xz'$

remaining two squares with 1's are also adjacent by the new definition and are shown in the diagram enclosed by half rectangles. These two squares when combined give the two-literal term xz'. The simplified function becomes

$$F = yz + xz'$$

Consider now any combination of four adjacent squares in the three-variable map. Any such combination represents the ORing of four adjacent minterms and results in an expression of only one literal. As an example, the sum of the four adjacent minterms m_0, m_2, m_4, and m_6 reduces to the single literal z' as shown:

$$\begin{aligned} x'y'z' + x'yz' + xy'z' + xyz' &= x'z'(y' + y) + xz'(y' + y) \\ &= x'z' + xz' = z'(x' + x) = z' \end{aligned}$$

EXAMPLE 3-3. Simplify the Boolean function

$$F = A'C + A'B + AB'C + BC$$

The map to simplify this function is shown in Fig. 3-6. Some of the terms in the function have less than three literals and are represented in the map by more than one square. For example, to find the squares corresponding

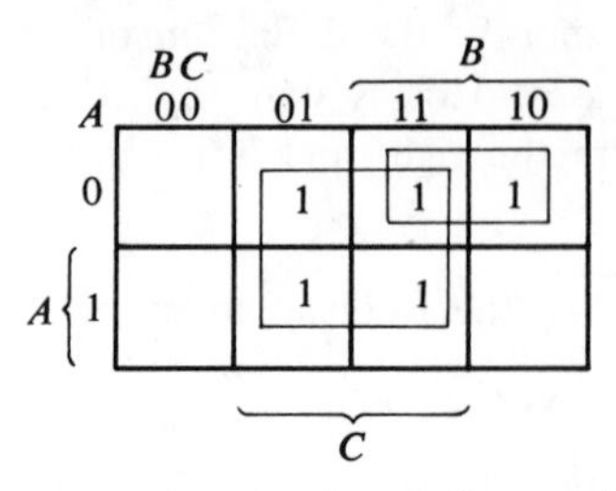

Figure 3-6 Map for Example 3-3; $A'C + A'B + AB'C + BC = C + A'B$

to $A'C$, we form the coincidence of A' (first row) and C (two middle columns) and obtain squares 001 and 011. Note that when marking 1's in the squares, it is possible to find a 1 already placed there by a preceding term. In this example, the second term $A'B$ has 1's in squares 011 and 010, but square 011 is common to the first term $A'C$ and only one 1 is marked in it. The function in this example has five minterms, as indicated by the five squares marked with 1's. It is simplified by combining four squares in the center to give the literal C. The remaining single square marked with a 1 in 010 is combined with an adjacent square that has already been used once. This is permissible and even desirable since the combination of the two squares gives the term $A'B$ while the single minterm represented by the square gives the three-variable term $A'BC'$. The simplified function is

$$F = C + A'B$$

EXAMPLE 3-4. Simplify the Boolean function

$$F(x, y, z) = \Sigma(0, 2, 4, 5, 6)$$

Here we are given the minterms by their decimal numbers. The corresponding squares are marked by 1's as shown in Fig. 3-7. From the map we obtain the simplified function

$$F = z' + xy'$$

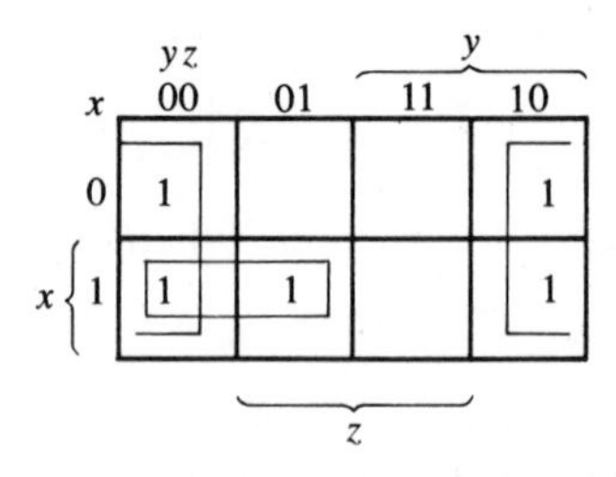

Figure 3-7 $f(x,y,z) = \Sigma$ **(0,2,4,5,6)** $= z' + xy'$

3-3 FOUR-VARIABLE MAP

The map for Boolean functions of four binary variables is shown in Fig. 3-8. In part (a) are listed the 16 minterms and the squares assigned to each. In (b) the map is redrawn to show the relation with the four variables. The rows and columns are numbered in a reflected code sequence with only one digit changing value between two adjacent rows or columns. The minterm corresponding to each square can be obtained from the concatenation of the row number with the column number. For example, the number of the third row (11) and the second column (01), when concatenated, give the binary number 1101, the binary equivalent of decimal 13. Thus, the square in the third row and second column represents minterm m_{13}.

The map minimization of four-variable Boolean functions is similar to the method used to minimize three-variable functions. Adjacent squares are defined to be squares next to each other. In addition, the map is considered to lie on a surface with the top and bottom edges, as well as the right and left edges, touching each other to form adjacent squares. For example, m_0 and m_2 form adjacent squares, as do m_3 and m_{11}. The combination of adjacent squares that is useful during the simplification process is easily determined from inspection of the four-variable map:

One square represents one minterm, giving a term of four literals.
Two adjacent squares represent a term of three literals.
Four adjacent squares represent a term of two literals.
Eight adjacent squares represent a term of one literal.
Sixteen adjacent squares represent the function equal to 1.

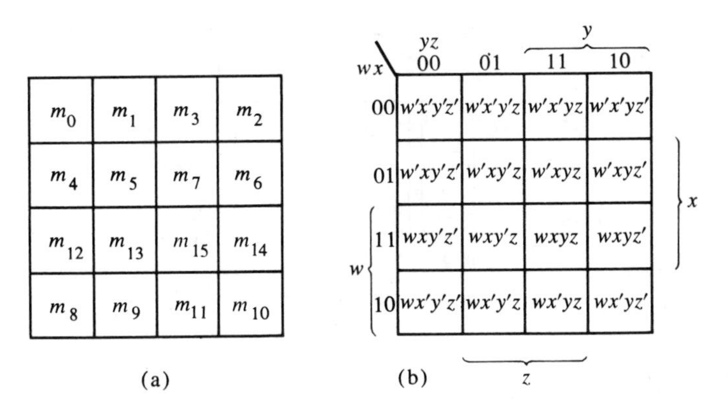

Figure 3-8 Four-variable map

No other combination of squares can simplify the function. The following two examples show the procedure used to simplify four-variable Boolean functions.

EXAMPLE 3-5. Simplify the Boolean function

$$F(w, x, y, z) = \Sigma(0, 1, 2, 4, 5, 6, 8, 9, 12, 13, 14)$$

Since the function has four variables, a four-variable map must be used. The minterms listed in the sum are marked by 1's in the map of Fig. 3-9. Eight adjacent squares marked with 1's can be combined to form the one literal term y'. The remaining three 1's on the right cannot be combined together to give a simplified term. They must be combined as two or four adjacent squares. The larger the number of squares combined, the less the number of literals in the term. In this example, the top two 1's on the right are combined with the top two 1's on the left to give the term $w'z'$. Note that it is permissible to use the same square more than once. We are now left with a square marked by 1 in the third row and fourth column (square 1110). Instead of taking this square alone (which will give a term of four literals), we combine it with squares already used to form an area of four adjacent squares. These squares comprise the two middle rows and the two end columns, giving the term xz'. The simplified function is

$$F = y' + w'z' + xz'$$

EXAMPLE 3-6. Simplify the Boolean function

$$F = A'B'C' + B'CD' + A'BCD' + AB'C'$$

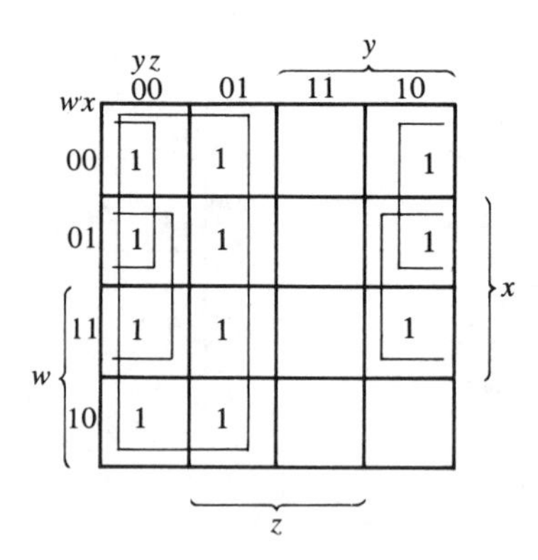

Figure 3-9 Map for Example 3-5; $F(w,x,y,z) = \Sigma(0,1,2,4,5,6,8,9,12,13,14) = y' + w'z' + xz'$

The area in the map covered by this function consists of the squares marked with 1's in Fig. 3-10. This function has four variables and, as expressed, consists of three terms, each with three literals, and one term of four literals. Each term of three literals is represented in the map by two squares. For example, $A'B'C'$ is represented in squares 0000 and 0001. The function can be simplified in the map by taking the 1's in the four corners to give the term $B'D'$. This is possible because these four squares are adjacent when the map is drawn in a surface with top and bottom or left and right edges touching one another. The two left-hand 1's in the top row are combined with the two 1's in the bottom row to give the term $B'C'$. The remaining 1 may be combined in a two-square area to give the term $A'CD'$. The simplified function is

$$F = B'D' + B'C' + A'CD'$$

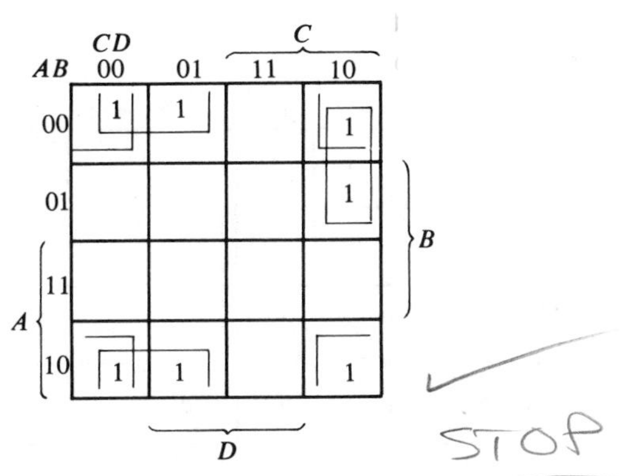

Figure 3-10 Map for Example 3-6; $A'B'C' + B'CD' + A'BCD' + AB'C' = B'D' + B'C' + A'CD'$

3-4 FIVE- AND SIX-VARIABLE MAPS

Maps of more than four variables are not as simple to use. The number of squares becomes excessively large and the geometry for combining adjacent squares becomes more involved. The number of squares is always equal to the number of minterms. For five-variable maps, we need 32 squares; for six-variable maps, we need 64 squares. Maps with seven or more variables need too many squares. They are impractical to use. The five- and six-variable maps are shown in Figs. 3-11 and 3-12, respectively. The rows and columns are numbered in a reflected code sequence; the minterm assigned to each square is read from these numbers. In this way, the square in the third row (11) and second column (001), in the five-variable map, is number 11001, the equivalent of decimal 25. Therefore, this square

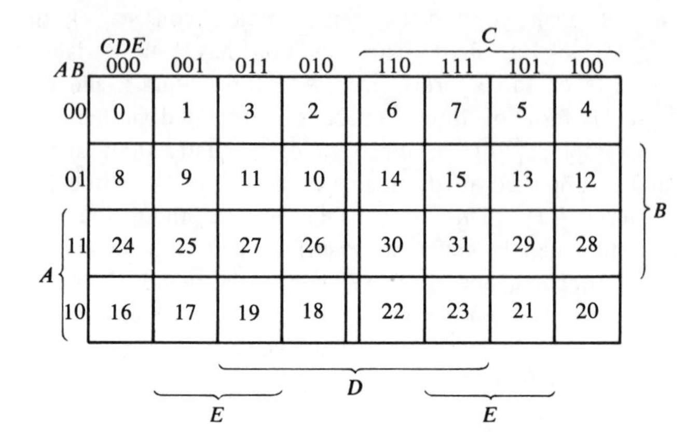

Figure 3-11 Five-variable map

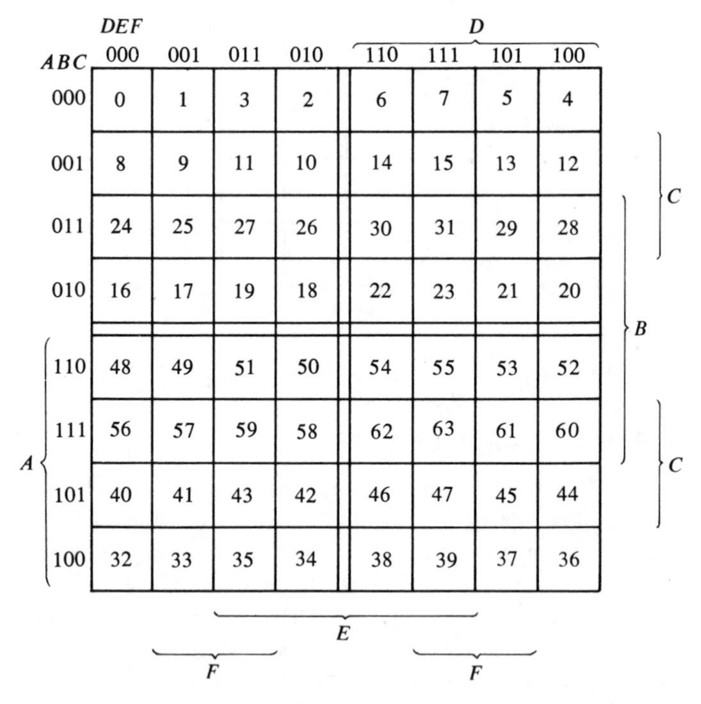

Figure 3-12 Six-variable map

represents minterm m_{25}. The letter symbol of each variable is marked along those squares where the corresponding bit value of the reflected code number is a 1. For example, in the five-variable map, the variable A is a 1 in the last two rows; B is a 1 in the middle two rows. The reflected numbers in the columns show variable C with a 1 in the rightmost four columns, variable D with a 1 in the middle four columns, and the 1's for variable E not physically adjacent but split in two parts. The variable assignment in the six-variable map is determined similarly.

The definition of adjacent squares for the maps of Figs. 3-11 and 3-12 must be modified again to take into account the fact that some variables are split in two parts. The five-variable map must be thought to consist of two four-variable maps, and the six-variable map to consist of four four-variable maps. Each of these four-variable maps is recognized from the double lines in the center of the map: each retains the previously defined adjacency when taken individually. In addition, the center double line must be considered as the center of a book, with each half of the map being a page. When the book is closed, two adjacent squares will fall one in each other. In other words, the center double line is like a mirror with each square being adjacent, not only to its four neighboring squares, but also to its mirror image. For example, minterm 31 in the five-variable map is adjacent to minterms 30, 15, 29, 23, *and* 27. The same minterm in the six-variable map is adjacent to all these minterms plus minterm 63.

From inspection, and taking into account the new definition of adjacent squares, it is possible to show that any 2^k adjacent squares, for $k = 0, 1, 2, \ldots, n$, in an n-variable map, will represent an area that gives a term of $n - k$ literals. For the above statement to have any meaning, n must be larger than k. When $n = k$, the entire area of the map is combined to give the identity function. Table 3-1 shows the relation between the number of adjacent squares and the number of literals in the term. For example, eight adjacent squares combine an area in the five-variable map to give a term of two literals.

Table 3-1. The Relation between the Number of Adjacent Squares and the Number of Literals in the Term

k	*Number of adjacent squares* 2^k	*Number of literals in a term in an* n-*variable map*					
		$n = 2$	$n = 3$	$n = 4$	$n = 5$	$n = 6$	$n = 7$
0	1	2	3	4	5	6	7
1	2	1	2	3	4	5	6
2	4	0	1	2	3	4	5
3	8		0	1	2	3	4
4	16			0	1	2	3
5	32				0	1	2
6	64					0	1

EXAMPLE 3-7. Simplify the Boolean function

$$F(A, B, C, D, E) = \Sigma(0, 2, 4, 6, 9, 11, 13, 15, 17, 21, 25, 27, 29, 31)$$

The five-variable map of this function is shown in Fig. 3-13. Each minterm is converted to its equivalent binary number and the 1's are marked in their corresponding squares. It is now necessary to find combinations of adjacent squares that will result in the largest possible area. The four squares in the center of the right-half map are reflected across the double line and are combined with the four squares in the center of the left-half map to give eight allowable adjacent squares equivalent to the term BE. The two 1's in the bottom row are the reflection of each other about the center double line. By combining them with the other two adjacent squares, the term $AD'E$ is obtained. The four 1's in the top row are all adjacent and can be combined to give the term $A'B'E'$. All the 1's are now included. The simplified function is

$$F = BE + AD'E + A'B'E'$$

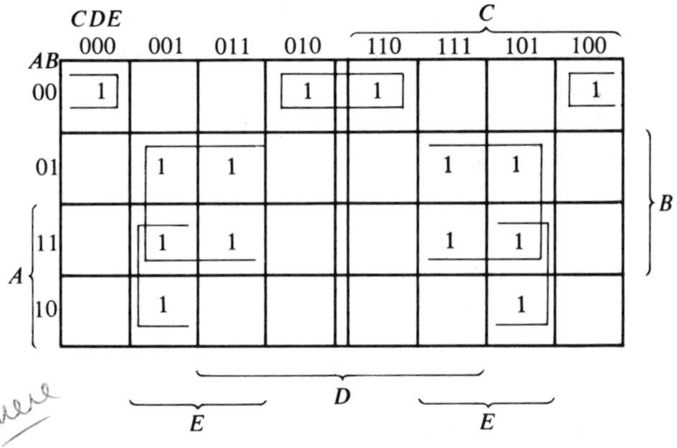

AB \ CDE	000	001	011	010	110	111	101	100
00	1			1	1			1
01		1	1			1	1	
11		1	1			1	1	
10		1					1	

Figure 3-13 Map for Example 3-7; $F(A,B,C,D,E) = \Sigma(0,2,4,6,9,11,13,15,17,21,25,27,29,31) = BE + AD'E + A'B'E'$

3-5 PRODUCT OF SUMS SIMPLIFICATION

The minimized Boolean functions derived from the map in all previous examples were expressed in the sum of products form. With a minor modification, the product of sums form can be obtained.

The procedure for obtaining a minimized function in product of sums follows from the basic properties of Boolean functions. The 1's placed in the squares of the map represent the minterms of the function. The minterms not included in the function denote the complement of the function. From this we see that the complement of a function is represented in the map by the squares not marked by 1's. If we mark the empty squares by 0's and combine them into valid adjacent squares, we shall obtain a simplified expression of the complement of the function; i.e., of F'. The complement of F' gives us back the function F. Because of the generalized De Morgan's theorem, the function so obtained is automatically in the product of sums form. The best way to show this is by example.

EXAMPLE 3-8. Simplify the following Boolean function in (a) sum of products and (b) product of sums.

$$F(A, B, C, D) = \Sigma(0, 1, 2, 5, 8, 9, 10)$$

The 1's marked in the map of Fig. 3-14 represent all the minterms of the function. The squares marked with 0's represent the minterms not included in F, and therefore, denote the complement of F. Combining the squares with 1's gives the simplified function in sum of products:

(a) $F = B'D' + B'C' + A'C'D$

If the squares marked with 0's are combined, as shown in the diagram, one obtains the simplified complemented function:

$$F' = AB + CD + BD'$$

Applying De Morgan's theorem (by taking the dual and complementing each literal as described in Sec. 2-4), we obtain the simplified function in product of sums:

(b) $F = (A' + B')(C' + D')(B' + D)$

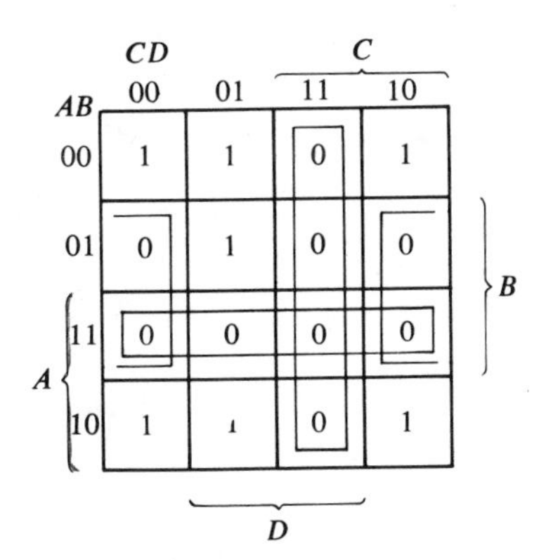

Figure 3-14 Map for Example 3-8; $F(A,B,C,D) = \Sigma(0,1,2,5,8,9,10)$
$B'D' + B'C' + A'C'D = (A' + B')(C' + D')(B' + D)$

The implementation of the simplified expressions obtained in Ex. 3-8 is shown in Fig. 3-15. The sum of products expression is implemented in (a) with a group of AND gates, one for each AND term. The outputs of the AND gates are connected to the inputs of a single OR gate. The same

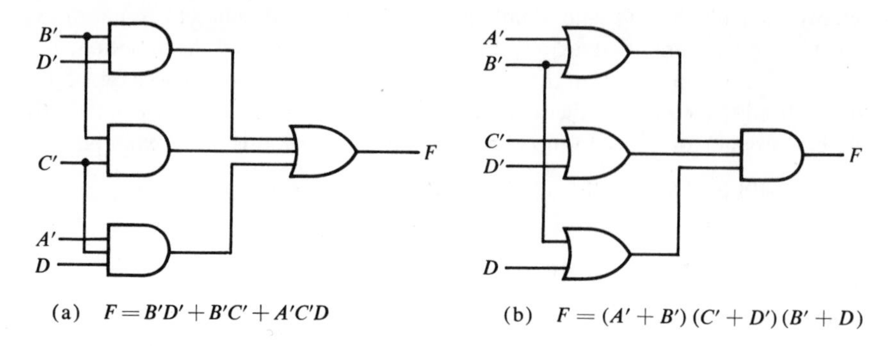

(a) $F = B'D' + B'C' + A'C'D$ (b) $F = (A' + B')(C' + D')(B' + D)$

Figure 3-15 Gate implementation of the function of Example 3-8

function is implemented in (b) in its product of sums form with a group of OR gates, one for each OR term. The outputs of the OR gates are connected to the inputs of a single AND gate. In each case it is assumed that the input variables are directly available in their complement so inverters are not needed. The configuration pattern established in Fig. 3-15 is the general form by which any Boolean function is implemented when expressed in one of the standard forms.  AND gates are connected to a single OR gate when in sum of products; OR gates are connected to a single AND gate when in product of sums. Either configuration forms two levels of gates. Thus, the implementation of a function in a standard form is said to be a two-level implementation. (2 levels of gates)

Example 3-8 has shown the procedure for obtaining the product of sums simplification when the function is originally expressed in the sum of minterms canonical form. The procedure is also valid when the function is originally expressed in the product of maxterm canonical form. Consider for example the truth table that defines the function F in Table 3-2. In sum of minterms, this function is expressed as

$$F(x, y, z) = \Sigma(1, 3, 4, 6)$$

Table 3-2 Truth Table of Function F

x	y	z	F
0	0	0	0
0	0	1	1
0	1	0	0
0	1	1	1
1	0	0	1
1	0	1	0
1	1	0	1
1	1	1	0

In product of maxterms, it is expressed as

$$F(x, y, z) = \Pi(0, 2, 5, 7)$$

In other words, the 1's of the function represent the minterms, while the 0's represent the maxterms. The map for this function is drawn in Fig. 3-16. One can start simplifying this function by first marking the 1's for each minterm that the function is a 1. The remaining squares are marked by 0's. If, on the other hand, the product of maxterms is initially given, one can start marking 0's in those squares listed in the function; the remaining squares are then marked by 1's. Once the 1's and 0's are marked, the function can be simplified in either one of the standard forms. For the sum of products, we combine the 1's to obtain

$$F = x'z + xz'$$

For the product of sums, we combine the 0's to obtain the simplified complemented function:

$$F' = xz + x'z'$$

which shows that the exclusive-or function is the complement of the equivalence function (Sec. 2-6). Taking the complement of F', we obtain the simplified function in product of sums

$$F = (x' + z')(x + z)$$

To enter a function expressed in product of sums in the map, take the complement of the function and from it find the squares to be marked by 0's. For example, the function

$$F = (A' + B' + C)(B + D)$$

can be entered in the map by first taking its complement

$$F' = ABC' + B'D'$$

and then marking 0's in the squares representing the minterms of F'. The remaining squares are marked with 1's.

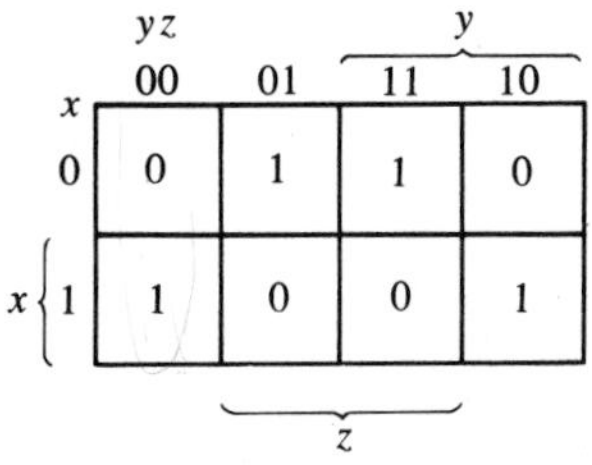

x \ yz	00	01	11	10
0	0	1	1	0
1	1	0	0	1

(y spans columns 11 and 10; z spans columns 01 and 11; x spans row 1)

Figure 3-16 Map for the function of Table 3-2

3-6 DON'T-CARE CONDITIONS

The 1's and 0's in the map signify the combination of variables that makes the function equal to 1 or 0, respectively. The combinations are usually obtained from a truth table that lists the conditions under which the function is a 1. The function is assumed equal to 0 under all other conditions. This assumption is not always true since there are applications where certain combinations of input variables never occur. A four-bit decimal code, for example, has six combinations which are not used. Any digital circuit using this code operates under the assumption that these unused combinations will never occur as long as the system is working properly. As a result, we don't care what the function output is to be for these combinations of the variables because they are guaranteed never to occur. These don't-care conditions can be used on a map to provide further simplification of the function.

It should be realized that a don't-care combination cannot be marked with a 1 on the map because it would require that the function always be a 1 for such input combination. Likewise, putting a 0 in the square requires the function to be 0. To distinguish the don't-care conditions from 1's and 0's, an X will be used.

When choosing adjacent squares to simplify the function in the map, the X's may be assumed to be either 0 or 1, whichever gives the simplest expression. In addition, an X need not be used at all if it does not contribute to covering a larger area. In each case, the choice depends only on the simplification that can be achieved.

EXAMPLE 3-9. Simplify the Boolean function

$$F(w, x, y, z) = \Sigma(1, 3, 7, 11, 15)$$

and the don't-care conditions

$$d(w, x, y, z) = \Sigma(0, 2, 5)$$

The minterms of F are the variable combinations that make the function equal to 1. The minterms of d are the don't-care combinations known never to occur. The minimization is shown in Fig. 3-17. The minterms of F are marked by 1's; those of d are marked by X's; and the remaining squares are filled with 0's. In (a), the 1's and X's are combined in any convenient manner so as to enclose the maximum number of adjacent squares. It is not necessary to include all or any of the X's, only those useful for simplifying a term. One combination that gives a minimum function encloses one X and leaves two out. This results in a simplified sum of products function

$$F = w'z + yz$$

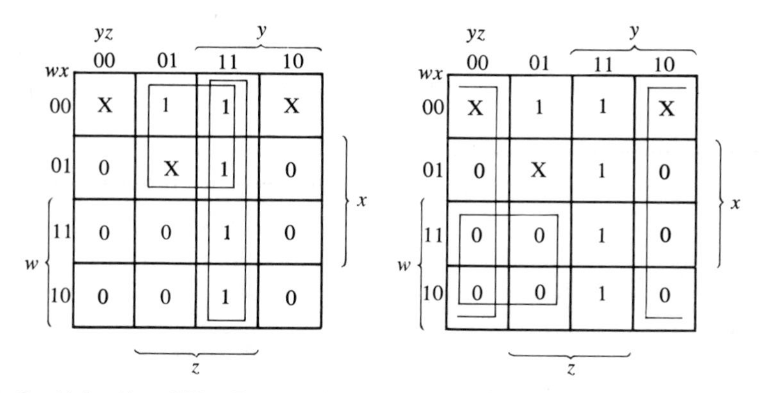

(a) Combining 1's and X's $F = w'z + yz$ (b) Combining 0's and X's $F = z(w' + y)$

Figure 3-17 Example with don't-care conditions

In (b), the 0's are combined with any X's convenient to simplify the complement of the function. The best results are obtained if we enclose the two X's as shown. The complement function is simplified to

$$F' = z' + wy'$$

Complementing again, we obtain a simplified product of sums function

$$F = z\,(w' + y)$$

The two expressions obtained in Ex. 3-9 give two functions which can be shown to be algebraically equal. This is not always the case when don't-care conditions are involved. As a matter of fact, if an X is used as a 1 when combining the 1's and again as a 0 when combining the 0's, the two resulting functions will not yield algebraically equal answers. The selection of the don't-care condition as a 1 in the first case and as a 0 in the second results in different minterm expressions and thus different functions. This can be seen from Ex. 3-9. In the solution of this example, the X chosen to be a 1 was not chosen to be a 0. Now, if in Fig. 3-17(a), we choose the term $w'x'$ instead of $w'z$, we still obtain a minimized function

$$F = w'x' + yz$$

But it is not algebraically equal to the one obtained in product of sums because the same X's are used as 1's in the first minimization and as 0's in the second.

This example also demonstrates that an expression with the minimum number of literals is not necessarily unique. Sometimes the designer is confronted with a choice between two terms with an equal number of literals, with either choice resulting in a minimized expression.

3-7 THE TABULATION METHOD

The map method of simplification is convenient as long as the number of variables does not exceed five or six. As the number of variables increases, the excessive number of squares prevents a reasonable selection of adjacent squares. The obvious disadvantage of the map is that it is essentially a trial-and-error procedure which relies on the ability of the human user to recognize certain patterns. For functions of six or more variables, it is difficult to be sure that the best selection has been made.

The tabulation method overcomes this difficulty. It is a specific step-by-step procedure that guarantees to produce a simplified standard form expression for a function. It can be applied to problems with many variables and has the advantage of being suitable for machine computation. However, it is quite tedious for human use and is prone to mistakes because of its routine, monotonous process. The tabulation method was first formulated by Quine (3) and later improved by McCluskey (4). It is also known as the Quine-McCluskey method.

The tabular method of simplification consists of two parts. The first is to find by an exhaustive search all the terms that are candidates for inclusion in the simplified function. These terms are called *prime-implicants*. The second operation is to choose among the prime-implicants those that give an expression with the least number of literals.

3-8 DETERMINATION OF PRIME-IMPLICANTS

The starting point of the tabulation method is the list of minterms that specify the function. The first tabular operation is to find the prime-implicants by using a matching process. This process compares each minterm with every other minterm. If two minterms differ in only one variable, that variable is removed and a term with one less literal is found. This process is repeated for every minterm until the exhaustive search is completed. The matching process cycle is repeated for those new terms just found. Third and further cycles are continued until a single pass through a cycle yields no further elimination of literals. The remaining terms and all the terms that did not match during the process comprise the prime-implicants. This tabulation method is illustrated by the following example.

EXAMPLE 3-10. Simplify the following Boolean function by using the tabulation method:

$$F = \Sigma\,(0, 1, 2, 8, 10, 11, 14, 15)$$

Step 1: Group binary representation of the minterms according to the number of 1's contained, as shown in Table 3-3, column (a). This is done by grouping the minterms into five sections separated by horizontal lines. The first section contains the number with no 1's in it. The second section contains those numbers that have only one 1. The third, fourth, and fifth sections contain those binary numbers with two, three, and four 1's, respectively. The decimal equivalents of the minterms are also carried along for identification.

Step 2: Any two minterms which differ from each other by only one variable can be combined, and the unmatched variable removed. Two minterm numbers fit in this category if they both have the same bit value in all positions except one. The minterms in one section are compared with those of the next section down only because two terms differing by more than one bit cannot match. The minterm in the first section is compared with each of the three minterms in the second section. If any two numbers are the same in every position but one, a check is placed to the right of both minterms to show that they have been used. The resulting term, together with the decimal equivalents, is listed in column (b) of the table. The variable eliminated during the matching is remembered by inserting a

Table 3-3 Determination of Prime-Implicants for Example 3-10

(a)

	w x y z	
0	0 0 0 0	√
1	0 0 0 1	√
2	0 0 1 0	√
8	1 0 0 0	√
10	1 0 1 0	√
11	1 0 1 1	√
14	1 1 1 0	√
15	1 1 1 1	√

(b)

	w x y z	
0, 1	0 0 0 –	
0, 2	0 0 – 0	√
0, 8	– 0 0 0	√
2, 10	– 0 1 0	√
8, 10	1 0 – 0	√
10, 11	1 0 1 –	√
10, 14	1 – 1 0	√
11, 15	1 – 1 1	√
14, 15	1 1 1 –	√

(c)

	w x y z
0, 2, 8, 10	– 0 – 0
0, 8, 2, 10	– 0 – 0
10, 11, 14, 15	1 – 1 –
10, 14, 11, 15	1 – 1 –

dash in its original position. In this case, m_0 (0000) combines with m_1 (0001) to form (000-). This combination is equivalent to the algebraic operation

$$m_0 + m_1 = w'x'y'z' + w'x'y'z = w'x'y'$$

Minterm m_0 also combines with m_2 to form (00-0) and with m_8 to form (-000). The result of this comparison is entered into the first section of column (b). The minterms of sections two and three of column (a) are next compared to produce the terms listed in the second section of column (b). All other sections of (a) are similarly compared and subsequent sections formed in (b). This exhaustive comparing process results in the four sections of (b).

Step 3: The terms of column (b) have only three variables. A 1 under the variable means it is unprimed. A 0 means it is primed. And a dash means the variable is not included in the term. The searching and comparing process is repeated for the terms in column (b) to form the two-variable terms of column (c). Again, terms in each section need to be compared only if they have dashes in the same position. Note that the term (000-) does not match with any other term. Therefore, it has no check mark on its right. The decimal equivalents are written on the left-hand side of each entry for identification purposes. The comparing process should be carried out again in column (c) and subsequent columns as long as proper matching is encountered. In the present example, the operation stops at the third column.

Step 4: The unchecked terms in the table form the prime-implicants. In this example we have the term $w'x'y'$ (000-) in column (b), and the terms $x'z'$(-0-0) and wy (1-1-) in column (c). Note that each term in column (c) appears twice in the table and as long as the term forms a prime-implicant, it is unnecessary to use the same term twice. The sum of the prime-implicants gives a simplified expression for the function. This is because each checked term in the table has been taken into account by an entry of a simpler term in a subsequent column. Therefore, the unchecked entries (prime-implicants) are the terms left to formulate the function. For the present example, the sum of prime-implicants gives the minimized function in sum of products:

$$F = w'x'y' + x'z' + wy$$

It is worth comparing this answer with that obtained by the map method. Figure 3-18 shows the map simplification of this function. The combinations of adjacent squares give the three prime-implicants of the function. The sum of these three terms is the simplified expression in sum of products.

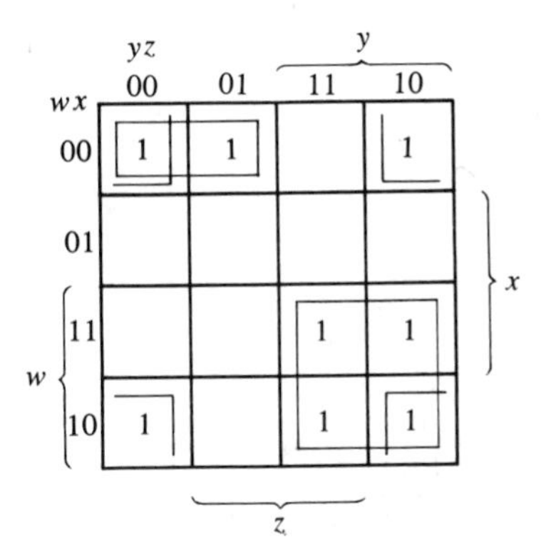

Figure 3-18 Map for the function of Example 3-10; $F = w'x'y' + x'z' + wy$

It is important to point out that Ex. 3-10 was purposely chosen to give the simplified function from the sum of prime-implicants. In most other cases, the sum of prime-implicants does not necessarily form the expression with the minimum number of terms. This is demonstrated in Ex. 3-11.

The tedious manipulation that one must undergo when using the tabulation method is reduced if the comparing is done with decimal numbers instead of binary. A method will now be shown that uses subtraction of decimal numbers instead of the comparing and matching of binary numbers. We note that each 1 in a binary number represents the coefficient multiplied by a power of two. When two minterms are the same in every position except one, the minterm with the extra 1 must be larger than the number of the other minterm by a power of two. Therefore, two minterms can be combined if the number of the first minterm differs by a power of two from a second larger number in the next section down the table. We shall illustrate this procedure by repeating Ex. 3-10.

As shown in Table 3-4, column (a), the minterms are arranged in sections as before except that now only the decimal equivalents of the minterms are listed. The process of comparing minterms is as follows: inspect every two decimal numbers in adjacent sections of the table. If the number in the section below is *greater* than the number in the section above by a power of two (that is, 1, 2, 4, 8, 16, etc.), check both numbers to show that they have been used, and write them down in column (b). The pair of numbers transferred to column (b) includes a third number in parentheses that designates the power of two by which the pair of numbers differ. The number in parentheses tells us the position of the dash in the binary notation. The result of all comparisons of column (a) is shown in column (b).

Table 3-4 Determination of Prime-Implicants of Example 3-10 With Decimal Notation

(a)	
0	√
1	√
2	√
8	√
10	√
11	√
14	√
15	√

(b)	
0, 1 (1)	
0, 2 (2)	√
0, 8 (8)	√
2, 10 (8)	√
8, 10 (2)	√
10, 11 (1)	√
10, 14 (4)	√
11, 15 (4)	√
14, 15 (1)	√

(c)
0, 2, 8, 10, (2, 8)
0, 2, 8, 10, (2, 8)
10, 11, 14, 15 (1, 4)
10, 11, 14, 15 (1, 4)

The comparison between adjacent sections in column (b) is carried out in a similar fashion except that only those terms with the same number in parentheses are compared. The pair of numbers in one section must differ by a power of two from the pair of numbers in the next section. And the numbers in the next section below must be *greater* for the combination to take place. In column (c), we write all four decimal numbers with the two numbers in parentheses designating the positions of the dashes. A comparison of Tables 3-3 and 3-4 may be helpful in understanding the derivations in Table 3-4.

The prime-implicants are those terms not checked in the table. These are the same as before except that they are given in decimal notation. To convert from decimal notation to binary, convert all decimal numbers in the term to binary and then insert a dash in those positions designated by the numbers in parentheses. Thus 0, 1 (1) is converted to binary as 0000, 0001; a dash in the first position of either number results in (000-). Similarly, 0, 2, 8, 10 (2, 8) is converted to the binary notation from 0000, 0010, 1000, and 1010, and a dash inserted in positions 2 and 8, to result in (-0-0)

EXAMPLE 3-11. Determine the prime-implicants of the function

$$F(w, x, y, z) = \Sigma(1, 4, 6, 7, 8, 9, 10, 11, 15)$$

The minterm numbers are grouped in sections as shown in Table 3-5, column (a). The binary equivalent of the minterm is included for the purpose of counting the number of 1's. The binary numbers in the first section have only one 1; the second section, two 1's, etc. The minterm numbers are compared by the decimal method and a match is found if the number in the section below is greater than that in the section above. If the number in the section below is smaller than the one above, a match is

Table 3-5 Determination of Prime-Implicants for Example 3-11

(a)			(b)			(c)
0001	1	√	1, 9	(8)		8, 9, 10, 11 (1, 2)
0100	4	√	4, 6	(2)		8, 9, 10, 11 (1, 2)
1000	8	√	8, 9	(1)	√	
			8, 10	(2)	√	
0110	6	√				
1001	9	√	6, 7	(1)		
1010	10	√	9, 11	(2)	√	
			10, 11	(1)	√	
0111	7	√				
1011	11	√	7, 15	(8)		
			11, 15	(4)		
1111	15	√				

Prime-Implicants:

Decimal	Binary wxyz	Term
1, 9 (8)	−001	$x'y'z$
4, 6 (2)	01−0	$w'xz'$
6, 7 (1)	011−	$w'xy$
7, 15 (8)	−111	xyz
11, 15 (4)	1−11	wyz
8, 9, 10, 11 (1, 2)	10−−	wx'

not recorded even if the two numbers differ by a power of two. The exhaustive search in column (a) results in the terms of column (b), with all minterms in column (a) being checked. There are only two matches of terms in column (b). Each gives the same two-literal term recorded in column (c). The prime-implicants consist of all the unchecked terms in the table. The conversion from the decimal to the binary notation is shown under the table. The prime-implicants are found to be $x'y'z$, $w'xz'$, $w'xy$, xyz, wyz, and wx'.

The sum of the prime-implicants gives a valid algebraic expression for the function. However, this expression is not necessarily the one with the minimum number of terms. This can be demonstrated from inspection of the map for the function of Ex. 3-11. As shown in Fig. 3-19, the minimized function is reorganized to be

$$F = x'y'z + w'xz' + xyz + wx'$$

which consists of the sum of four out of the six prime-implicants derived in Ex. 3-11. The tabular procedure for selecting the prime-implicants that give the minimized function is the subject of the next section.

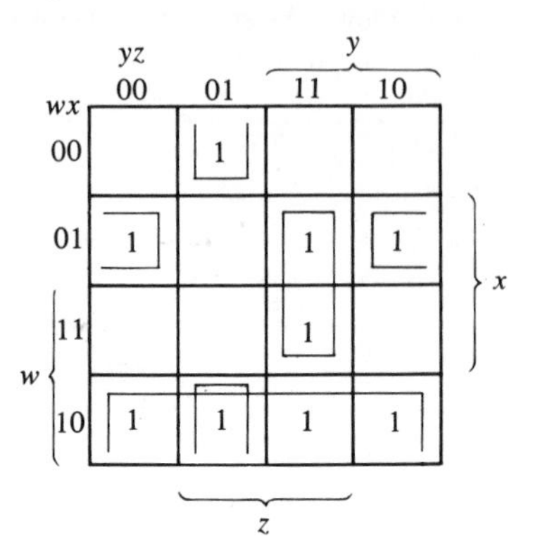

Figure 3-19 Map for the function of Example 3-11; $F = x'y'z + w'xz' + xyz + wx'$

3-9 SELECTION OF PRIME-IMPLICANTS

The selection of prime-implicants that form the minimized function is made from a prime-implicant table. In this table, each prime-implicant is represented in a row and each minterm in a column. Crosses are placed in each row to show the composition of minterms that make the prime-implicants. A minimum set of prime-implicants is then chosen that covers all the minterms in the function. This procedure is illustrated in Ex. 3-12.

EXAMPLE 3-12. Minimize the function of Ex. 3-11.

The prime-implicant table for this example is shown in Table 3-6. There are six rows, one for each prime-implicant (derived in Ex. 3-11), and nine columns, each representing one minterm of the function. Crosses are placed in each row to indicate the minterms contained in the prime-implicant of that row. For example, the two crosses in the first row indicate that minterms 1 and 9 are contained in the prime-implicant $x'y'z$. It is advisable to include the decimal equivalent of the prime-implicant in each row, as it conveniently gives the minterms contained in it. After all the crosses have been marked, we proceed to select a minimum number of prime-implicants.

The completed prime-implicant table is inspected for columns containing only a single cross. In this example, there are four minterms whose columns have a single cross: 1, 4, 8, and 10. Minterm 1 is covered by prime-implicant $x'y'z$: i.e., the selection of prime-implicant $x'y'z$ guarantees that minterm 1 is included in the function. Similarly, minterm 4 is covered by prime-implicant $w'xz'$; and minterms 8 and 10, by prime-implicant wx'. Prime-implicants that cover minterms with a single cross in their column are called *essential prime-implicants*. To enable the final simplified expression to

Table 3-6 Prime-Implicant Table for Example 3-12

		1	4	6	7	8	9	10	11	15
√ $x'y'z$	1, 9	X					X			
√ $w'xz'$	4, 6		X	X						
$w'xy$	6, 7			X	X					
xyz	7, 15				X					X
wyz	11, 15								X	X
√ wx'	8, 9, 10, 11					X	X	X	X	
		√	√	√		√	√	√	√	

contain all the minterms, we have no alternative but to include essential prime-implicants. A check mark is placed in the table next to the essential prime-implicants to indicate that they have been selected.

Next we check each column whose minterm is covered by the selected essential prime-implicants. For example, the selected prime-implicant $x'y'z$ covers minterms 1 and 9. A check is inserted in the bottom of the columns. Similarly, prime-implicant $w'xz'$ covers minterms 4 and 6 and wx' covers minterms 8, 9, 10, and 11. Inspection of the prime-implicant table shows that the selection of the essential prime-implicants covers all the minterms of the function except 7 and 15. These two minterms must be included by the selection of one or more prime-implicants. In this example, it is clear that prime-implicant xyz covers both minterms and is, therefore, the one to be selected. We have thus found the minimum set of prime-implicants whose sum gives the required minimized function:

$$F = x'y'z + w'xz' + wx' + xyz$$

The simplified expressions derived in the preceeding examples were all in the sum of products form. The tabulation method can be adapted to give a simplified expression in product of sums. As in the map method, we have to start with the complement of the function by taking the 0's as the initial list of minterms. This list contains those minterms not included in the original function which are numerically equal to the maxterms of the function. The tabulation process is carried out with the 0's of the function and terminates with a simplified expression in sum of products of the complement of the function. By taking the complement again, one obtains the simplified product of sums expression.

A function with don't-care conditions can be simplified by the tabulation method after a slight modification. The don't-care terms are included in the list of minterms when the prime-implicants are determined. This allows the derivation of prime-implicants with the least number of literals. The don't-care terms are not included in the list of minterms when setting up the prime-implicant table. This is because don't-care terms do not have to be covered by the selected prime-implicants.

3-10 CONCLUDING REMARKS

Two methods of Boolean function simplification were introduced in this chapter. The criterion for simplification was taken to be the minimization of the number of literals in sum of products or product of sums expressions. Both the map and the tabulation methods are restricted in their capabilities since they are useful for simplifying only Boolean functions expressed in the standard forms. Although this is a disadvantage of the methods, it is not very critical. Most applications prefer the standard forms over any other form. We have seen from Fig. 3-15 that the gate implementation of expressions in standard form consists of no more than two levels of gates. Expressions not in the standard form are implemented with more than two levels. Humphrey (5) shows an extension of the map method that produces simplified multilevel expressions.

One should recognize that the reflected code sequence chosen for the maps is not unique. It is possible to draw a map and assign a binary reflected code sequence to the rows and columns different from the sequence employed here. As long as the binary sequence chosen produces a change in only one bit between adjacent squares, it will produce a valid and useful map. Any map that looks different from the one used in this book, or is called by a different name, should be recognized as merely a variation of reflected code assignment.

As evident from Exs. 3-10 and 3-11, the tabulation method has the drawback that errors will inevitably occur in trying to compare numbers over long lists. The map method would seem to be preferable, but for more than five variables, we cannot be certain that the best simplified expression has been found. The real advantage of the tabulation method lies in the fact that it consists of specific step-by-step procedures that guarantee an answer. Moreover, this formal procedure is suitable for computer mechanization.

It was stated in Sec. 3-8 that the tabulation method always starts with the minterm list of the function. If the function is not in this form, it must be converted. In most applications, the function to be simplified comes from a truth table, from which the minterm list is readily available. Otherwise, the conversion to minterms adds considerable manipulative work to the problem. However, an extension of the tabulation method exists for finding prime-implicants from arbitrary sum of products expressions. See, for example, McCluskey (7).

In this chapter, we have considered the simplification of functions with many input variables and a single output variable. However, some digital circuits have more than one output. Such circuits are described by a set of Boolean functions, one for each output variable. A circuit with multiple outputs may sometimes have common terms among the various functions which can be utilized to form common gates during the implementation. This results in further simplification not taken into consideration when each function is simplified separately. There exists an extension of the tabulation method for multiple output circuits (6, 7). However, this method is too specialized and very tedious for human manipulation. It is of practical importance only if a computer program based on this method is available to the user.

PROBLEMS

3-1. Obtain the simplified expressions in sum of products for the following Boolean functions.

(a) $F(x, y, z) = \Sigma(2, 3, 6, 7)$

(b) $F(A, B, C, D) = \Sigma(7, 13, 14, 15)$

(c) $F(A, B, C, D) = \Sigma(4, 6, 7, 15)$

(d) $F(w, x, y, z) = \Sigma(2, 3, 12, 13, 14, 15)$

3-2. Obtain the simplified expressions in sum of products for the following Boolean functions.

(a) $xy + x'y'z' + x'yz'$

(b) $A'B + BC' + B'C'$

(c) $a'b' + bc + a'bc'$

(d) $xy'z + xyz' + x'yz + xyz$

3-3. Obtain the simplified expressions in sum of products for the following Boolean functions.

(a) $D(A' + B) + B'(C + AD)$

(b) $ABD + A'C'D' + A'B + A'CD' + AB'D'$

(c) $k'lm' + k'm'n + klm'n' + lmn'$

(d) $A'B'C'D' + AC'D' + B'CD' + A'BCD + BC'D$

(e) $x'z + w'xy' + w(x'y + xy')$

3-4. Obtain the simplified expressions in sum of products for the following Boolean functions.

(a) $F(A, B, C, D, E) = \Sigma(0, 1, 4, 5, 16, 17, 21, 25, 29)$

(b) $BDE + B'C'D + CDE + A'B'CE + A'B'C + B'C'D'E'$

(c) $A'B'CE' + A'B'C'D' + B'D'E' + B'CD' + CDE' + BDE'$

3-5. Given the following truth table:

x	y	z	F_1	F_2
0	0	0	0	0
0	0	1	1	0
0	1	0	1	0
0	1	1	0	1
1	0	0	1	0
1	0	1	0	1
1	1	0	0	1
1	1	1	1	1

(a) express F_1 and F_2 in product of maxterms.

(b) obtain the simplified functions in sum of products.

(c) obtain the simplified functions in product of sums.

3-6. Obtain the simplified expressions in product of sums:

(a) $F(x,y,z) = \Pi(0, 1, 4, 5)$

(b) $F(A,B,C,D) = \Pi(0, 1, 2, 3, 4, 10, 11)$

(c) $F(w,x,y,z) = \Pi(1, 3, 5, 7, 13, 15)$

3-7. Obtain the simplified expressions in (1) sum of products and (2) product of sums:

(a) $x'z' + y'z' + yz' + xyz$

(b) $(A + B' + D)(A' + B + D)(C + D)(C' + D')$

(c) $(A' + B' + D')(A + B' + C')(A' + B + D')(B + C' + D')$

(d) $(A' + B' + D)(A' + D')(A + B + D')(A + B' + C + D)$

(e) $w'yz' + vw'z' + vw'x + v'wz + v'w'y'z'$

3-8. Draw the gate implementation of the simplified Boolean functions obtained in Prob. 3-7.

3-9. Simplify the Boolean function F in sum of products using the don't-care conditions d:

(a) $F = y' + x'z'$

$d = yz + xy$

(b) $F = B'C'D' + BCD' + ABC'D$

$d = B'CD' + A'BC'D'$

3-10. Simplify the Boolean function F, using the don't-care conditions d, in (1) sum of products and (2) product of sums:

(a) $F = A'B'D' + A'CD + A'BC$

$d = A'BC'D + ACD + AB'D'$

(b) $F = w'(x'y + x'y' + xyz) + x'z'(y + w)$

$d = w'x(y'z + yz') + wyz$

(c) $F = ACE + A'CD'E' + A'C'DE$

$d = DE' + A'D'E + AD'E'$

(d) $F = B'DE' + A'BE + B'C'E' + A'BC'D'$

$d = BDE' + CD'E'$

3-11. Given Boolean functions F_1 and F_2:

(a) Show that the Boolean function G obtained from ANDing F_1 and F_2 (i.e., $G = F_1 \cdot F_2$) contains those minterms common to both F_1 and F_2.

(b) Show that the Boolean function H obtained from ORing F_1 and F_2 (i.e., $H = F_1 + F_2$) contains all the minterms of both F_1 and F_2.

(c) Explain how the maps of F_1 and F_2 can be used to find G and H.

3-12. The following Boolean expression

$$BE + B'DE'$$

is a simplified version of the expression

$$A'BE + BCDE + BC'D'E + A'B'DE' + B'C'DE'$$

Are there any don't-care conditions? If so, what are they?

3-13. Give three possible ways to express the function

$$F = A'B'D' + AB'CD' + A'BD + ABC'D$$

with eight or less literals.

3-14. With the use of maps, find the simplest form in sum of products of the function $F = fg$, where f and g are given by:

$$f = wxy' + y'z + w'yz' + x'yz'$$

$$g = (w + x + y' + z')(x' + y' + z)(w' + y + z')$$

Hint: See Prob. 3-11.

3-15. Simplify the following Boolean functions by means of the tabulation method.

(a) $F(A, B, C, D, E, F, G) = \Sigma(20, 28, 52, 60)$

(b) $F(A, B, C, D, E, F, G) = \Sigma(20, 28, 38, 39, 52, 60, 102, 103, 127)$

(c) $F(A, B, C, D, E, F) = \Sigma(6, 9, 13, 18, 19, 25, 27, 29, 41, 45, 57, 61)$

3-16. Repeat Prob. 3-6 using the tabulation method.

3-17. Repeat Prob. 3-10(c) and (d) using the tabulation method.

REFERENCES

1. Veitch, E. W., "A Chart Method for Simplifying Truth Functions," *Proc. of the ACM* (May 1952), 127-33.

2. Karnaugh, M., "A Map Method for Synthesis of Combinational Logic Circuits," *Trans. AIEE, Comm. and Electronics,* Vol. 72, Part I (November 1953), 593-99.

3. Quine, W. V., "The Problem of Simplifying Truth Functions," *Am. Math. Monthly*, Vol. 59, No. 8 (October 1952), 521-31.

4. McCluskey, E. J., Jr., "Minimization of Boolean Functions," *Bell System Tech. J.*, Vol. 35, No. 6 (November 1956), 1417-44.

5. Humphrey, W. S., Jr., *Switching Circuits with Computer Applications.* New York: McGraw-Hill Book Co., 1958. Ch. 4.

6. Hill, F. J., and G. R. Peterson, *Introduction to Switching Theory and Logical Design*. New York: John Wiley & Sons, Inc., 1968. Chs. 6 and 7.

7. McCluskey, E. J., Jr., *Introduction to the Theory of Switching Circuits.* New York: McGraw-Hill Book Co., 1965. Ch. 4.

4 COMBINATIONAL LOGIC

4-1 INTRODUCTION TO COMBINATIONAL LOGIC CIRCUITS

Logic circuits for digital systems may be combinational or sequential. A combinational circuit consists of logic gates whose outputs at any time are determined directly from the present combination of inputs without regard for previous inputs. A combinational circuit performs a specific information processing operation fully specified logically by a set of Boolean functions. Sequential circuits employ memory elements (binary cells) in addition to logic gates. Their outputs are a function of the inputs and the state of the memory elements. The state of memory elements, in turn, is a function of previous inputs. As a consequence, the outputs of a sequential circuit depend not only on present, but also on past, inputs, and the circuit behavior must be specified by a time sequence of inputs and internal states. Sequential circuits are discussed in Ch. 6.

In Ch. 1 we learned to recognize binary numbers and binary codes that represent discrete quantities of information. These binary variables are represented by electric voltages or by some other signal. The signals can be manipulated in digital logic gates to perform required functions. In Ch. 2 we introduced Boolean algebra as a way to express logic functions algebraically. In Ch. 3 we learned how to simplify Boolean functions to achieve economical gate implementations. The purpose of this chapter is to utilize the knowledge acquired in previous chapters and formulate various systematic design and analysis procedures of combinational circuits. The solution of some typical examples will provide a useful catalog of elementary functions important for the understanding of digital computers and systems.

A combinational circuit consists of input variables, logic gates, and output variables. The logic gates accept signals from the inputs and generate signals to the outputs. This process transforms binary information from the given input data to the required output data. Obviously, both input and output data are represented by binary signals; i.e., they exist in two possible values, one representing logic–1 and the other logic–0. A block diagram of a combinational circuit is shown in Fig. 4-1. The n input binary variables come from an external source; the m output variables go to an external destination. In many applications, the source and/or destination are storage registers (Sec. 1-7) located either in the vicinity of the combinational circuit or in a remote external device. By definition, an external register does not influence the behavior of the combinational circuit, because if it does, the total system becomes a sequential circuit.

For n input variables, there are 2^n possible combinations of binary input values. For each possible input combination, there is one and only one possible output combination. A combinational circuit can be described by m Boolean functions, one for each output variable. Each output function is expressed in terms of the n input variables.

Each input variable to a combinational circuit may have one or two wires. When only one wire is available, it may represent the variable either in the normal form (unprimed) or in the complement form (primed). Since a variable in a Boolean expression may appear primed and/or unprimed, it is necessary to provide an inverter for each literal not available in the input wire. On the other hand, an input variable may appear in two wires, supplying both the normal and complement forms to the input of the circuit. If so, it is unnecessary to include inverters for the inputs. The type

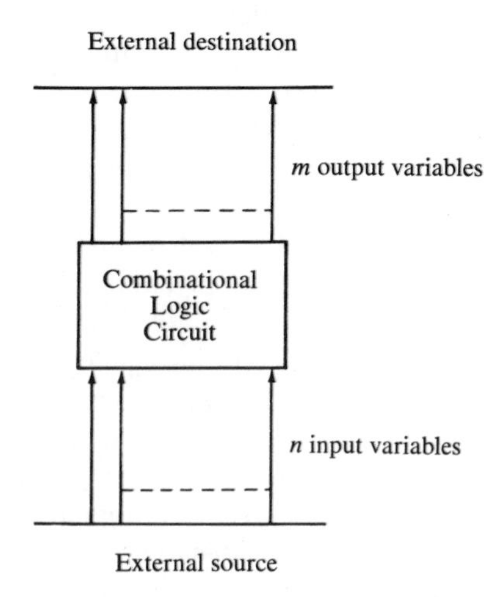

Figure 4-1 Block diagram of a combinational circuit

of binary cells used in most digital systems are flip-flop circuits (Ch. 6) that have outputs for both the normal and complement values of the stored binary variable. In our subsequent work, we shall assume that each input variable appears in two wires, supplying both the normal and complement values simultaneously. We must also realize that an inverter circuit can always supply the complement of the variable if only one wire is available.

4-2 DESIGN PROCEDURE

The design of combinational circuits starts from the verbal outline of the problem and ends in a logic circuit diagram, or a set of Boolean functions from which the logic diagram can be easily obtained. The procedure involves the following steps:

1. The problem is stated.
2. The number of available input variables and required output variables is determined.
3. Each input and output variable is assigned a letter symbol.
4. The truth table that defines the required relations between inputs and outputs is derived.
5. The simplified Boolean function for each output is obtained.
6. The logic diagram is drawn.

A truth table for a combinational circuit consists of input columns and output columns. The 1's and 0's in the input columns are obtained from the 2^n binary combinations available for n input variables. The binary values for the outputs are determined from examination of the stated problem. An output can be equal to either 0 or 1 for every valid input combination. However, the specifications may indicate that some input combinations will not occur. These combinations become don't-care conditions.

The output functions specified in the truth table give the exact definition of the combinational circuit. It is important that the verbal specifications are interpreted correctly into a truth table. Sometimes the designer must use his intuition and experience to arrive at the correct interpretation. Word specifications are very seldom complete and exact. Any wrong interpretation which results in an incorrect truth table produces a combinational circuit that will not fulfil the stated requirements.

The output Boolean functions from the truth table are simplified by any available method, such as algebraic manipulation, the map method, or the tabulation procedure. Usually there will be a variety of simplified expressions from which to choose. However, in any particular application, certain restrictions, limitations, and criteria will serve as a guide in the process of choosing a particular algebraic expression. A practical design method would have to consider such constraints as (a) minimum number of gates, (b) minimum number of inputs to a gate, (c) minimum propagation time of the signal through the circuit, (d) minimum number of interconnections, and (e) limitations of the driving capabilities of each gate. Since all these criteria cannot be satisfied simultaneously, and since the importance of each constraint is dictated by the particular application, it is difficult to make a general statement as to what constitutes an acceptable simplification. In most cases, the simplification begins by satisfying an elementary objective such as producing a simplified Boolean function in a standard form and from that proceeds to meet any other performance criteria.

In practice, designers tend to go from the Boolean functions to a wiring list that shows the interconnections among various standard logic gates. In that case, the design need not go any further than the required simplified output Boolean functions. However, a logic diagram is helpful for visualizing the gate implementation of the expressions.

4-3 ADDERS

Digital computers perform a variety of information processing tasks. Among the basic functions encountered are the various arithmetic operations. The most basic arithmetic operation, no doubt, is the addition of two binary digits. This simple addition consists of four possible elementary operations, namely: $0 + 0 = 0$, $0 + 1 = 1$, $1 + 0 = 1$, and $1 + 1 = 10$. The first three operations produce a sum whose length is one digit, but when both augend and addend bits are equal to 1, the binary sum consists of two digits. The higher significant bit of this result is called a *carry*. When the augend and addend numbers contain more significant digits, the carry obtained from the addition of two bits is added to the next higher order pair of significant bits. A combinational circuit that performs the addition of two bits is called a *half-adder*. One that performs the addition of three bits (two significant bits and a previous carry) is a *full-adder*. The name of the former stems from the fact that two half-adders can be employed to implement a full-adder. The two adder circuits are the first combinational circuits we shall design.

Half-Adder

From the verbal explanation of a half-adder, we find that this circuit needs two binary inputs and two binary outputs. The input variables designate the augend and addend bits; the output variables produce the sum and carry. It is necessary to specify two output variables because the result may consist of two binary digits. We arbitrarily assign symbols x and y to the two inputs and S (for sum) and C (for carry) to the outputs.

Now that we have established the number and name of the input and output variables, we are ready to formulate a truth table to identify exactly the function of the half-adder. This truth table is shown below:

x	y	C	S
0	0	0	0
0	1	0	1
1	0	0	1
1	1	1	0

The carry output is 0 unless both inputs are 1. The S output represents the least significant bit of the sum.

The simplified Boolean functions for the two outputs can be obtained directly from the truth table. The simplified sum of products expressions are:

$$S = x'y + xy'$$
$$C = xy$$

The logic diagram for this implementation is shown in Fig. 4-2 (a), as are four other implementations for a half-adder. They all achieve the same result as far as the input-output behavior is concerned. They illustrate the flexibility available to the designer when implementing even a simple combinational logic function such as this.

Figure 4-2(a), as mentioned above, is the implementation of the half-adder in sum of products. Figure 4-2(b) shows the implementation in product of sums:

$$S = (x + y)(x' + y')$$
$$C = xy$$

To obtain the implementation of Fig. 4-2(c), we note that S is the exclusive-or of x and y. The complement of S is the equivalence of x and y (Sec. 2-6):

$$S' = xy + x'y'$$

but $C = xy$, and therefore we have

$$S = (C + x'y')'$$

In Fig. 4-2(d) we use the product of sums implementation with C derived as follows:

$$C = xy = (x' + y')'$$

In Fig. 4-2(e) we obtain the product term $(x' + y')$ from the complement of C as follows:

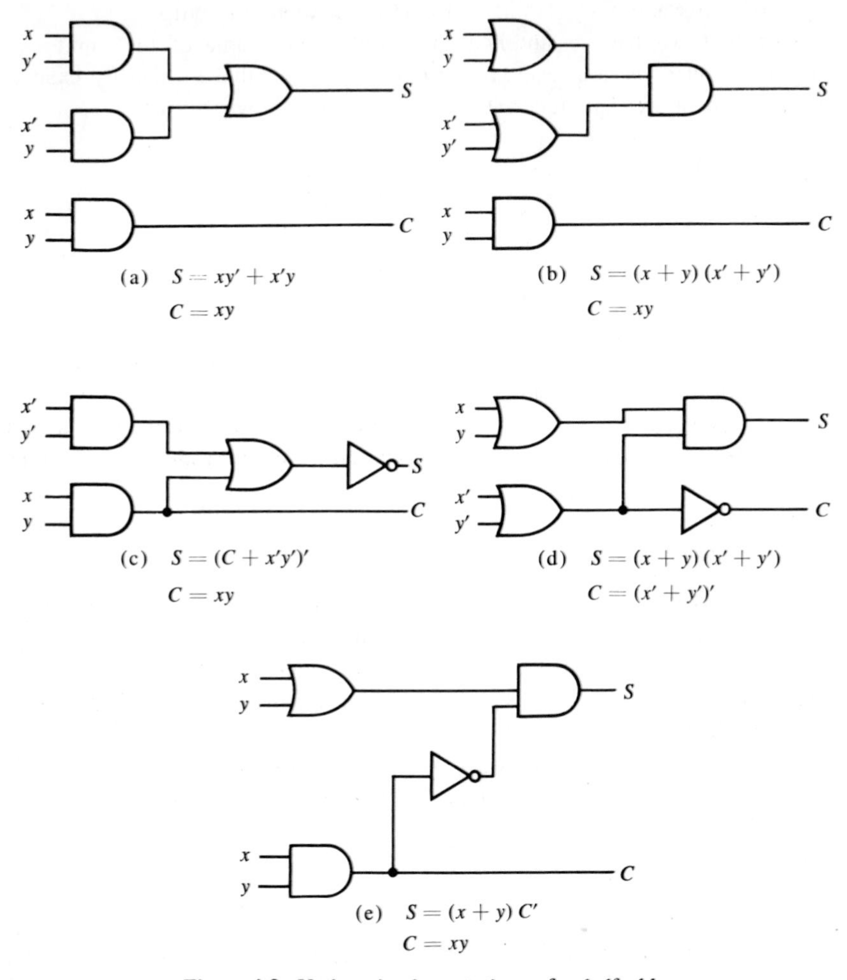

Figure 4-2 Various implementations of a half-adder

$$S = (x + y)(x' + y') = (x + y)(xy)' = (x + y)C'$$

The half-adder is limited; it adds only two single bits. Although it generates a carry for the next higher pair of significant bits, it cannot accept a carry generated from the previous pair of lower significant bits. A full-adder solves this problem.

Full-Adder

A full-adder is a combinational circuit that forms the arithmetic sum of three input bits. It consists of three inputs and two outputs. Two of the input variables, denoted by x and y, represent the two significant bits to be added. The third input, z, represents the carry from the previous lower significant position. Two outputs are necessary because the arithmetic sum of three binary digits ranges in value from 0 to 3, and binary 2 or 3 needs two digits. The two outputs are designated by the symbols S for sum, and C for carry. The binary variable S gives the value of the least significant bit of the sum. The binary variable C gives the output carry. The truth table of the full-adder is as follows:

x	y	z	C	S
0	0	0	0	0
0	0	1	0	1
0	1	0	0	1
0	1	1	1	0
1	0	0	0	1
1	0	1	1	0
1	1	0	1	0
1	1	1	1	1

The eight rows under the input variables designate all possible combinations of 1's and 0's that these variables may have. The 1's and 0's for the ouput variables are determined from the arithmetic sum of the input bits. When all input bits are 0's, the output is 0. The S output is equal to 1 when only one input is equal to 1 or when all three inputs are equal to 1. The C output has a carry of 1 if two or three inputs are equal to 1.

The input and output bits of the combinational circuit have different interpretations at various stages of the problem. Physically, the binary signals of the input wires are considered binary digits added arithmetically to form a two-digit sum at the output wires. On the other hand, the same binary values are considered variables of Boolean functions when expressed in the truth table or when the circuit is implemented with logic gates. It is important to realize that two different interpretations are given to the values of the bits encountered in this circuit.

The input-output logical relation of the full-adder circuit may be expressed in two Boolean functions, one for each output variable. Each output Boolean function requires a unique map for its simplification. Each map must have eight squares since each output is a function of three input variables. The maps of Fig. 4-3 are used for simplifying the two output functions. The 1's in the squares for the maps of S and C are determined directly from the truth table. The squares with 1's for the S output do not combine in adjacent squares to give a simplified expression in sum of products. The C output can be simplified to a six-literal expression. The logic diagram for the full-adder implemented in sum of products is shown

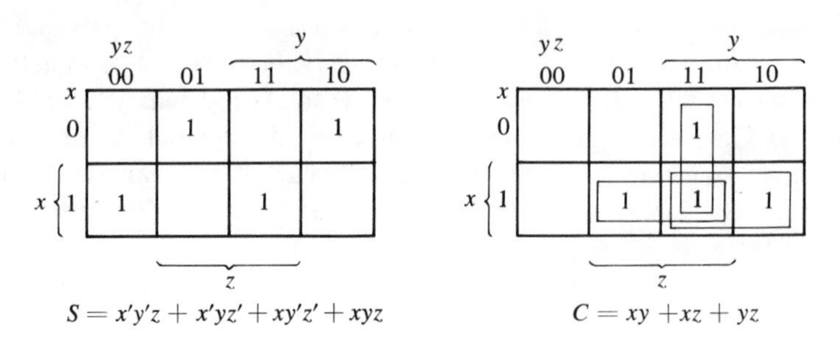

Figure 4-3 Maps for full-adder

in Fig. 4-4. This implementation uses the following Boolean expressions:

$$S = x'y'z + x'yz' + xy'z' + xyz$$
$$C = xy + xz + yz$$

Other configurations for a full-adder may be developed. The product of sums implementation requires the same number of gates as in Fig. 4-4, with the number of AND and OR gates interchanged. A full-adder can be

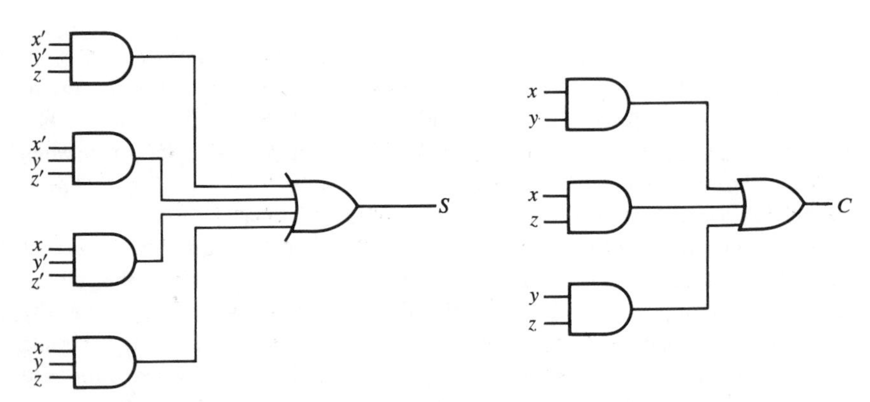

Figure 4-4 Implementation of full-adder in sum of products

implemented with two half-adders and one OR gate as shown in Fig. 4-5. The S output from the second half-adder is the exclusive-or of z and the output of the first half-adder giving:

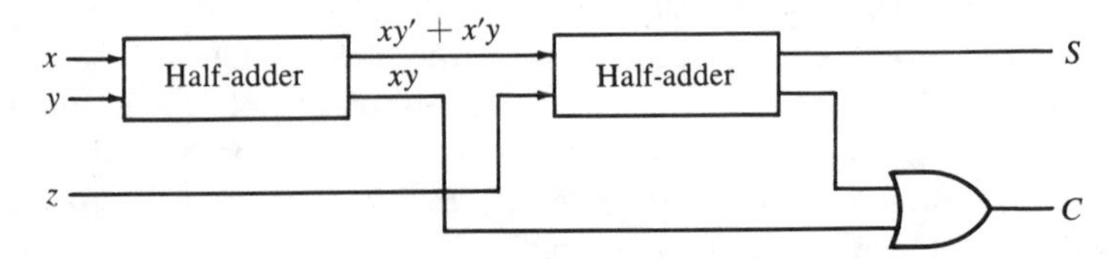

Figure 4-5 Implementation of full-adder with two half-adders and one OR gate

$$
\begin{aligned}
S &= z'(xy' + x'y) + z(xy' + x'y)' \\
&= z'(xy' + x'y) + z(xy + x'y') \\
&= xy'z' + x'yz' + xyz + x'y'z
\end{aligned}
$$

and the carry output is

$$C = z(xy' + x'y) + xy = xy'z + x'yz + xy$$

4-4 SUBTRACTORS

The subtraction of two binary numbers may be accomplished by taking the complement of the subtrahend and adding it to the minuend (Sec. 1-5). By this method, the subtraction operation becomes an addition operation requiring full-adders for its machine implementation. It is possible to implement subtraction with logic circuits in a direct manner as done with paper and pencil. By this method, each subtrahend bit of the number is subtracted from its corresponding significant minuend bit to form a difference bit. If the minuend bit is smaller than the subtrahend bit, a 1 is borrowed from the next higher significant position. The fact that a 1 has been borrowed must be conveyed to the next higher pair of bits by means of a binary signal coming out (output) of a given stage and going into (input) the next higher stage. Just as there are half- and full-adders, there are half- and full-subtractors.

Half-Subtractor

A half-subtractor is a combinational circuit that subtracts two bits and produces their difference. It also has an output to specify if a 1 has been borrowed. Designate the minuend bit by x and the subtrahend bit by y. To perform $x - y$ we have to check the relative magnitudes of x and y. If $x \geqslant y$ we have three possibilities: $0 - 0 = 0$, $1 - 0 = 1$, $1 - 1 = 0$. The result is called the difference bit. If $x < y$, we have $0 - 1$, and it is necessary to borrow a 1 from the next higher stage. The 1 borrowed from the next higher stage adds two to the minuend bit, just as in the decimal system a borrow adds 10 to a minuend digit. With the minuend equal to 2, the difference becomes $2 - 1 = 1$. The half-subtractor needs two outputs. One output generates the difference and will be designated by the symbol D. The second output, designated B for borrow, generates the binary signal that informs the next stage if a 1 has been borrowed. The truth table for the input-output relations of a half-subtractor can now be derived as follows:

x	y	B	D
0	0	0	0
0	1	1	1
1	0	0	1
1	1	0	0

The output borrow B is a 0 as long as $x \geqslant y$. It is a 1 for $x = 0$ and $y = 1$. The D output is the result of the arithmetic operation $2B + x - y$.

The Boolean functions for the two outputs of the half-subtractor are derived directly from the truth table:

$$D = x'y + xy'$$
$$B = x'y$$

It is interesting to note that the logic for D is exactly the same as the output S in the half-adder.

Full-Subtractor

A full-subtractor is a combinational circuit that performs a subtraction between two bits, taking into account that a 1 may have been borrowed by a lower significant stage. This circuit has three inputs and two outputs. The three inputs, x, y, and z, denote the minuend, subtrahend, and previous borrow, respectively. The two outputs, D and B, represent the difference and output borrow, respectively. The truth table for the circuit is as follows:

x	y	z	B	D
0	0	0	0	0
0	0	1	1	1
0	1	0	1	1
0	1	1	1	0
1	0	0	0	1
1	0	1	0	0
1	1	0	0	0
1	1	1	1	1

The eight rows under the input variables designate all possible combinations of 1's and 0's that the binary variables may take. The 1's and 0's for the output variables are determined from the subtraction of $x - y - z$. The combinations having input borrow $z = 0$ reduce to the same four conditions of the half-adder. For $x = 0$, $y = 0$, and $z = 1$, we have to borrow a 1 from the next stage, which makes $B = 1$ and adds 2 to x. Since $2 - 0 - 1 = 1$, $D = 1$. For $x = 0$ and $yz = 11$, we need to borrow again, making $B = 1$ and $x = 2$. Since $2 - 1 - 1 = 0$, $D = 0$. For $x = 1$ and $yz = 01$, we have $x - y - z = 0$, which makes $B = 0$ and $D = 0$. Finally, for $x = 1$, $y = 1$, $z = 1$, we have to borrow 1, making $B = 1$ and $x = 3$, and $3 - 1 - 1 = 1$, making $D = 1$.

The simplified Boolean functions for the two outputs of the full-subtractor are derived in the maps of Fig. 4-6. The simplified sum of product output functions are:

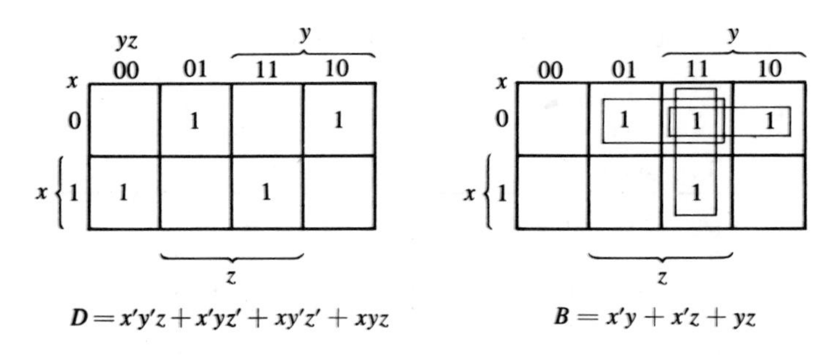

Figure 4-6 Maps for full-subtractor

$$D = x'y'z + x'yz' + xy'z' + xyz$$
$$B = x'y + x'z + yz$$

Again we note that the logic function for output D in the full-subtractor is exactly the same as output S in the full-adder. Moreover, the output B resembles the function for C in the full-adder except that the input variable x is complemented. Because of these similarities, it is possible to convert a full-adder into a full-subtractor by merely complementing input x prior to its application to the gates that form the carry output.

4-5 CODE CONVERSION

The availability of a large variety of codes for the same discrete elements of information results in the use of different codes by different digital systems. It is sometimes necessary to use the output of one system as input to another. A conversion circuit must be inserted between the two systems if each uses different codes for the same information. Thus, a code converter is a circuit that makes the two systems compatible even though each uses a different binary code.

To convert from binary code A to binary code B, the input lines must supply the bit combination of elements as specified by code A and the output lines must generate the corresponding bit combination of code B. A combinational circuit performs this transformation by means of logic gates. The design procedure of code converters will be illustrated by means of a specific example of conversion from the BCD to the excess-3 code.

The bit combinations for the BCD and excess-3 codes are listed in Table 1-2, Sec. 1-6. Since each code uses four bits to represent a decimal digit, there must be four input variables and four output variables. Let us designate the four input binary variables by the symbols A, B, C, D, and the four output variables by w, x, y, z. The truth table relating the input-output variables is shown in Table 4-1. The bit combinations for the

Table 4-1 Truth Table for Code Conversion Example

Input BCD				*Output Excess-3 Code*			
A	B	C	D	w	x	y	z
0	0	0	0	0	0	1	1
0	0	0	1	0	1	0	0
0	0	1	0	0	1	0	1
0	0	1	1	0	1	1	0
0	1	0	0	0	1	1	1
0	1	0	1	1	0	0	0
0	1	1	0	1	0	0	1
0	1	1	1	1	0	1	0
1	0	0	0	1	0	1	1
1	0	0	1	1	1	0	0

inputs and their corresponding outputs are obtained directly from Table 1-2. We note that four binary variables may have 16 bit combinations, only 10 of which are listed in the truth table. The six bit combinations not listed for the *input* variables are don't-care combinations. Since they will never occur, we are at liberty to assign to the output variables either a 1 or a 0, whichever gives a simpler circuit.

The maps in Fig. 4-7 are drawn to obtain a simplified Boolean function for each output. Each of the four maps of Fig. 4-7 represents one of the four outputs of this circuit as a function of the four input variables. The 1's marked inside the squares are obtained from the minterms that make the output equal to 1. The 1's are obtained from the truth table by going over the output columns one at a time. For example, the column under output z has five 1's; therefore, the map for z must have five 1's, each being in a square corresponding to the minterm that makes z equal to 1. The six don't-care combinations are marked by X's. One possible way of simplifying the functions in sum of products is listed under the map of each variable.

A two-level logic diagram may be obtained directly from the Boolean expressions derived by the maps. There are various other possibilities for a logic diagram that implements this circuit. The expressions obtained in Fig. 4-7 may be manipulated algebraically for the purpose of using common gates for two or more outputs. This manipulation, shown below, illustrates the flexibility obtained with multiple output systems when implemented with three or more levels of gates.

$$z = D'$$

$$y = CD + C'D' = CD + (C + D)'$$

$$\begin{aligned} x &= B'C + B'D + BC'D' = B'(C + D) + BC'D' \\ &= B'(C + D) + B(C + D)' \end{aligned}$$

$$w = A + BC + BD = A + B(C + D)$$

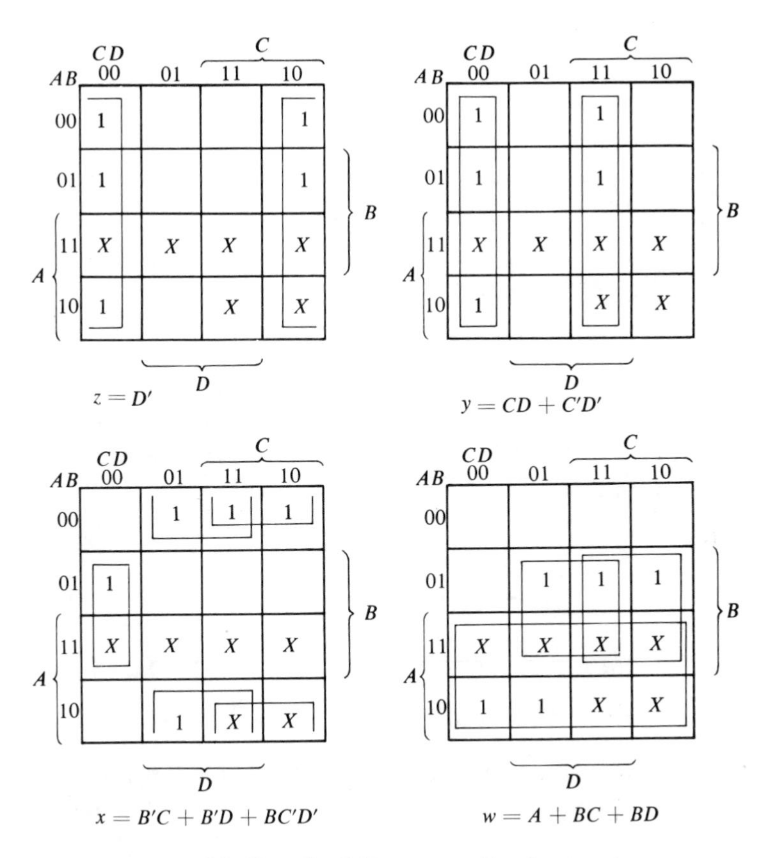

Figure 4-7 Maps for BCD to excess-3 code converter

The logic diagram that implements the above expressions is shown in Fig. 4-8. In it we see that the OR gate whose output is $C + D$ has been used to implement partially each of three outputs.

Not counting input inverters, the implementation in sum of products requires seven AND gates and three OR gates. The implementation of Fig. 4-8 requires four AND gates, four OR gates, and one inverter. If only the normal inputs are available, the first implementation will require inverters for variables B, C, and D, while the second implementation requires inverters for variables B and D.

4-6 COMPARATORS

A comparator is a combinational circuit that compares two numbers A and B and determines their relative magnitude. The outcome of the comparison is displayed in three outputs that indicate whether $A > B$, $A = B$, or $A < B$.

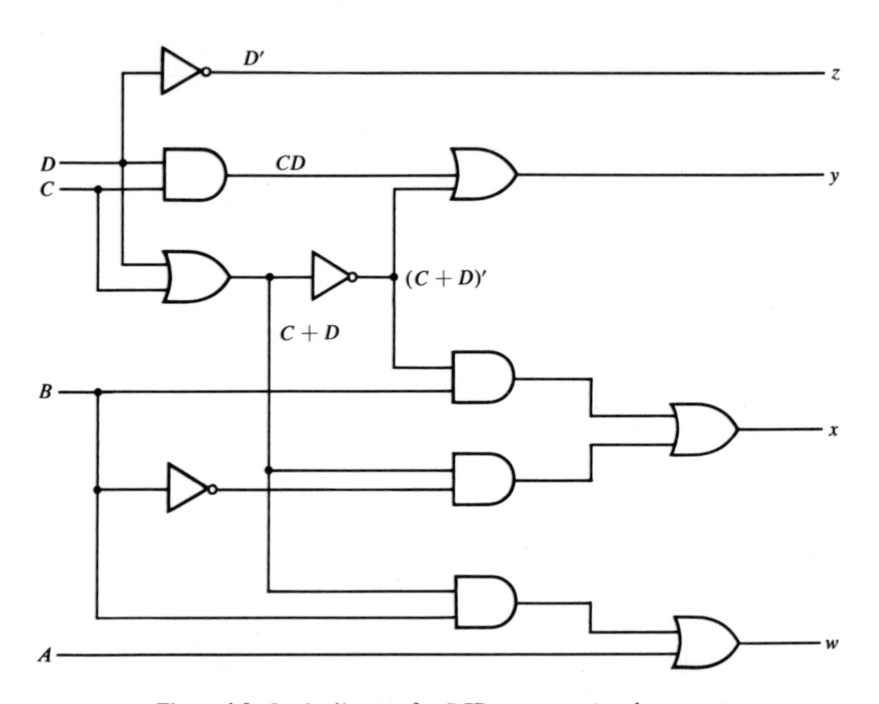

Figure 4-8 Logic diagram for BCD to excess-3 code converter

Possible types of digital data that one may wish to compare are: binary numbers, decimal numbers represented in a binary code, or any other ordered set of discrete elements of information. By following the design procedure outlined in Sec. 4-2, we shall now demonstrate the design of a circuit that compares the relative magnitude of two binary numbers each of two bits in length. We shall then proceed to formulate a general scheme for the design of comparators that compare any ordered set of data of any bit length.

A circuit that compares two binary numbers A and B, each consisting of two bits, must have two inputs for each number. Label the four input variables A_1, A_0, B_1, and B_0, with the subscript 0 denoting the least significant bit. The circuit contains three outputs, one for each of the possibilities $A > B$, $A = B$, $A < B$, labeled x, y, and z, respectively. The truth table for the input-output relations is shown in Table 4-2. We note that the outputs are mutually exclusive; only one output is equal to 1 for each input combination. The four input combinations that make the "equality" output y equal to 1 are those with $A_1A_0 = B_1B_0$. The six input combinations that make the "greater than" output x equal to 1 are those with $A_1A_0 > B_1B_0$. The remaining six input combinations for the "less than" output z are those with $A_1A_0 < B_1B_0$. The three maps for the

Table 4-2 Truth Table for Comparator

Inputs				*Outputs*		
				$A>B$	$A=B$	$A<B$
A_1	A_0	B_1	B_0	x	y	z
0	0	0	0	0	1	0
0	0	0	1	0	0	1
0	0	1	0	0	0	1
0	0	1	1	0	0	1
0	1	0	0	1	0	0
0	1	0	1	0	1	0
0	1	1	0	0	0	1
0	1	1	1	0	0	1
1	0	0	0	1	0	0
1	0	0	1	1	0	0
1	0	1	0	0	1	0
1	0	1	1	0	0	1
1	1	0	0	1	0	0
1	1	0	1	1	0	0
1	1	1	0	1	0	0
1	1	1	1	0	1	0

outputs derived from the truth table are shown in Fig. 4-9. The simplified output Boolean functions in sum of products are obtained from the maps:

$$x = A_1B_1' + A_1A_0B_0' + A_0B_1'B_0'$$

$$y(A_1, A_0, B_1, B_0) = \Sigma(0, 5, 10, 15)$$

$$z = B_1A_1' + B_1B_0A_0' + B_0A_1'A_0'$$

The function for y cannot be simplified and is expressed in its sum of minterms form for convenience. Note the symmetry of the maps and the Boolean functions: the 1's of y lie in the squares forming the diagonal, and the functions for x and z are similar with corresponding A's and B's interchanged. The logic diagram implementation of this circuit can be easily derived from the Boolean functions and is not drawn here.

The truth table for comparing two n-bit binary numbers requires 2^{2n} rows and becomes unmanageable even with $n = 3$. Combinational circuits with greater than six inputs are difficult to design on paper and require a certain amount of machine computation assistance. However, circuits that possess a certain amount of symmetry may sometimes be designed by means of an algorithmic procedure if one is found to exist. An *algorithm* is a procedure that specifies a set of rules that gives the solution to a problem. This method will now be illustrated by deriving an algorithm for the design of comparators.

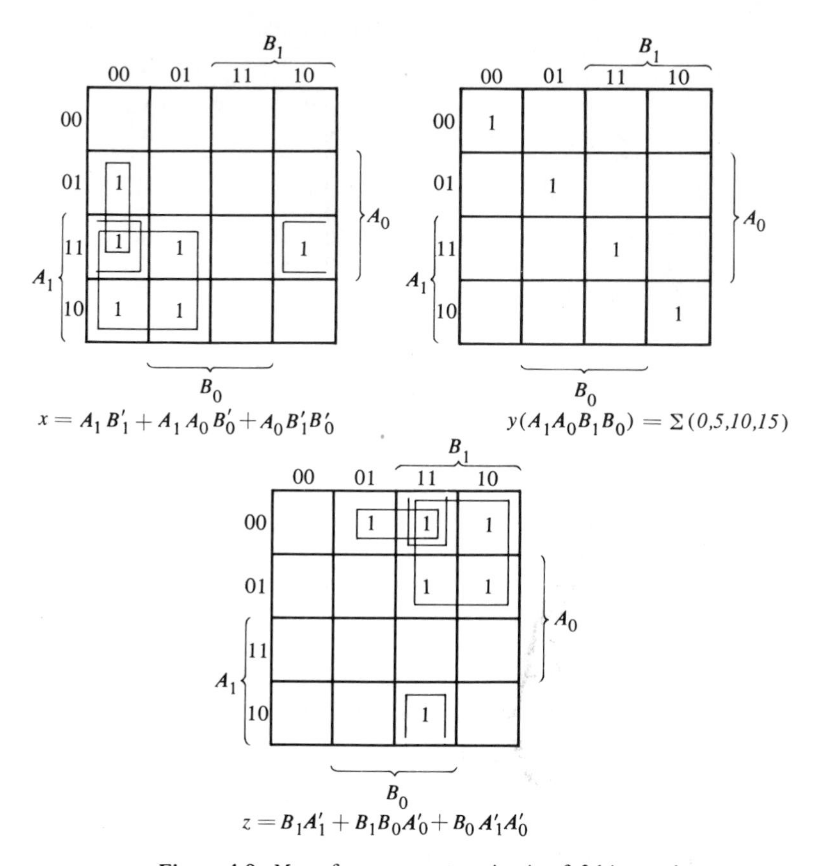

Figure 4-9 Maps for comparator circuit of 2-bit numbers

The algorithm is a direct application of the procedure a human uses to compare the relative magnitude of numbers, or for that matter, the collating order of the alphabet. Consider two numbers A and B with three digits each. Write the coefficients of the numbers with descending significance as follows:

$$A_2 A_1 A_0$$
$$B_2 B_1 B_0$$

The two numbers are equal if $A_2 = B_2$ and $A_1 = B_1$ and $A_0 = B_0$. When the digits are binary, they are equal if each pair of significant bits forms either two 1's or two 0's. This is expressed logically by the following Boolean function:

$$(A = B) = (A_2 B_2 + A_2' B_2')(A_1 B_1 + A_1' B_1')(A_0 B_0 + A_0' B_0')$$

where $(A = B)$ is an output binary variable equal to logic-1 if A is arithmetically equal to B.

To determine if A is greater than or less than B, we inspect the relative magnitude of pairs of significant digits starting from the most significant position. If the two digits are equal, we compare the next lower significant pair of digits. This comparison is continued until a pair of unequal digits is reached. If the corresponding digit of A is greater than that of B, $A > B$, otherwise $A < B$. In the case of binary digits, this sequential comparison can be logically expressed by the following Boolean functions:

$$(A > B) = A_2B_2' + A_1B_1'(A_2B_2 + A_2'B_2') + A_0B_0'(A_2B_2 + A_2'B_2')(A_1B_1 + A_1'B_1')$$

$$(A < B) = A_2'B_2 + A_1'B_1(A_2B_2 + A_2'B_2') + A_0'B_0(A_2B_2 + A_2'B_2')(A_1B_1 + A_1'B_1')$$

where again $(A > B)$ and $(A < B)$ are binary output variables which are equal to logic-1 when $A > B$ or $A < B$, respectively. In words, the first Boolean function states that binary output $(A > B)$ is equal to logic-1 if $A_2 = 1$ and $B_2 = 0$, or if $A_1 = 1$ and $B_1 = 0$ (provided that $A_2 = B_2$) or if $A_0 = 1$ and $B_0 = 0$ (provided that both $A_2 = B_2$ and $A_1 = B_1$).

The gate implementation of the three outputs just derived is simpler than it seems because the "unequal" outputs can use portions of the outputs generated by the "equality" output. The logic diagram for a comparator of two three-bit numbers is shown in Fig. 4-10. The procedure for obtaining comparator circuits for binary numbers with more than three bits is obvious. Comparators of decimal numbers will use the same algorithm except that two four-bit numbers must be compared for each decimal digit.

4-7 ANALYSIS PROCEDURE

The design of a combinational circuit starts from the verbal specifications of a required function and culminates with a set of output Boolean functions or a logic diagram. The *analysis* of a combinational circuit is a somewhat reverse process: it starts with a given logic diagram and culminates with a set of Boolean functions, a truth table, or a verbal explanation of the circuit operation. If the logic diagram to be analyzed is accompanied with a function name or with an explanation of what it is assumed to accomplish, then the analysis problem reduces to a verification of the stated function.

The first step in the analysis is to make sure that the given circuit is combinational and not sequential. The diagram of a combinational circuit has logic gates with no feedback paths or memory elements. A feedback path is a connection from the output of one gate to the input of a second gate that forms part of the input to the first gate. Feedback paths or memory elements in a digital circuit define a sequential circuit and must be analyzed according to procedures outline in Ch. 6.

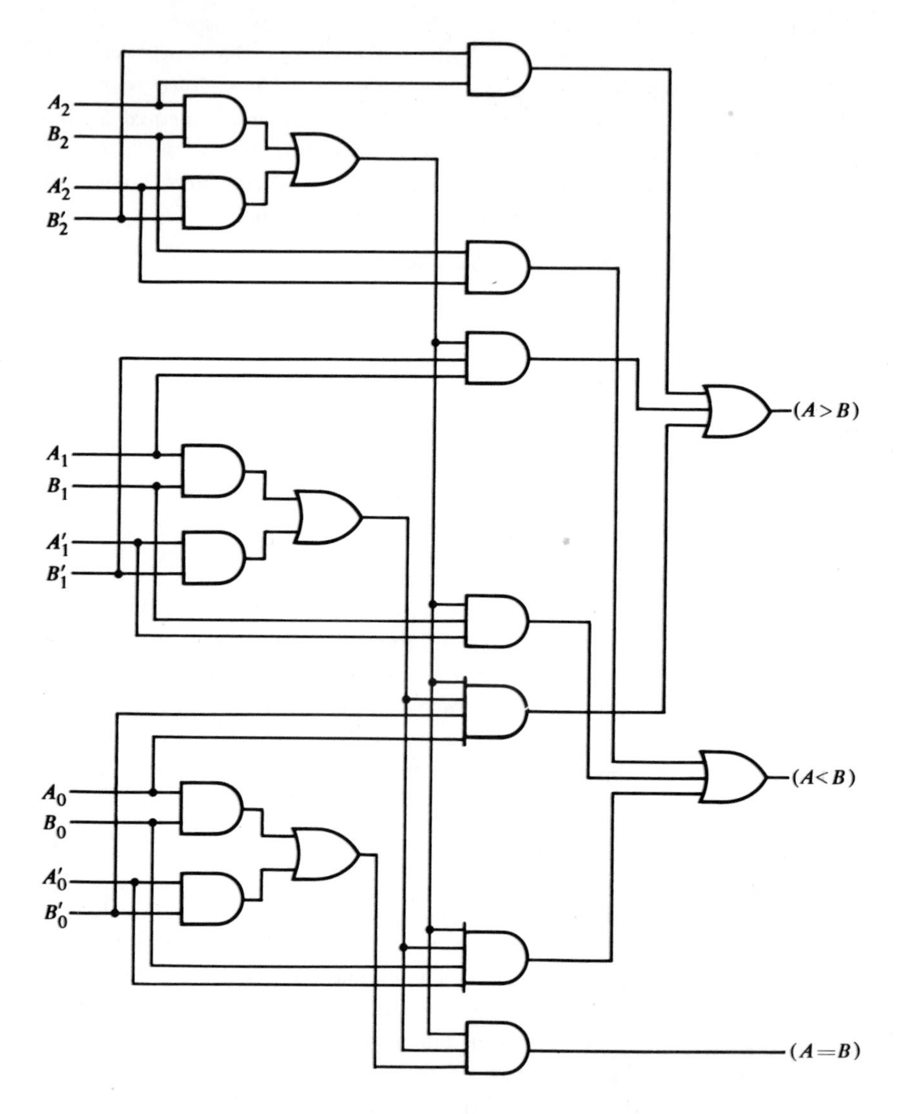

Figure 4-10 Logic diagram of a 3-bit comparator

Once the logic diagram is verified as a combinationl circuit, one can proceed to obtain the output Boolean functions and/or the truth table. If the circuit is accompanied with a verbal explanation of its function, then the Boolean functions or the truth table is sufficient for verification. If the function of the circuit is under investigation, then it is necessary to interpret the operation of the circuit from the derived truth table. The success of such investigation is enhanced if one has previous experience and familiarity with a wide variety of digital circuits. The ability to correlate a truth table with an information processing task is an art one acquires with experience.

To obtain the output Boolean functions from a logic diagram, proceed as follows:

1. Label with arbitrary symbols all gate outputs that are a function of the input variables. Obtain the Boolean functions for each gate.
2. Label with other arbitrary symbols those gates which are a function of input variables and/or previously labeled gates. Find the Boolean functions for these gates.
3. Repeat the process outlined in step 2 until the outputs of the circuit are obtained.
4. By repeated substitution of previously defined functions, obtain the output Boolean functions in terms of input variables only.

The analysis of the combinational circuit in Fig. 4-11 will illustrate the proposed procedure. We note that the circuit has three binary inputs, A, B, and C, and two binary outputs, F_1 and F_2. The outputs of various gates are labeled with intermediate symbols. The output of gates that are a function of input variables only are F_2, T_1, and T_2. The Boolean functions for these three outputs are:

$$\begin{aligned} F_2 &= AB + AC + BC \\ T_1 &= A + B + C \\ T_2 &= ABC \end{aligned}$$

Next we consider outputs of gates which are a function of already defined symbols.

$$\begin{aligned} T_3 &= F_2'T_1 \\ F_1 &= T_3 + T_2 \end{aligned}$$

The output Boolean function F_2 expressed above is already given as a function of the inputs only. To obtain F_1 as a function of A, B, and C, form a series of substitutions as follows:

$$\begin{aligned} F_1 = T_3 + T_2 &= F_2'T_1 + ABC = (AB + AC + BC)'(A + B + C) + ABC \\ &= (A' + B')(A' + C')(B' + C')(A + B + C) + ABC \\ &= (A' + B'C')(AB' + AC' + BC' + B'C) + ABC \\ &= A'BC' + A'B'C + AB'C' + ABC \end{aligned}$$

If we want to pursue the investigation and determine the information transformation task achieved by this circuit, we can derive the truth table

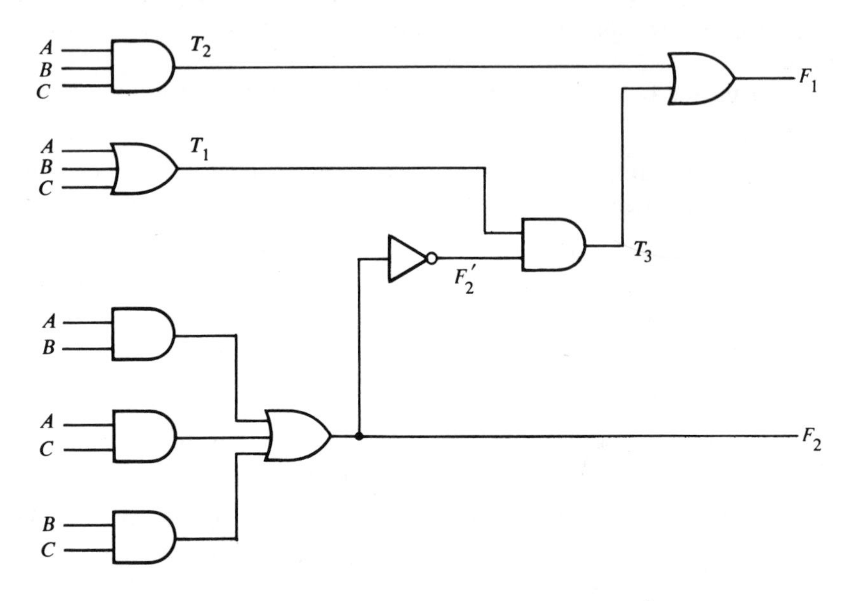

Figure 4-11 Logic diagram for analysis example

directly from the Boolean functions and try to recognize a familiar operation. For this example, we note that the circuit is a full-adder with F_1 being the sum output and F_2 the carry output. A, B, and C are the three inputs added arithmetically.

The derivation of the truth table for the circuit is a straightforward process once the output Boolean functions are known. To obtain the truth table directly from the logic diagram without going through the derivations of the Boolean functions, proceed as follows:

1. Determine the number of input variables to the circuit. For n inputs, form the 2^n possible input combinations of 1's and 0's by listing the binary numbers from 0 to $2^n - 1$.
2. Label the outputs of selected gates with arbitrary symbols.
3. Obtain the truth table for the outputs of those gates that are a function of the input variables only.
4. Proceed to obtain the truth table for the outputs of those gates that are a function of previously defined values until the columns for all outputs are determined.

This process can be illustrated using the circuit of Fig. 4-11. In Table 4-3 we form the eight possible combinations for the three input variables. The truth table for F_2 is determined directly from the values of

Table 4-3 Truth Table for Logic Diagram of Figure 4-11

A	B	C	F_2	F'_2	T_1	T_3	T_2	F_1
0	0	0	0	1	0	0	0	0
0	0	1	0	1	1	1	0	1
0	1	0	0	1	1	1	0	1
0	1	1	1	0	1	0	0	0
1	0	0	0	1	1	1	0	1
1	0	1	1	0	1	0	0	0
1	1	0	1	0	1	0	0	0
1	1	1	1	0	1	0	1	1

A, B, and C, with F_2 equal to 1 for any combination that has two or three inputs equal to 1. The truth table for F'_2 is the complement of F_2. The truth tables for T_1 and T_2 are the OR and AND function of the input variables, respectively. The values for T_3 are derived from T_1 and F'_2: T_3 is equal to 1 when both T_1 and F'_2 are equal to 1 and to 0 otherwise. Finally, F_1 is equal to 1 for those combinations in which either T_2 or T_3 or both are equal to 1. Inspection of the truth table combinations for A, B, C, F_1, and F_2 of Table 4-3 shows that it is identical to the truth table of the full-adder listed in Sec. 4-3 for x, y, z, S, and C, respectively.

Consider now a combinational circuit that has don't-care input combinations. When such a circuit is designed, the don't-care combinations are marked by X's in the map and assigned an output of either a 1 or a 0, whichever is more convenient for the simplification of the output Boolean function. When a circuit with don't-care combinations is being analyzed, the situation is entirely different. Even though we assume that the don't-care input combinations will never occur, the fact of the matter is that if any one of these combinations is applied to the inputs (intentionally or in error), a binary output will be present. The value of the output will depend on the choice for the X's taken during the design. Part of the analysis of such a circuit may involve the determination of the output values for the don't-care input combinations. As an example, consider the BCD to excess-3 code converter designed in Sec. 4-5. The outputs obtained when the six unused combinations of the BCD code are applied to the inputs are:

Unused BCD Inputs				*Outputs*			
A	B	C	D	w	x	y	z
1	0	1	0	1	1	0	1
1	0	1	1	1	1	1	0
1	1	0	0	1	1	1	1
1	1	0	1	1	0	0	0
1	1	1	0	1	0	0	1
1	1	1	1	1	0	1	0

These outputs may be derived by means of the truth table analysis method as outlined in this section. In this particular case, the outputs may be obtained directly from the maps of Fig. 4-7. From inspection of the maps, we determine whether the X's in the corresponding minterm squares for each output have been included with the 1's or the 0's. For example, the square for minterm m_{10} (1010) has been included with the 1's for outputs w, x, and z but not for y. Therefore, the outputs for m_{10} are $wxyz = 1101$ as listed in the above table. We also note that the first three outputs in the table have no meaning in the excess-3 code, while the last three outputs correspond to decimals 5, 6, and 7, respectively. This coincidence is entirely a function of the choice for the X's taken during the design.

4-8 DECODERS AND ENCODERS

Discrete elements of information are represented in digital systems by binary numbers or binary codes. Consider, for example, a binary code or a binary number of n bits capable of representing $m \leqslant 2^n$ discrete elements of information. A *decoder* is a combinational circuit that converts a binary code of n variables into m output lines, one for each discrete element of information. An *encoder* is a combinational circuit that accepts m input lines, one for each element of information, and generates a binary code of n output lines. A decoder is a construction of AND gates with n inputs and 2^n (or less) outputs. An encoder has OR gates with 2^n (or less) inputs and n outputs. A binary code of n bits with don't-care combinations represents, by definition, less than 2^n elements of information. Its decoder or encoder will use less outputs or inputs, respectively.

As an example, consider the binary to octal decoder circuit of Fig. 4-12. The three inputs (x,y, and z) represent a binary number of three bits. The eight outputs (D_0 to D_7) represent octal digits 0 to 7. The decoder consists of a group of AND gates that decode the input binary number. It supplies as many outputs as there are possible input binary number combinations. In this particular example, the elements of information are the eight octal digits. The code for this discrete information consists of the binary numbers represented by three bits. The operation of the decoder may be further clarified from its input-output relations shown in Table 4-4. Observe that the output variables are mutually exclusive because only one output can be equal to 1 at one time. The output line whose value is equal to 1 represents the octal digit equivalent of the binary number in the input lines.

An example of an encoder is shown in Fig. 4-13. The octal to binary encoder consists of eight inputs, one for each of the eight digits, and three outputs that generate the corresponding binary number. It is constructed with OR gates. Its truth table is shown in Table 4-5. It is assumed that only one input line can be equal to 1 at any time, otherwise the circuit has

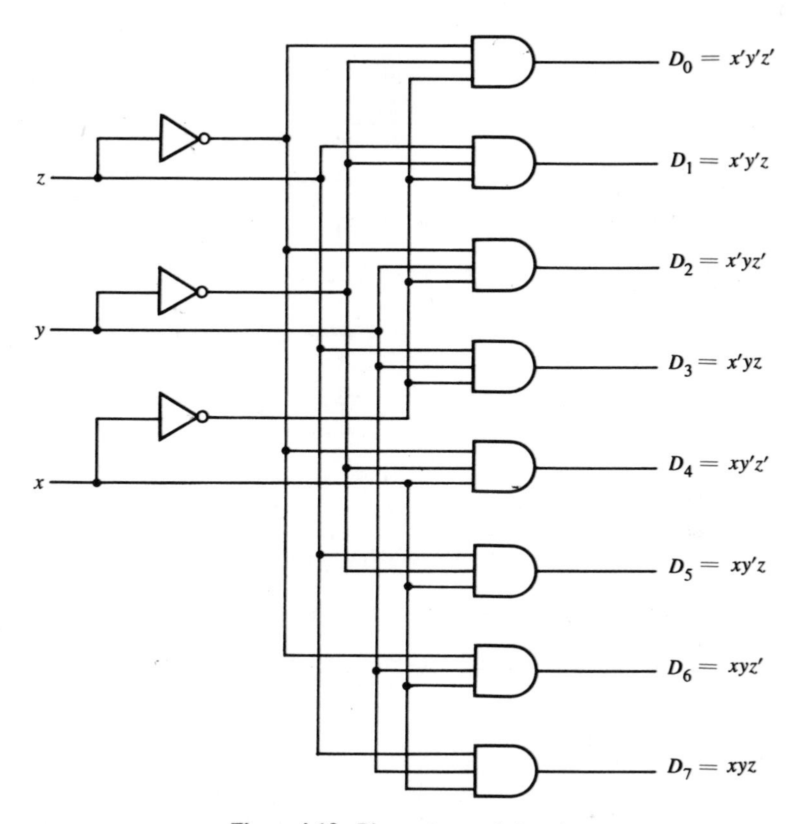

Figure 4-12 Binary to octal decoder

Table 4-4 Truth Table of Binary to Octal Decoder

Inputs			*Outputs*							
x	y	z	D_0	D_1	D_2	D_3	D_4	D_5	D_6	D_7
0	0	0	1	0	0	0	0	0	0	0
0	0	1	0	1	0	0	0	0	0	0
0	1	0	0	0	1	0	0	0	0	0
0	1	1	0	0	0	1	0	0	0	0
1	0	0	0	0	0	0	1	0	0	0
1	0	1	0	0	0	0	0	1	0	0
1	1	0	0	0	0	0	0	0	1	0
1	1	1	0	0	0	0	0	0	0	1

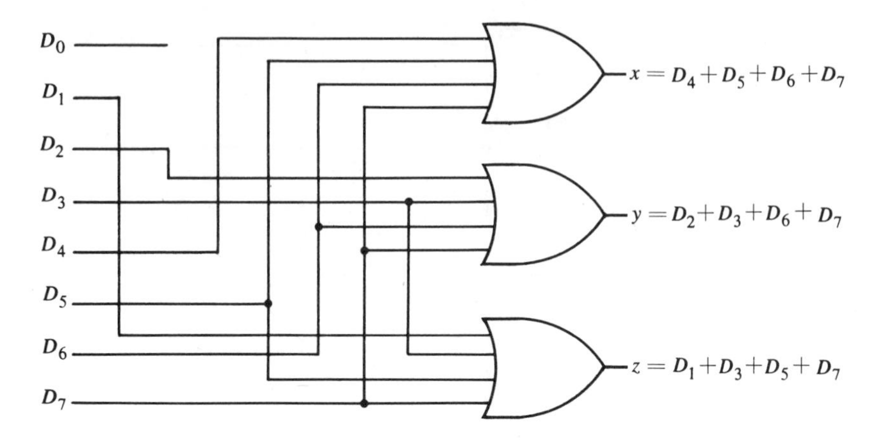

Figure 4-13 Octal to binary encoder

Table 4-5 Truth Table of Octal to Binary Encoder

Inputs								*Outputs*		
D_0	D_1	D_2	D_3	D_4	D_5	D_6	D_7	x	y	z
1	0	0	0	0	0	0	0	0	0	0
0	1	0	0	0	0	0	0	0	0	1
0	0	1	0	0	0	0	0	0	1	0
0	0	0	1	0	0	0	0	0	1	1
0	0	0	0	1	0	0	0	1	0	0
0	0	0	0	0	1	0	0	1	0	1
0	0	0	0	0	0	1	0	1	1	0
0	0	0	0	0	0	0	1	1	1	1

no meaning. Note that the circuit has eight inputs, which can give 2^8 possible input combinations, but that only eight of these combinations have any meaning. The other $2^8 - 8$ input combinations are don't-care conditions.

A BCD to decimal decoder is shown in Fig. 4-14. The elements of information in this case are the 10 decimal digits represented by the BCD code. The function of the decoder is to supply one output for each decimal digit. Each output is equal to 1 only when the input variables constitute a bit combination corresponding to the decimal digit as represented in BCD.

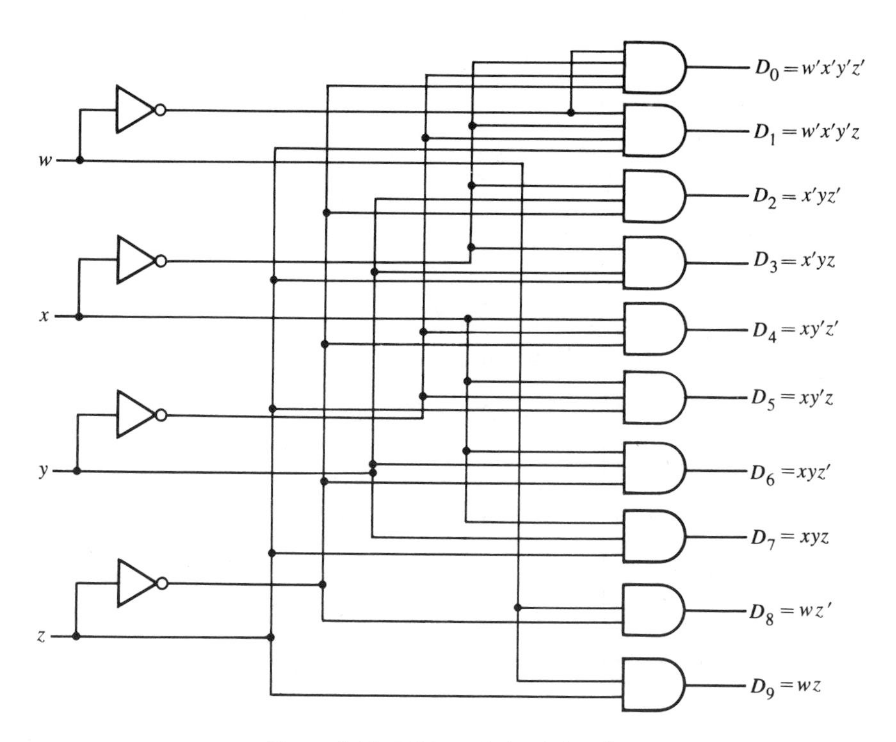

Figure 4-14 BCD to decimal decoder

Table 4-6 shows the input-output relations of the decoder. Only the first 10 input combinations are valid code assignments, the last six are not used and are, by definition, don't-care conditions. It is obvious that the don't-care conditions were used initially to simplify the output functions because, otherwise, each gate would have required four inputs. For the benefit of a complete analysis, Table 4-6 lists the outputs for the six not used combinations of the BCD code, but these combinations obviously have no meaning in this circuit.

Decoders and encoders have many applications in digital systems. Decoders are useful for displaying discrete elements of information stored in registers. For example, a decimal digit represented in BCD and stored in a four-cell register may be displayed with the help of a BCD to decimal decoder, with the outputs of the four binary cells feeding the inputs of the decoder and the outputs of the decoder driving 10 indicator lights. The indicator lights may be in the form of display digits, such that one decimal digit lights up when the corresponding decoder output is a logic-1. Decoder circuits are also useful in applications where the contents of registers need to be determined for the purpose of decision making. Another application

Table 4-6 Truth Table of BCD to Decimal Decoder of Figure 4-14

Inputs				*Outputs*									
w	x	y	z	D_0	D_1	D_2	D_3	D_4	D_5	D_6	D_7	D_8	D_9
0	0	0	0	1	0	0	0	0	0	0	0	0	0
0	0	0	1	0	1	0	0	0	0	0	0	0	0
0	0	1	0	0	0	1	0	0	0	0	0	0	0
0	0	1	1	0	0	0	1	0	0	0	0	0	0
0	1	0	0	0	0	0	0	1	0	0	0	0	0
0	1	0	1	0	0	0	0	0	1	0	0	0	0
0	1	1	0	0	0	0	0	0	0	1	0	0	0
0	1	1	1	0	0	0	0	0	0	0	1	0	0
1	0	0	0	0	0	0	0	0	0	0	0	1	0
1	0	0	1	0	0	0	0	0	0	0	0	0	1
1	0	1	0	0	0	1	0	0	0	0	0	1	0
1	0	1	1	0	0	0	1	0	0	0	0	0	1
1	1	0	0	0	0	0	0	1	0	0	0	1	0
1	1	0	1	0	0	0	0	0	1	0	0	0	1
1	1	1	0	0	0	0	0	0	0	1	0	1	0
1	1	1	1	0	0	0	0	0	0	0	1	0	1

of decoder circuits is in the generation of timing and sequencing signals for control purposes. These applications are introduced in their appropriate places in the following chapters.

Encoder circuits are useful for binary code formation when the discrete elements of information are each available from a single line. An example of an encoder circuit is demonstrated in Fig. 1-2. The box labeled "control" contains an encoder circuit that accepts an input from each key of the keyboard and generates the corresponding eight-bit code of the letter whose key is struck.

4-9 MULTIPLEXERS AND DEMULTIPLEXERS

Multiplexing means transmitting a large number of information units over a smaller number of channels or lines. *Demultiplexing* is a reverse operation and denotes receiving information from a small number of channels and distributing it over a larger number of destinations. A digital multiplexer is a combinational circuit that selects data from 2^n input lines and directs it to a single output line. The selection of input-output transfer paths is controlled by a set of input selection lines. An example of a multiplexer is shown in Fig. 4-15. The eight input lines I_0 to I_7 are applied to eight AND gates whose outputs go to a single OR gate. Only one input line has a path to the output at any particular time. The selection lines S_0, S_1, and S_2

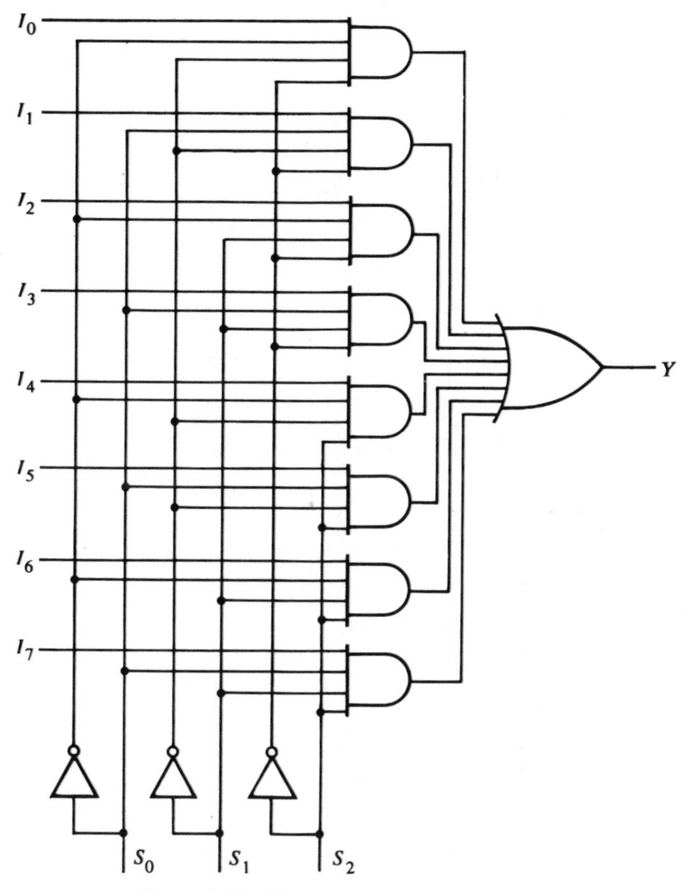

Figure 4-15 Eight-input digital multiplexer

determine which input is selected to have a direct path to the output. The eight AND gates resemble a decoder circuit and indeed decode the three input selection lines. The output Boolean function of the eight-input multiplexer shows clearly how the selection is accomplished.

$$Y = I_0S_2'S_1'S_0' + I_1S_2'S_1'S_0 + I_2S_2'S_1S_0' + I_3S_2'S_1S_0 + I_4S_2S_1'S_0' + I_5S_2S_1'S_0 + I_6S_2S_1S_0' + I_7S_2S_1S_0$$

Figure 4-16 is an example of a multiplexer that selects one of two data inputs A and B, each input data consisting of three bits A_2, A_1, A_0 and B_2, B_1, B_0. A one-bit selection line determines which input, A or B, is to be applied to the outputs. The Boolean functions for the outputs are:

$$Y_0 = A_0S' + B_0S$$

$$Y_1 = A_1S' + B_1S$$

$$Y_2 = A_2S' + B_2S$$

In general, an m-input, k-bit multiplexer requires n selection lines (with m =

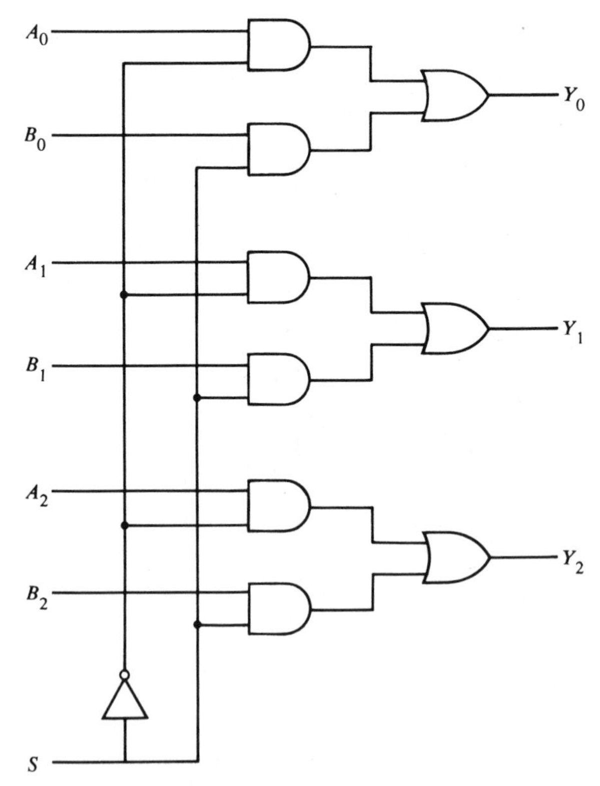

Figure 4-16 Two-input, 3-bit digital multiplexer

2^n) for decoding the input data. There are k outputs, each with an OR gate, and mk inputs, each with an AND gate. The decoding scheme is repeated k times.

An example of a demultiplexer is shown in Fig. 4-17. A single input line is steered to any of four identical outputs under control of two selection lines. It consists of four three-input AND gates, each receiving the data input along with one of the four possible combinations of the selection variables. The single input variable has a path to all four outputs, but the information is directed to the one output specified by the two selection lines. A demultiplexer can function as a decoder circuit if the single input is connected permanently to a signal that corresponds to logic-1. Multiplexer and demultiplexer devices, used in conjunction, are ideal in systems where it is desired to multiplex many data lines, transmit on one line, and convert back to the original data form at the receiving end for processing.

An interesting application for a multiplexer circuit is its use as a universal logic element; i.e., a circuit that implements any Boolean function of n variables. A multiplexer circuit with k selection lines and 2^k input wires is a universal logic element for Boolean functions of k + 1 variables. For example, the circuit of Fig. 4-15 with three selection lines S_2, S_1, S_0 and eight input lines I_0 through I_7 can be used to implement *any* Boolean function of four variables $Y(I, S_2, S_1, S_0)$, where S_2, S_1, and S_0 are three of the input variables and I is the fourth input variable. By expanding the

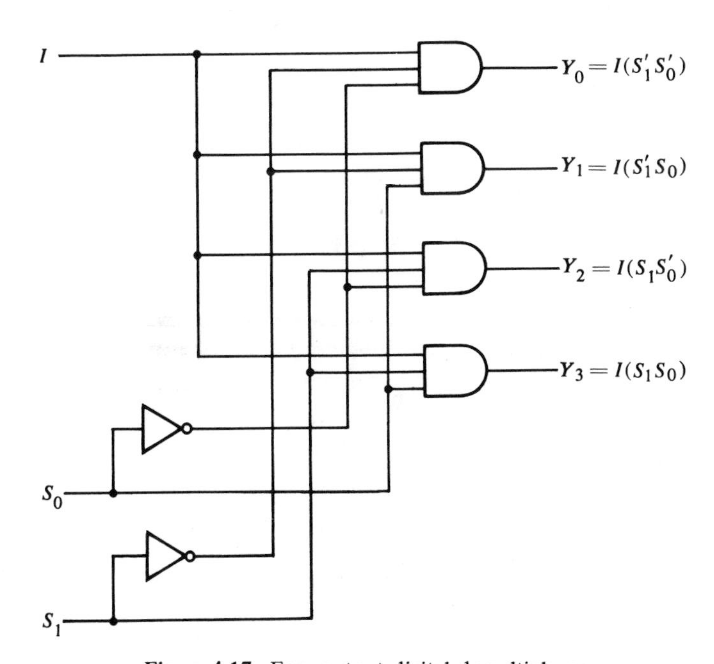

Figure 4-17 Four-output digital demultiplexer

function into its sum of minterms, it is possible to determine values for inputs I_0 through I_7. These values are either 0 or 1 or I or I'. Inputs 0 and 1 are fixed logic signals. Inputs I and I' are the normal and complement values of the fourth variable.

As an example, consider the following Boolean function of four variables expressed in sum of minterms:

$$Y(I, S_2, S_1, S_0) = \Sigma(0, 4, 8, 13)$$

To implement this function with the multiplexer circuit of Fig. 4-15 we need to determine values for I_0 through I_7. The function is first expressed in terms of the four variables:

$$Y = I'S_2'S_1'S_0' + I'S_2S_1'S_0' + IS_2'S_1'S_0' + IS_2S_1'S_0$$

Two minterms are combined if and only if they cause the elimination of the variable I. The first and third minterms in this example are combined and Y can be expressed as follows:

$$Y = S_2'S_1'S_0' + I'S_2S_1'S_0' + IS_2S_1'S_0$$

Comparing this expression with the Boolean function for the multiplexer given above, we determine the following input values:

$$I_0 = 1$$

$$I_4 = I'$$

$$I_5 = I$$

$$I_1 = I_2 = I_3 = I_6 = I_7 = 0$$

This example demonstrates the procedure for determining the values of I_0 to I_7 from the minterm expression of the Boolean function.

PROBLEMS

4-1. A combinational circuit has four inputs and one output. The output is equal to 1 when: (1) all the inputs are equal to 1 or (2) none of the inputs are equal to 1 or (3) an odd number of inputs are equal to 1.

(a) Obtain the truth table.

(b) Find the simplified output function in sum of products.

(c) Find the simplified output function in product of sums.

(d) Draw the two logic diagrams.

4-2. A three-digit binary number is represented by the Boolean variables w, x, y, and z. w represents the sign of the number so that $w = 1$ if the number is negative and $w = 0$ if the number is non-negative (positive or zero). x, y, and z represent the magnitude of the number, with z being the least significant digit. Design a combinational circuit that fulfills the following requirements: (1) if the input number is non-negative, the output number is equal to the input number minus two; (2) if the input number is negative, the output number is equal to the input number plus two.

4-3. Design a combinational circuit that accepts a three-bit number and generates an output binary number equal to the square of the input number.

4-4. It is necessary to multiply two binary numbers, each two bits long, in order to form their product in binary. Let the two numbers be represented by a_1,a_0 and b_1,b_0, where subscript 0 denotes the least significant bit.

(a) Determine the number of output lines required.

(b) Find the simplified Boolean expressions for each output.

4-5. Repeat Prob. 4-4 to form the sum (instead of the product) of the two binary numbers.

4-6. Obtain the simplified Boolean functions of the full-adder and full-subtractor in product of sums and draw the logic diagrams.

4-7. The half-adder is a circuit that adds two bits. The full-adder is a circuit that adds three bits. Now design a circuit that adds four bits. Determine the number of outputs necessary to form the sum and carries and find their simplified Boolean functions.

4-8. Design a combinational circuit with four input lines that represent a decimal digit in BCD and four output lines that generate the 9's complement of the input digit.

4-9. Design a combinational circuit whose input is a four-bit number and whose output is the 2's complement of the input number.

4-10. Design a combinational circuit that multiplies by 5 an input decimal digit represented in BCD. The output is also in BCD. Show that the outputs can be obtained from the input lines without using any logic gates.

4-11. Design a combinational circuit that detects an error in the representation of a decimal digit in BCD. In other words, obtain a logic diagram whose output is logic-1 when the inputs contain an unused combination in the code.

4-12. Implement a full-subtractor with two half-subtractors and an OR gate.

4-13. Show how a full-adder can be converted to a full-subtractor with the addition of one inverter circuit.

4-14. Design a combinational circuit that converts a decimal digit from the 8,4,-2,-1 code to BCD.

4-15. Design a combinational circuit that converts a decimal digit from the 2,4,2,1 code to the 8,4,-2,-1 code.

4-16. Obtain the logic diagram that converts a four-digit binary number to a decimal number in BCD. Note that two decimal digits are needed since the binary numbers range from 0 to 15.

4-17. Obtain the logic diagram of a circuit that compares two four-bit numbers. Use the algorithm given in Sec. 4-6.

4-18. Show the external connections between two three-bit comparators (Fig. 4-10) that will form a comparator circuit for two numbers of five bits each.

4-19. Analyze the two output combinational circuits shown in Fig. P4-19. Obtain the Boolean functions for the two outputs and explain the circuit operation.

4-20. Derive the truth table of the circuit shown in Fig. P4-19.

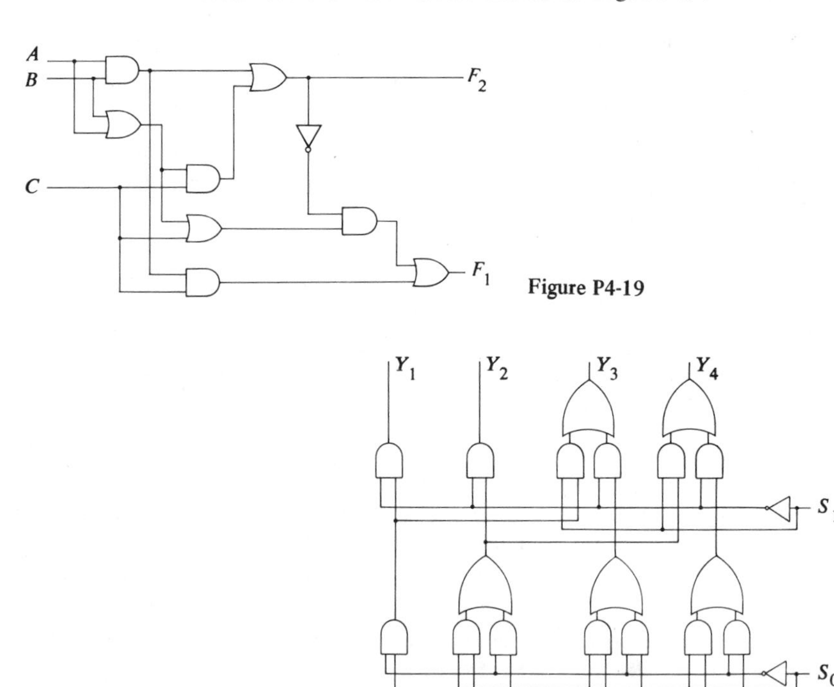

Figure P4-19

Figure P4-21

4-21. Analyze the circuit shown in Fig. P4-21 and explain the circuit operation.

4-22. The circuit of Fig. P4-21 establishes a given relation between the inputs and outputs depending on the value of the selection lines S_1 and S_0. Extend the circuit to seven outputs; that is, add outputs Y_5, Y_6, and Y_7 and the needed gates to accomplish an equivalent relation.

4-23. Obtain the logic diagram of an excess-3 to decimal decoder.

4-24. A BCD to seven-segment decoder is a combinational circuit that accepts a decimal digit in BCD and generates the appropriate outputs for selection of segments in a display indicator used for displaying the decimal digit. The seven outputs of the decoder (a, b, c, d, e, f, g) select the corresponding segments in the display as shown in Fig. P4-24(a). The numeric designation chosen to represent the decimal digit is shown in Fig. P4-24(b).

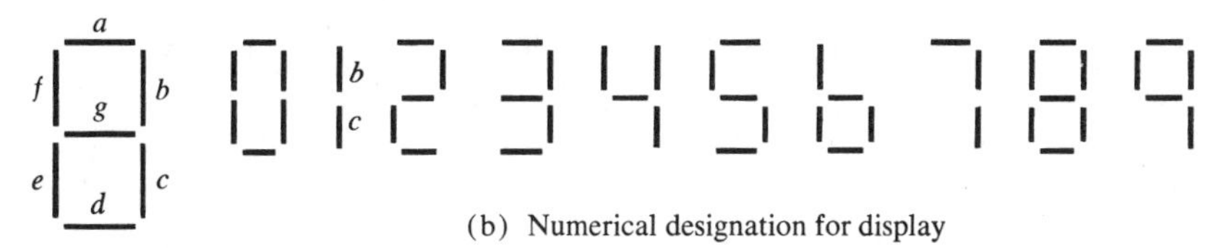

(b) Numerical designation for display

(a) Segment designation

Figure P4-24

(a) Design a seven-segment decoder using all don't-care conditions.

(b) Show the resulting displays generated when the unused bit combinations of the BCD code are applied to the inputs.

(c) Redesign the decoder so that unused combinations give meaningless displays.

4-25. Obtain the logic diagram of a decimal to BCD encoder.

4-26. Design an encoder whose inputs are the letters A to I and whose outputs represent the corresponding character in the six-bit internal code given in Table 1-5.

4-27. Six input lines must be decoded to obtain 64 output lines. AND gates are available with up to six inputs maximum. Each output of an AND gate can be connected to no more than 10 other AND gate inputs. Each external input can be connected to an inverter and no more than 10 AND gate inputs. Each inverter output can be connected to no more than 10 AND gate inputs. Determine a gate configuration for the decoder.

4-28. Show the logic diagram of a three-input, four-bit digital multiplexer.

4-29. Show the logic diagram of an eight-output digital demultiplexer. Show how this circuit in conjunction with the multiplexer of Fig. 4-15 can be used to multiplex eight data lines, transmit one line, and convert back to the original line at the receiving end.

4-30. Implement the following Boolean function with an eight-input digital multiplexer.

$$Y(I, S_2, S_1, S_0) = \Sigma(2, 4, 7, 10, 12, 14)$$

4-31. Implement a full-adder circuit with two four-input digital multiplexers.

5 GATE IMPLEMENTATION

5-1 INTEGRATED CIRCUITS

An integrated circuit (IC) is a small silicon crystal called a "chip" containing electronic components such as transistors, diodes, resistors, and capacitors. The various components are interconnected inside the chip to form an electronic circuit. The chip is mounted in a metal or plastic package and connections are welded to external pins. Integrated digital circuits with several logic gates are available in small packages like those shown in Fig. 5-1. Integrated circuits differ from conventional circuits in that individual components cannot be separated or disconnected and the circuit inside the package is accessed only through external pins. The benefits derived from ICs are: (1) substantial reduction in size, (2) substantial reduction in cost, (3) high reliability against failures, (4) increase in operating speed, and (5) reduction of externally wired connections.

As the technology of ICs has improved, the number of devices and components which can be put on a single silicon chip has increased considerably. The differentiation between those chips that have a few internal gates and those having tens or hundreds of gates is made by a customary reference to a package as being either a small-, medium-, or large-scale integration device. Several logic gates in a single package make it a small-scale integration (SSI) device. To qualify as medium-scale (MSI), a device must perform a complete logic function and have a complexity of approximately 10 gates or more. A large-scale integration (LSI) device performs a logic function with more than a hundred gates.

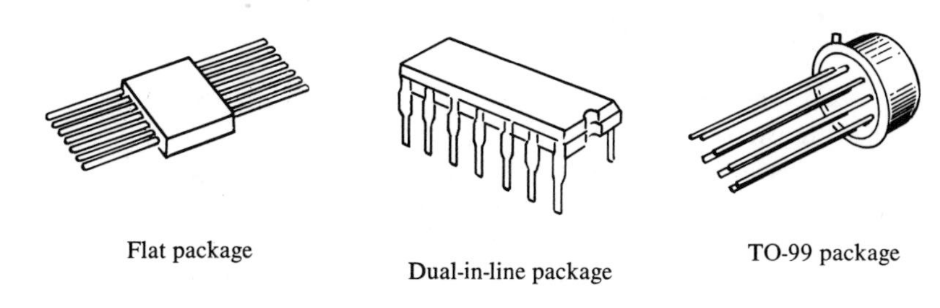

Figure 5-1 Integrated circuit packages

Examples of MSI functions are those introduced in Ch. 4: adders, subtractors, code converters, comparators, decoders, encoders, multiplexers, and demultiplexers. These functions are used extensively in the design of digital systems and are classified as standard items. These and similar digital functions are available in one IC package like the one shown in Fig. 5-1. Digital integrated circuits in an MSI package provide the logic designer with standard functional building blocks that can be very efficiently connected to satisfy a large variety of logic systems requirements.

An example of an LSI function is the general purpose accumulator register introduced in Ch. 9. LSI devices provide greater circuit complexity and may, at times, encompass the entire digital system in one chip. A digital computer may be constructed with only a few LSI devices to provide an extremely efficient method of packaging.

MSI and LSI devices provide a considerable decrease in package count over the same design using individual gates from SSI devices. This reduction in package count is significantly advantageous because the reliability of a system is a function of the number of packages used. In addition, fewer packages also mean fewer connections and less wiring. Higher speed of operation is achieved since interpackage delays are eliminated. Other advantages of MSI and LSI are reduction in power consumption, improvement of noise immunity, and a considerable savings in costs.

Logic diagrams of digital systems considered throughout the book are shown in detail up to the individual gates and their interconnections. Such logic diagrams are useful for demonstrating the logical construction of a particular function. However, it must be realized that, in practice, the function may be obtained from an MSI or LSI device and the user has access only to external inputs and outputs but not to inputs or outputs of intermediate gates. For example, a designer who wants to incorporate a decoder in his system is more likely to choose such a function from an available MSI integrated circuit instead of designing an individual gate structure.

5-2 DIGITAL LOGIC GATES

In Sec. 2-6 we defined eight binary operators and gave them the following characteristic names: AND, OR, inhibition, implication, exclusive-or, equivalence, NOR, and NAND. Each operator defines a logical function of two variables. The AND and OR operators together with the unary NOT operator express Boolean functions. For this reason, logic gates that perform these functions have been employed to implement Boolean functions. The possibility of constructing logic gates for the other six binary operators is of practical interest. Factors to be weighed when considering the construction of other types of logic gates are: (1) the feasibility and economy of producing the gate with physical components, (2) the possibility of extending the gate to more than two inputs, (3) the basic properties of the binary operator such as commutativity and associativity, and (4) the ability of the gate to implement Boolean functions alone or in conjunction with other gates.

The binary operators "inhibition" and "implication" are not commutative and are thus impractical for use as standard logic gates. The exclusive-or and equivalence operations have many excellent characteristics as candidates for logic gates but are expensive to construct with physical components. They are available as standard logic gates in IC packages but are usually constructed internally with other standard gates (see Fig. 5-28). The NAND and NOR functions are extensively used as standard logic gates and are in fact more popular than the AND and OR gates. This is because NAND and NOR gates are easily constructed with transistor circuits and because Boolean functions can be easily implemented with them.

The block diagram symbols for the digital logic gates of exclusive-or and equivalence are shown in Fig. 5-2. The symbol for the exclusive-or gate is similar to the OR gate except for the curved line on the input side. Since equivalence is the complement of exclusive-or, it is customary to use identical symbols for both except for the small circle in the output of the equivalence gate. Henceforth, a small circle in the output (or input) of a gate will designate complementation.

The NAND function is the complement of the AND function. For this reason we use identical symbols with the exception of the small circle in the output of the NAND gate as shown in Fig. 5-3(a). Similarly, the NOR and OR functions, being the complement of each other, use the same symbol except for the circle in the output of the NOR gate as shown in Fig. 5-3(b). It should be clear now why the symbol for the inverter circuit has been drawn with a small circle. It differentiates it from a non-inverting amplifier as shown in Fig. 5-4(a). A non-inverting amplifier is used for signal amplitude restoration or for signal amplification when a large amount of power is needed to drive heavy loads.

The exclusive-or and equivalence gates are rarely extended to more than

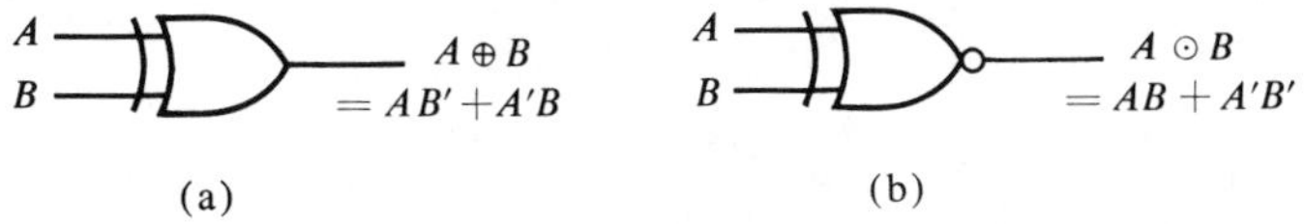

A
B
A ⊕ B
= AB′ + A′B
(a)
A
B
A ⊙ B
= AB + A′B′
(b)

Figure 5-2 Symbols for logic gates (a) exclusive-or, (b) equivalence

A
B
A↑B
=(AB)′
(a)
A
B
A↓B
= (A+B)′
(b)

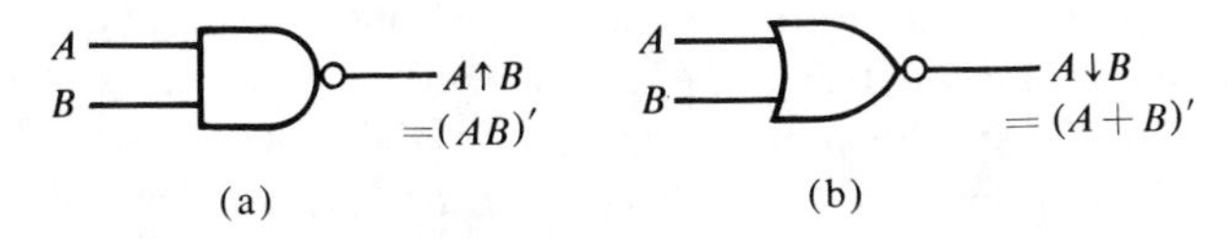

Figure 5-3 Symbols for logic gates (a) NAND, (b) NOR

A
A
(a)
A
A′
(b)

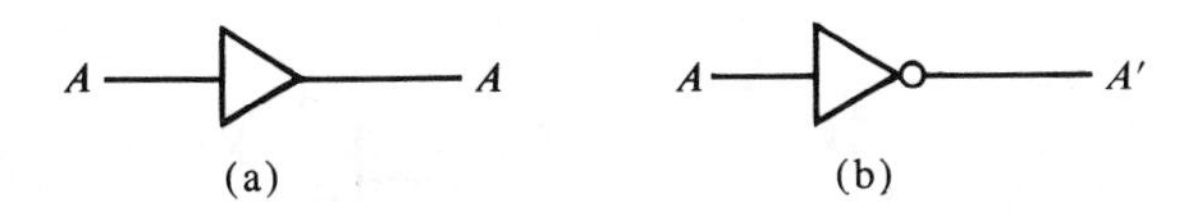

Figure 5-4 Symbols for (a) non-inverting amplifier, (b) inverter

two inputs. The NOR and NAND gates can be extended to many inputs provided the definition of the operation is slightly modified. The difficulty is that the NOR and NAND operators are not associative; i.e., $(x \downarrow y) \downarrow z \neq x \downarrow (y \downarrow z)$ as shown below and in Fig. 5-5.

$$(x \downarrow y) \downarrow z = [(x + y)' + z]' = (x + y) z'$$

$$x \downarrow (y \downarrow z) = [x + (y + z)']' = x' (y + z)$$

To overcome this difficulty, we define the multiple-input NOR (or NAND) gate as a complemented OR (or AND) gate.

x
y
z
(x↓y)↓z = (x + y) z′
x
y
z
x↓(y↓z) = x′(y+z)

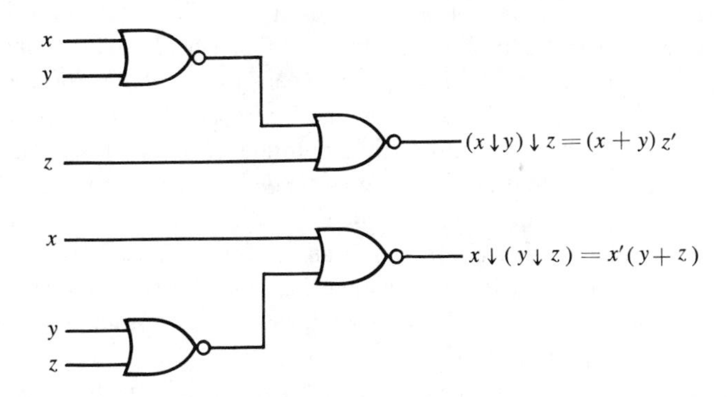

Figure 5-5 Demonstrating the non-associativity of the NOR operator; $(x \downarrow y) \downarrow z \neq x\ (y \downarrow z)$

Thus by definition we have:

$$x \downarrow y \downarrow z = (x + y + z)'$$

$$x \uparrow y \uparrow z = (xyz)'$$

The symbols for the three-input gates are shown in Fig. 5-6. NOR and NAND are sometimes called OR-invert and AND-invert to conform with the generalized definition. In writing cascaded NOR and NAND operations using the symbolic operators (↓) and (↑) as in Fig. 5-5, one must use the correct parentheses to signify the proper sequence of these operations. The operation signifies a multiple-input gate when no parentheses are included. NAND and NOR gates are discussed in more detail in Secs. 5-5 and 5-6. Functions suitable for implementation with exclusive-or and equivalence gates are presented in Sec. 5-8.

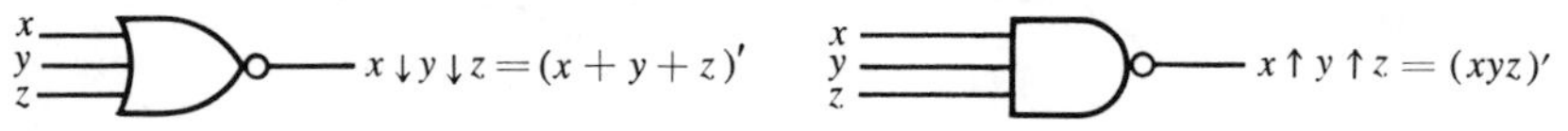

x y z — $x \downarrow y \downarrow z = (x + y + z)'$

(a) NOR

x y z — $x \uparrow y \uparrow z = (xyz)'$

(b) NAND

Figure 5-6 Three-input NOR and NAND gates

5-3 POSITIVE AND NEGATIVE LOGIC

The relation between a binary signal and a binary variable was discussed in Sec. 1-8. A binary signal exists in one of two values except during transition. One signal value represents logic-1 and the other, logic-0. Since two signal values are assigned to two logic values, there exists two different assignments of signals to logic. Because of the principle of duality of Boolean algebra, an interchange of signal value assignment results in a dual function implementation. The consequences of an interchange of signal assignment is investigated in this section.

Consider the two values of a binary signal as shown in Fig. 5-7. One value must be higher in amplitude than the other since the two values must be different in order to distinguish between them. Designate the higher amplitude by H and and the lower by L. There are two choices for signal value assignment. Choosing the high amplitude H to represent logic-1 as shown in Fig. 5-7(a) defines a *positive logic* system. Choosing the low amplitude L to represent logic-1 as in Fig. 5-7(b) defines a *negative logic* system. The terms *positive* and *negative* are somewhat misleading since both signal values may be positive or both may be negative. It is not the polarity of the signal that determines the type of logic but the assignment according to their relative magnitude.

Consider for example, a two-input AND gate whose truth table is given in Table 5-1(a). Assume that we are operating with a positive logic system so that logic-1 is equivalent to signal H and logic-0 to signal L. By a direct

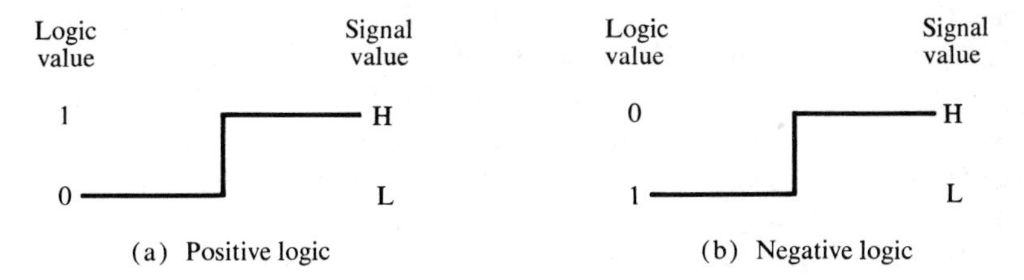

Figure 5-7 Signal amplitude assignment and type of logic

Table 5-1 Positive Logic AND, Negative Logic OR

(a)

x	*y*	*z*
0	0	0
0	1	0
1	0	0
1	1	1

(b)

x	*y*	*z*
L	*L*	L
L	*H*	L
H	*L*	L
H	*H*	H

(c)

x	*y*	*z*
1	1	1
1	0	1
0	1	1
0	0	0

substitution, we obtain the input-output signal amplitude relations of the circuit as shown in Table 5-1(b). Similarly, the truth table of a two-input OR gate in Table 5-2(a) translates into the positive logic assignment of Table 5-2(b). It is important to realize that part (b) of the two tables defines the input-output behavior of the physical circuit and that the logic assignment of 1 and 0 is chosen arbitrarily by the user. Therefore, if we take the same piece of hardware whose physical behavior is specified in Table 5-1(b) and employ negative logic, the *L* signal becomes logic-1 and the *H* signal logic-0. A direct substitution of logic values for signal values

Table 5-2 Positive Logic OR, Negative Logic AND

(a)

x	*y*	*z*
0	0	0
0	1	1
1	0	1
1	1	1

(b)

x	*y*	*z*
L	*L*	*L*
L	*H*	*H*
H	*L*	*H*
H	*H*	*H*

(c)

x	*y*	*z*
1	1	1
1	0	0
0	1	0
0	0	0

gives the relations listed in Table 5-1(c), from which we obtain the OR function. Similarly, the AND function of Table 5-2(c) is obtained from the negative logic choice of the circuit specified by Table 5-2(b). From these observations we can easily conclude that a positive logic AND gate and a negative logic OR gate are the same piece of hardware. A similar statement can be made for a positive logic OR gate and a negative logic AND gate. The type of logic gate we choose to call the physical circuit depends entirely on whether we choose positive or negative logic.

What has been said with respect to AND and OR in conjunction with positive and negative logic also applies to NOR and NAND because these two functions are the dual of each other. Thus, a circuit that performs the logic function of NOR when positive logic symbols are used will perform the function of NAND for negative logic and vice versa.

The choice of positive or negative logic for a given set of standard logic circuits is arbitrary. For this reason, standard functions in MSI devices are sometimes described in tables in terms of H and L signals instead of 1 and 0. The choice of positive or negative logic in many instances has been dictated by the type of transistors employed. Circuits using NPN-type transistors use positive signals and are usually assigned positive logic values, while circuits using PNP-type transistors use negative signals and are usually assigned negative logic values. This is the reason for the terminology of positive and negative logic.

5-4 SIMPLIFICATION CRITERIA

The purpose of Boolean function simplification is to obtain an algebraic expression that, when implemented, results in a low-cost circuit. However, the criteria that determine a low-cost circuit or system must be defined if we are to evaluate the success of the achieved simplification. It has previously been mentioned that criteria for simplification depend on the constraints imposed by the particular application. The subject of *switching theory* is concerned, among other things, in finding algorithms for simplifying Boolean functions. The best known algorithm for simplifying Boolean functions is the tabulation (or map) method presented in Ch. 3. This method gives an answer in one of the two standard forms and achieves a minimization of the number of terms and the number of literals of the function. Other algorithms may be found in the literature of switching theory, but they usually satisfy a narrow set of criteria and are not general enough for inclusion in a textbook.* The simplest approach available to a logic designer is to use the map method in conjunction with algebraic manipulation and his skill, experience, and ingenuity. The labor involved

*The best reference for articles on switching theory is the *IEEE Transactions on Computers.*

may be reduced if a computer program for simplifying Boolean functions is available.

Although algorithms that satisfy all possible criteria for simplification are not available, it is important that the designer be familiar with the different requirements, restrictions, and limitations of a practical situation. Criteria for Boolean function simplification may be divided into three broad categories: (1) reduction in the number of components and wires, (2) reduction in propagation delay, and (3) loading constraints. Each of these categories is discussed below.

Number of Components and Wires

It seems reasonable that we would try to minimize the following items if we wanted to reduce the cost of implementing digital systems:

1. The number of logic gates.
2. The number of inputs to a gate.
3. The number of integrated circuit packages.
4. The number of printed-circuit boards (IC packages are sometimes mounted on printed circuit boards).
5. The number of connecting wires.

The map (or tabulation) method minimizes the number of logic gates and inputs to a gate provided we are satisfied with a standard form implementation. Given two circuits that perform the same function, the one that requires less gates is more likely to be preferable because it will cost less. This is not necessarily true when integrated circuits are used. Since several logic gates (or an entire function) are included in a single IC package, it becomes economical to use as many of the gates from an already used package as possible even if, by doing so, we increase the total number of gates. Moreover, some of the interconnections among gates in an IC are internal to the chip and it is more economical to use as many internal interconnections as possible in order to minimize the number of wires between external pins. With integrated circuits, it is not the count of logic gates that determines the cost but the number and type of ICs used to implement the given function. In order to minimize the number of ICs and the number of external interconnections, we must specify the physical layout and construction of the various components. Computer programs that consider the pertinent physical variables are sometimes used for optimizing the cost of digital systems.

Propagation Delay and Logic Levels

The signals through a logic gate take a certain amount of time to propagate from the inputs of the circuit to its output. This interval of time is defined as the *propagation delay* of the circuit. The signals that travel from the inputs of a combinational circuit to its outputs pass through a series of gates. The sum of the propagation delays through the gates is the total propagation delay of the circuit. A reduced propagation delay means a faster operation. When speed of operation is important, logic gates must have small propagation delays and combinational circuits must have a minimum number of series gates between inputs and outputs.

The input signals in most combinational circuits are applied simultaneously to more than one gate. All those gates that receive their inputs exclusively from external inputs constitute the first logic level of the circuit. Gates that receive at least one input from an output of a first logic level gate constitute the second level and similarly for third and higher levels. The total propagation delay through the combinational circuit is equal to the propagation delay through a gate times the number of logic levels in the circuit. Thus, a reduction in the number of levels results in a reduction of signal delay and a faster circuit. The reduction of the propagation delay in circuits may be more important than the reduction in the number of gates if speed of operation is a major factor.

A Boolean function expressed in sum of products or product of sums can be implemented with two levels of gates provided both normal and complement inputs are available. A Boolean function expressed in one of the standard forms is said to provide a two-level implementation. Obviously, if only the normal input variables are available, the inverters that generate the complements constitute a third level in the implementation.

Because any function can be written as a sum of products, any function can be implemented with a two-level implementation. The use of two-levels provides the least propagation delay and thus the fastest circuit. If speed is unimportant, a multilevel implementation of less gates is normally more desirable. General algorithms for obtaining a simplified multilevel implementation are not available. The designer must resort to algebraic manipulation or to any available computer program that searches for a minimum gate implementation under certain given conditions. Examples of multilevel implementation can be found in Ch. 4. These are: the full-adder of Fig. 4-11 (five levels), the code converter of Fig. 4-8 (four levels), and the comparator of Fig. 4-10 (four levels).

Fan-In, Fan-Out, and Loading

Fan-in specifies the number of inputs to a gate. For example, a four-input AND gate has a fan-in of four. For logic gates constructed with discrete components, an increase in fan-in usually means adding more inputs by inserting more diodes or resistors. The fan-in limitation in this case is determined by the ability of the circuit to support additional inputs and still function within the allowable tolerance. The fan-in of IC gates is determined mostly from the limitation of external pins in the package. For example, the dual-in-line package shown in Fig. 5-1 has 14 external pins. One pin is needed for supply voltage and one for ground connection, leaving 12 for inputs and outputs. Suppose it is desirable to include one gate in a package. The largest fan-in that can be achieved is 11 since one pin must be available for the output. If it is desired to include two gates in one IC package with equal number of inputs per gate, then the largest fan-in for each is five (two gates, each with five inputs and one output, require a total of 12 pins). When Boolean functions are implemented with IC gates, the fan-in of available gates must be considered for minimizing the number of IC packages.

Fan-out specifies the number of "standard loads" an output of a gate can drive without impairing its normal operation. A standard load may be an input to a gate, to an inverter, or to any other circuit which must be specified in defining fan-out. Sometimes the term *loading* is used instead of fan-out. This term is derived from the fact that the output of a gate can supply a limited amount of power, above which it ceases to operate properly. Each circuit connected to the output consumes a certain amount of power, so that each additional circuit adds to the "load" of the gate. "Loading rules" are usually listed for a family of standard digital circuits. These rules specify the maximum amount of loading allowed for each output of each circuit. Exceeding the specified maximum load may cause a malfunction because the circuit cannot supply the power demanded from it. The loading (or fan-out) capabilities of a gate must be considered when simplifying Boolean functions. Care must be taken not to develop expressions that result in an overloaded gate. Non-inverting amplifiers (whose symbol is shown in Fig. 5-4) are sometimes employed to provide additional driving capabilities for heavy loads.

5-5 NAND LOGIC

Combinational circuits are more frequently constructed with NAND or NOR gates than with AND, OR, and NOT gates. NAND and NOR circuits are superior to AND and OR gates from the hardware point of view, as they supply outputs that maintain the signal value without loss of amplitude. OR and AND gates sometimes need amplitude restoration after the signal travels through a few levels of gates. Because of the prominence of NAND and NOR gates in the design of combinational circuits, rules and procedures have been developed for the conversion from Boolean functions given in terms of AND, OR, and NOT into equivalent NAND or NOR logic diagrams. The procedures for NAND logic are presented in this section and for NOR logic in the next section. Since the NOR operation is the dual of the NAND, the rules and procedures for NOR logic are the dual of those for NAND logic.

Universal Gate

The NAND gate is said to be a universal gate because any digital system can be implemented with it. Not only combinational circuits but sequential circuits as well can be constructed with this gate. This is because the flip-flop circuit (the memory element most frequently used in sequential circuits) can be constructed from two NAND gates connected back to back as shown in Sec. 6-2.

To show that any Boolean function can be implemented with NAND gates, we need only show that the logical operations AND, OR, and NOT can be implemented with NAND gates. The implementation of the AND, OR, and NOT operations with NAND gates is shown in Fig. 5-8. The NOT operation is obtained from a one-input NAND gate, actually another symbol for an inverter circuit. The AND operation requires two NAND gates. The first produces the inverted AND and the second acts as an inverter to obtain the normal output. The OR operation is achieved through a NAND gate with additional inverters in each input.

A convenient way to implement a combinational circuit with NAND gates is to obtain the simplified Boolean functions in terms of AND, OR, and NOT and convert the functions to NAND logic. The conversion of the algebraic expression from AND, OR, and NOT operations to NAND opera-

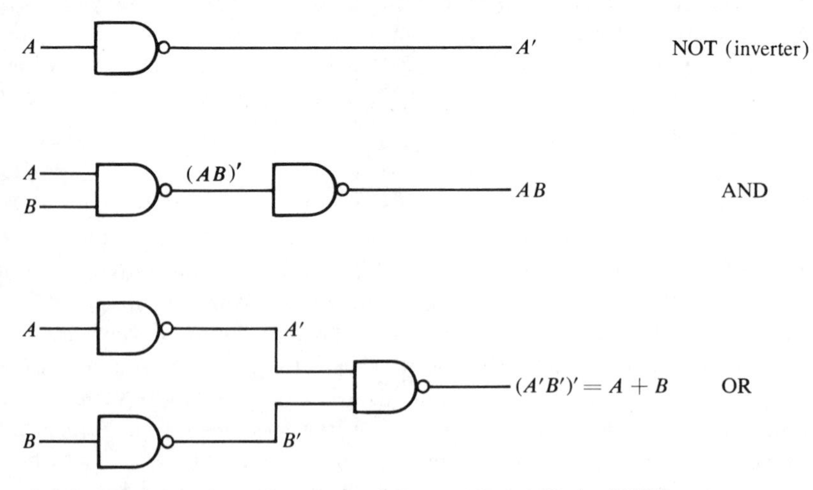

Figure 5-8 Implementation of NOT, AND or OR by NAND gates

tions (using ↑ symbol) is usually quite complicated because it involves a large number of applications of De Morgan's theorem. This difficulty is avoided by the use of simple circuit manipulations and simple rules as outlined below.

Boolean Function Implementation—Block Diagram Method

The implementation of Boolean functions with NAND gates may be obtained by means of a simple block diagram manipulation technique. This method requires that two other logic diagrams be drawn prior to obtaining the NAND logic diagram. Nevertheless, the procedure is very simple and straightforward:

1. From the given algebraic expression, draw the logic diagram with AND, OR, and NOT gates. Assume that both the normal and complement inputs are available.
2. Draw a second logic diagram with the equivalent NAND logic as given in Fig. 5-8 substituted for each AND, OR, and NOT gate.
3. Remove any two cascaded inverters from the diagram since double inversion does not perform a logic function. Remove inverters connected to single external inputs and complement the corresponding input variable. The new logic diagram obtained is the required NAND gate implementation.

This procedure is illustrated in Fig. 5-9 for the function:

$$F = A\ (B + CD) + BC'$$

The AND/OR implementation of this function is drawn in the logic diagram of Fig. 5-9(a). For each AND gate we substitute a NAND gate followed by an inverter; for each OR gate we substitute input inverters followed by a NAND gate. This substitution follows directly from the logic equivalences of Fig. 5-8 and is drawn in the diagram of Fig. 5-9(b). This diagram has seven inverters and five two-input NAND gates listed with numbers inside the gate symbol. Pairs of inverters connected in cascade (from each AND box to each OR box) are removed since they form double inversion. The inverter connected to input B is removed and the input variable is designated by B'. The result is the NAND logic diagram shown in Fig. 5-9(c), with the number inside each symbol identifying the gate from Fig. 5-9(b).

This example demonstrates that the number of NAND gates required to implement the Boolean function is equal to the number of AND/OR gates, provided both the normal and complement inputs are available. If only the normal inputs are available, inverters must be used to generate any required complemented inputs.

Don't worry about this

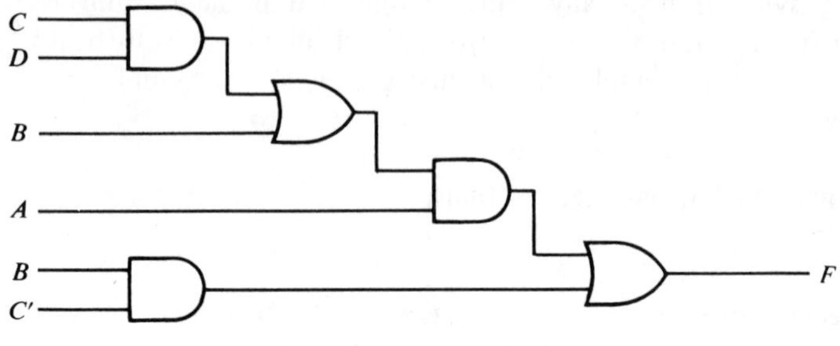

(a) AND/OR implementation

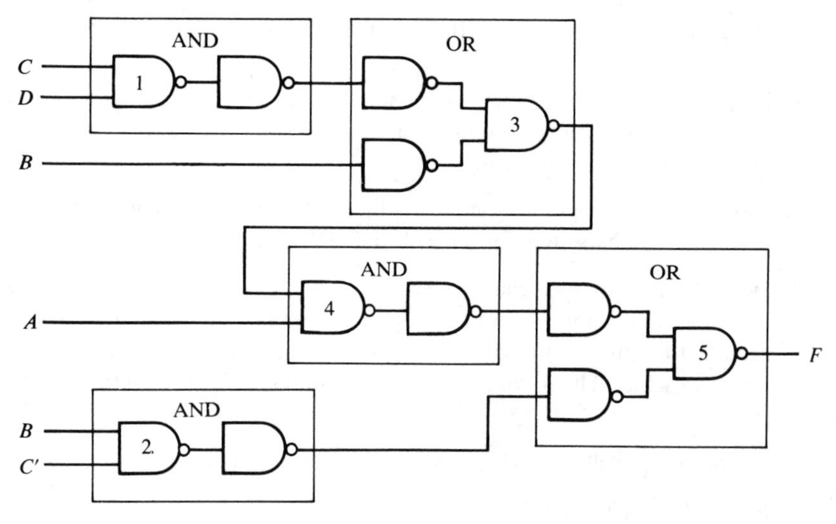

(b) Substituting equivalent NAND functions from Fig. 5-8

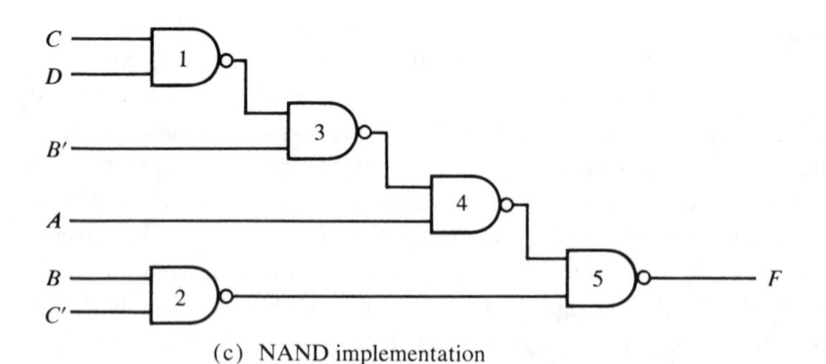

(c) NAND implementation

Figure 5-9 Implementation of $F = A(B + CD) + BC'$ with NAND gates

A second example of NAND implementation is shown in Fig. 5-10. The Boolean function to be implemented is:

$$F = (A + B')\,(CD + E)$$

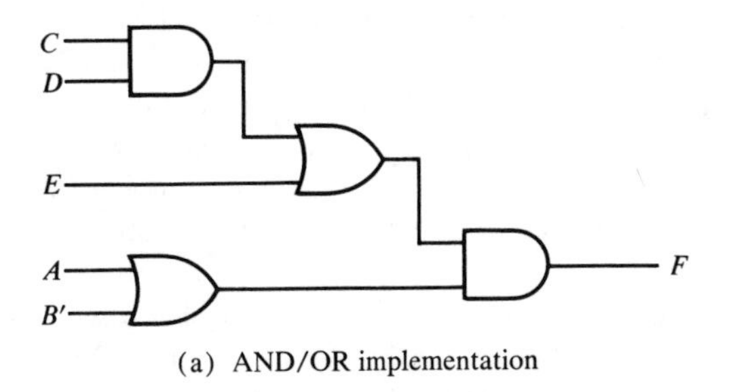

(a) AND/OR implementation

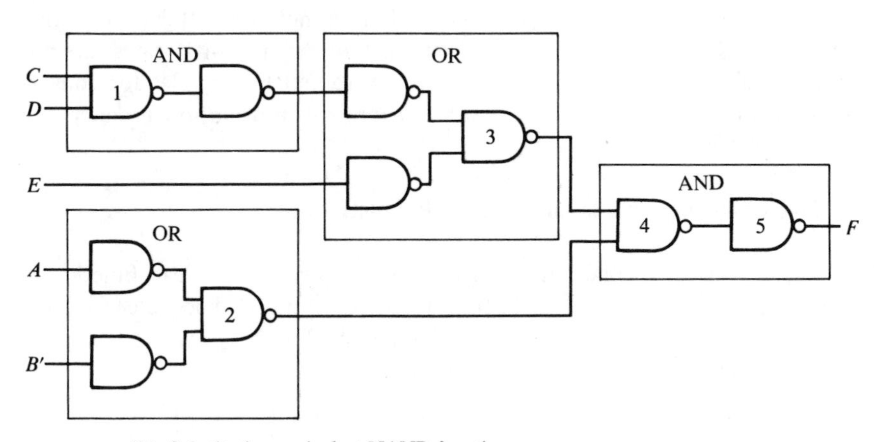

(b) Substituting equivalent NAND functions

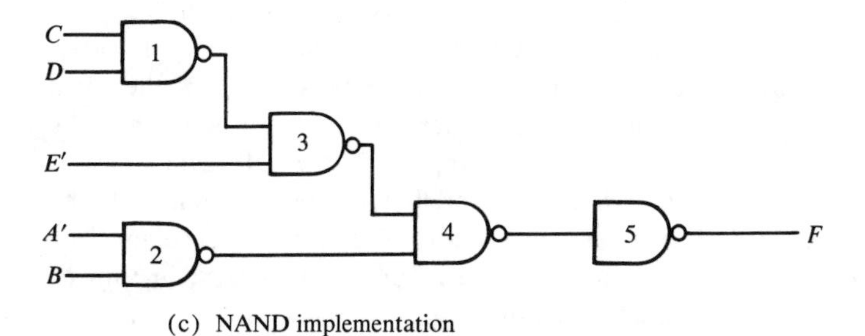

(c) NAND implementation

Figure 5-10 Implementation of $(A + B')(CD + E)$ with NAND gates

The AND/OR implementation is shown in Fig. 5-10(a) and its NAND logic substitution in Fig. 5-10(b). One pair of cascaded inverters may be removed. The three external inputs E, A, and B', which go directly to inverters, are complemented and the corresponding inverters removed. The final NAND gate implementation is in Fig. 5-10(c).

The number of NAND gates for the second example is equal to the number of AND/OR gates plus an additional inverter in the output (NAND gate number 5). In general, the number of NAND gates required to implement a function equals the number of AND/OR gates, except for an occasional inverter. This is true provided both normal and complement inputs are available because the conversion forces certain input variables to be complemented.

The block diagram method is somewhat tiresome to use because it requires the drawing of two logic diagrams to obtain the answer in a third. With some experience, it is possible to reduce the amount of labor by anticipating the pairs of cascaded inverters and the inverters in the inputs. Starting from the procedure just outlined, it is not too difficult to derive general rules for implementing Boolean functions with NAND gates directly from an algebraic expression.* We shall now formulate the rules for conversion when the Boolean functions are expressed in sum of products or product of sums.

Two-Level Implementation—Sum of Products

The derivation of the NAND logic diagram when the Boolean function is expressed in sum of products is simple and direct. The rule for this conversion can be deduced from inspection of Fig. 5-11, where a typical sum of product expression is implemented:

$$F = AB + CD + E$$

A sum of products expression is always implemented with a group of AND gates in the first logic level and a single OR gate in the second level, as shown in Fig. 5-11(a). The substitution of the NAND equivalent logic as drawn in Fig. 5-11(b) clearly shows that each first-level AND produces an output inverter and each input of the OR gate produces an input inverter. The pairs of cascaded inverters can be removed as shown in the final diagram of Fig. 5-11(c). A term with one literal is an exception, since the input variable is applied directly to the input of the OR gate. This is clearly shown in the diagram where the variable E, being a term of one literal, is complemented during the conversion. The rule for obtaining the NAND logic diagram directly from a Boolean function expressed in sum of products should be clear from this example:

1. Draw a NAND gate for each AND term of the function that has at least two literals. The inputs to each NAND gate are the literals of the term. This constitutes a group of first-level gates.

*A comprehensive set of rules for both NAND and NOR logic can be found in reference (1) at the end of this chapter.

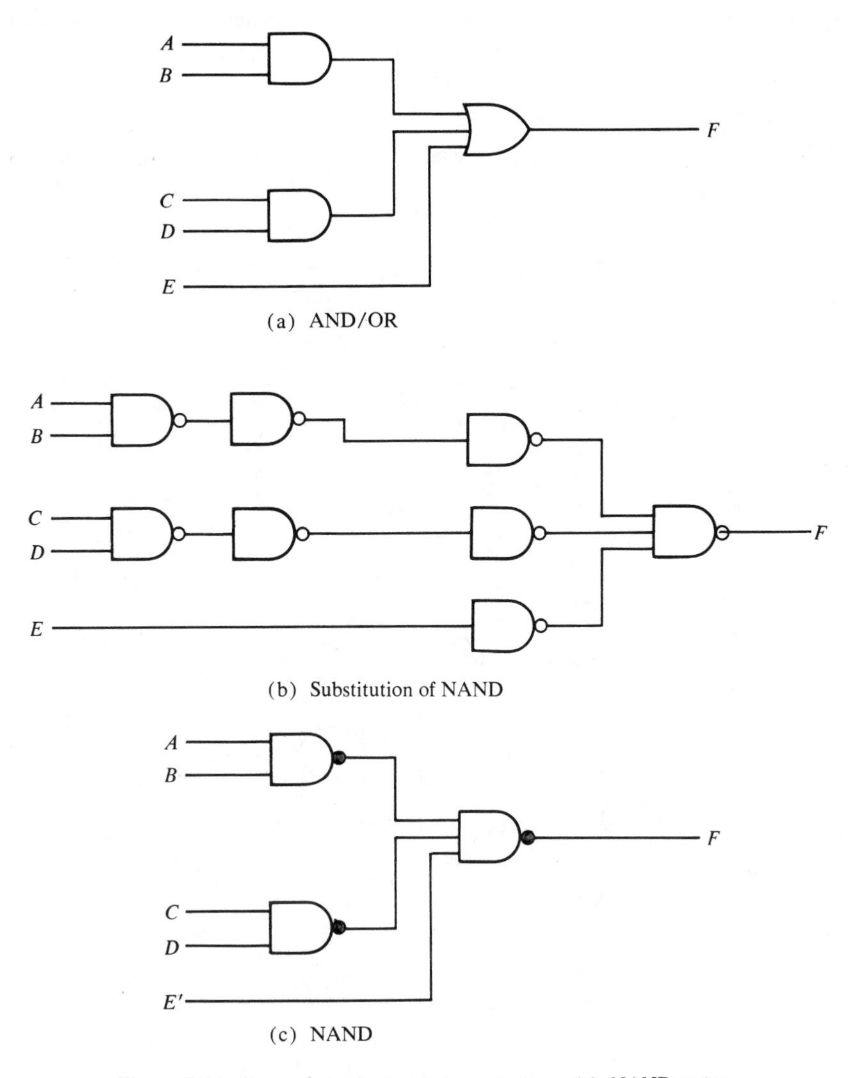

Figure 5-11 Sum of products implementation with NAND gates

2. Draw a single NAND gate in the second level with inputs coming from outputs of first-level gates.

3. Any literal that appears as a term by itself is complemented and applied as input to the second-level NAND gate.

Three-Level Implementation—Product of Sums

A Boolean function expressed in product of sums when implemented with NAND gates requires three levels of gates, the third and last level being an inverter. The rules of conversion from the algebraic expression to the logic diagram can be deduced from the example shown in Fig. 5-12, where the following function is implemented:

$$F = (A + B)(C + D)E$$

A product of sums expression is always implemented with a group of OR

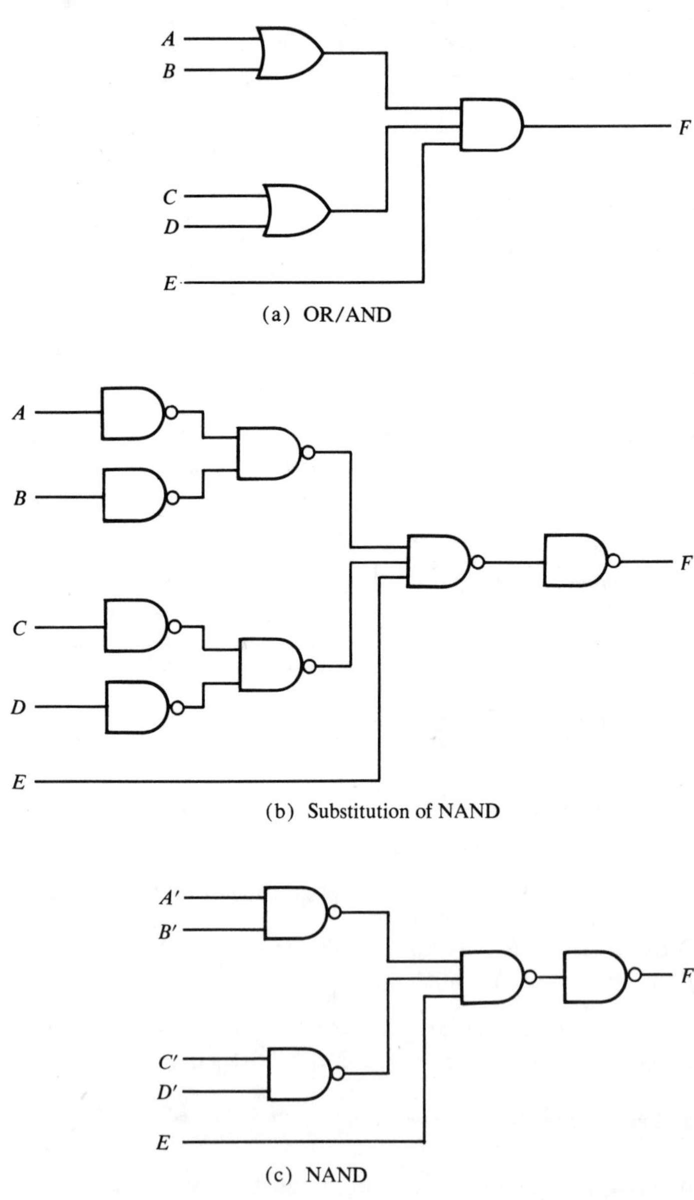

Figure 5-12 Product of sums implementation with NAND gates

gates in the first level and a single AND gate in the second level as shown in Fig. 5-12(a). The substitution of NAND equivalent logic, as in Fig. 5-12(b), clearly shows that each OR gate produces an input inverter and that the AND gate produces a single output inverter. There are no pairs of inverters that can be removed, but inverters connected to external inputs can be removed provided the input variables are complemented. The literal that forms a single term goes directly to the input of the second level and is not complemented during the conversion. The final NAND logic diagram is shown in Fig. 5-12(c). From it we can deduce the rule for obtaining the NAND logic diagram directly from a Boolean function expressed in product of sums:

1. Draw a NAND gate for each OR term of the function that has at least two literals. The inputs to each NAND gate are the *complements* of the literals in the term. This constitutes a group of first-level gates.
2. Draw a single NAND gate in the second level with inputs coming from outputs of first-level gates.
3. Any literal that appears as a term by itself is applied as input to the second-level NAND gate.
4. Draw a third-level NAND gate with a single input connected to the output of the second-level NAND gate.

In stating the rules for the sum of products and product of sums implementations it was assumed that both the normal and complement inputs were available. If only the normal inputs are available, the inverters that generate the complements constitute one additional logic level.

The following example demonstrates the derivation of the NAND logic diagrams for a Boolean function initially expressed in canonical form.

EXAMPLE 5-1. Implement the following Boolean function with NAND gates.

$$F(A, B, C, D) = \Sigma(0, 1, 3, 7, 8, 9, 10, 14, 15)$$

First, the expression must be simplified. This is done by using the map shown in Fig. 5-13. In (a) we combine the 1's to obtain a simplified expression in sum of products:

$$F = B'C' + A'CD + BCD + ACD'$$

Following the rules for implementing a sum of products expression, we obtain the NAND logic diagram of Fig. 5-14(a).

A different NAND gate structure is obtained when the function is simplified in a product of sums form. This is achieved by combining the 0's of the function as shown in the map of Fig. 5-13(b) and then complementing to obtain:

$$F = (B' + C)(A + C' + D)(A' + B + C' + D')$$

Following the rules for implementing a product of sums expression, we obtain the NAND logic diagram of Fig. 5-14(b). Both implementations

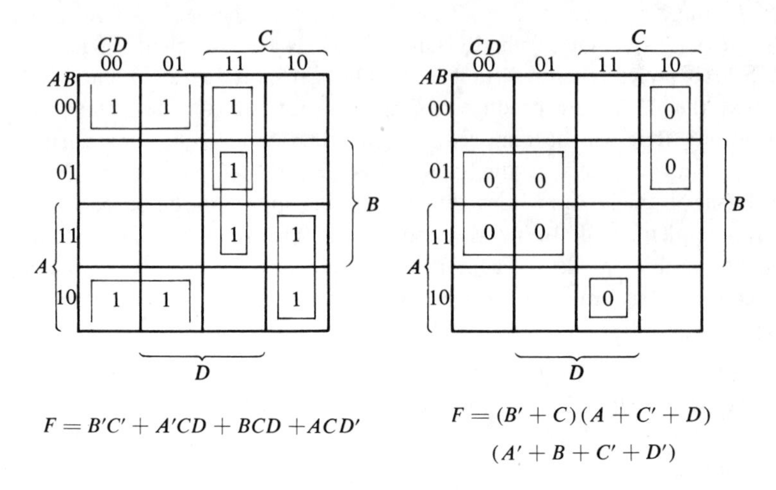

Figure 5-13 Map for Example 5-1

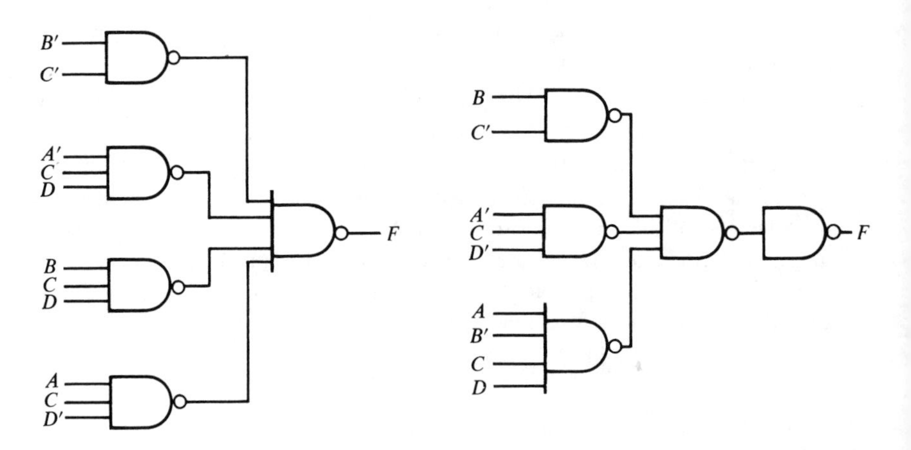

(a) Two-level from sum of products

(b) Three-level from product of sums

Figure 5-14 NAND implementation of the Boolean function of Example of 5-1

require five NAND gates, but one has two logic levels and the other, three. If for any reason only the complemented output of this function is required, the output inverter in Fig. 5-14(b) can be removed and the circuit becomes more economical than the one of Fig. 5-14(a).

Analysis Procedure

Up to now we have considered the problem of deriving a NAND logic diagram from a given Boolean function. The reverse process is the analysis problem which starts with a given NAND logic diagram and culminates with a Boolean expression or a truth table. The analysis of NAND logic diagrams follows the same procedures presented in Sec. 4-7 for the analysis of combinational circuits. The only difference is that NAND logic requires a repeated application of De Morgan's theorem. We shall now demonstrate the derivation of the Boolean function from a logic diagram. We shall then proceed to show the derivation of the truth table directly from the NAND logic diagram. Finally, a method will be presented for converting a NAND logic diagram to AND/OR/NOT logic diagram by means of block diagram manipulation.

Derivation of the Boolean Function by Algebraic Manipulation

The procedure for deriving the Boolean function from a logic diagram is outlined in Sec. 4-7. This procedure is demonstrated for the NAND logic diagram shown in Fig. 5-15, which is the same as Fig. 5-9(c). First, all gate outputs are labeled with arbitrary symbols. Second, the Boolean functions for the outputs of gates that receive only external inputs are derived:

$$T_1 = (CD)' = C' + D'$$
$$T_2 = (BC')' = B' + C$$

The second form follows directly from De Morgan's theorem and may, at times, be more convenient to use. Third, Boolean functions of gates which have inputs from previously derived functions are determined in consecutive order until the output is expressed in terms of input variables:

$$T_3 = (B'T_1)' = (B'C' + B'D')'$$
$$= (B + C)\ (B + D) = B + CD$$
$$T_4 = (AT_3)' = [A\ (B + CD)]'$$
$$F = (T_2T_4)' = \{(BC')'\ [A\ (B + CD)]'\}'$$
$$= BC' + A\ (B + CD)$$

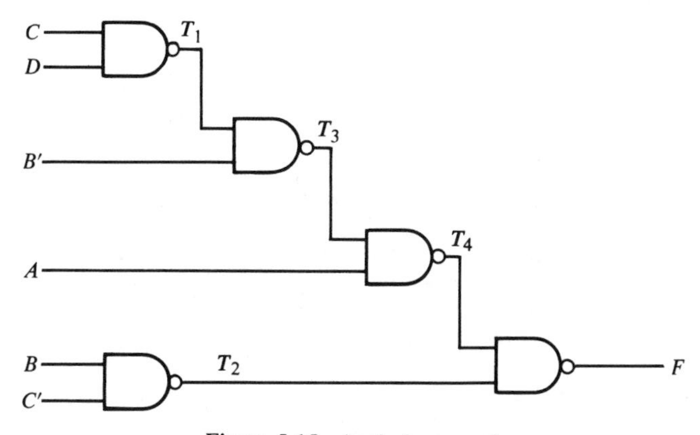

Figure 5-15 Analysis example

Derivation of the Truth Table

The procedure for obtaining the truth table directly from a logic diagram is also outlined in Sec. 4-7. This procedure is demonstrated for the NAND logic diagram of Fig. 5-15. First, the four input variables, together with their 16 combinations of 1's and 0's, are listed as in Table 5-3. Second, the outputs of all gates are labeled with arbitrary symbols as in Fig. 5-15. Third, we obtain the truth table for the outputs of those gates that are a function of the input variables only. These are T_1 and T_2. $T_1 = (CD)'$, so we mark 0's in those rows where both C and D are equal to 1 and fill the rest of the rows of T_1 with 1's. Also $T_2 = (BC')'$, so we mark 0's in those rows where $B = 1$ and $C = 0$, and fill the rest of the rows of T_2 with 1's. We then proceed to obtain the truth table for the outputs of those gates

Table 5-3 Truth Table for the Circuit of Figure 5-15

A	B	C	D	T_1	T_2	T_3	T_4	F
0	0	0	0	1	1	0	1	0
0	0	0	1	1	1	0	1	0
0	0	1	0	1	1	0	1	0
0	0	1	1	0	1	1	1	0
0	1	0	0	1	0	1	1	1
0	1	0	1	1	0	1	1	1
0	1	1	0	1	1	1	1	0
0	1	1	1	0	1	1	1	0
1	0	0	0	1	1	0	1	0
1	0	0	1	1	1	0	1	0
1	0	1	0	1	1	0	1	0
1	0	1	1	0	1	1	0	1
1	1	0	0	1	0	1	0	1
1	1	0	1	1	0	1	0	1
1	1	1	0	1	1	1	0	1
1	1	1	1	0	1	1	0	1

that are a function of previously defined outputs until the column for the output F is determined. It is now possible to obtain an algebraic expression for the output from the derived truth table. The map drawn in Fig. 5-16 is obtained directly from Table 5-3 and has 1's in the squares of those minterms for which F is equal to 1. The simplified expression obtained from the map is

$$F = AB + ACD + BC' = A(B + CD) + BC'$$

which is the same as the one of Fig. 5-9, thus verifying the correct answer.

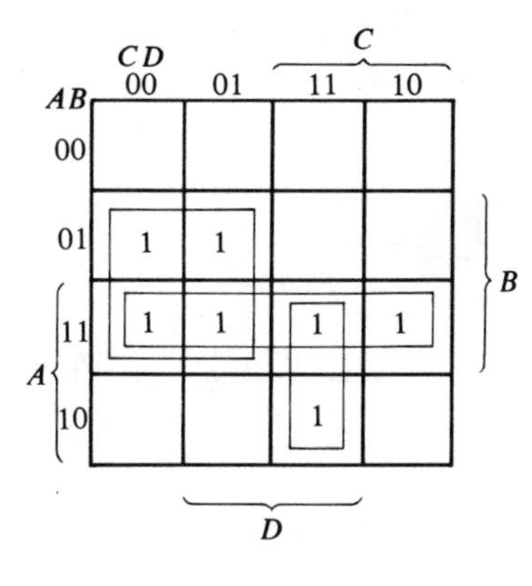

$$F = AB + BC' + ACD$$

Figure 5-16 Derivation of F from Table 5-2

Block Diagram Transformation

It is sometimes convenient to convert a NAND logic diagram to its equivalent AND/OR/NOT logic diagram to facilitate the analysis procedure. By doing so, the Boolean function can be derived more easily without employing De Morgan's theorem. The conversion of logic diagrams is accomplished through a process reverse from that used for implementation. A convenient equivalent symbol for a NAND gate for the conversion is shown in Fig. 5-17(b). Instead of representing a NAND gate with an AND symbol followed by a circle, we can represent it by an OR gate preceded by circles in all inputs. The invert-OR symbol for a NAND gate follows from De Morgan's theorem and from the convention that small circles denote complementation.

The conversion of a NAND logic diagram to an AND/OR/NOT diagram is achieved through a change in symbols from AND-invert to invert-OR in *alternate* levels of gates. The first level to be changed to an invert-OR symbol should be the last level. These changes produce pairs of circles along

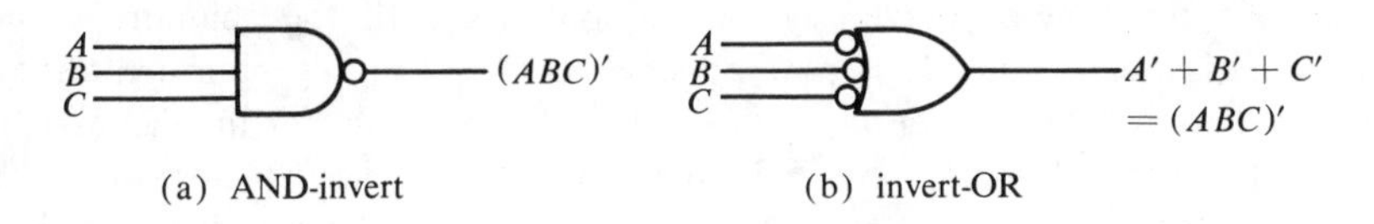

Figure 5-17 Two symbols for NAND gate

the same line which can be removed since they represent double complementation. Moreover, a one-input AND or OR gate can be removed since it does not perform a logical function. A one-input AND or OR with a circle in the input or output is changed to an inverter circuit.

This procedure is demonstrated in Fig. 5-18. The NAND logic diagram of

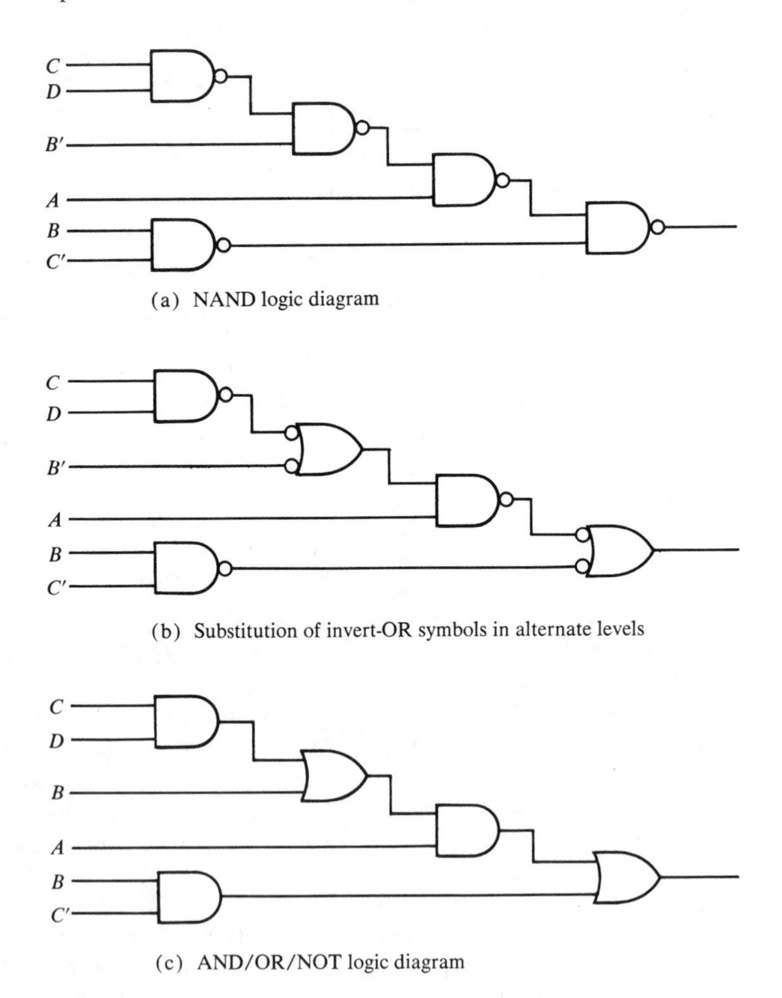

Figure 5-18 Conversion of NAND logic diagram to AND/OR/NOT

Fig. 5-18(a) is to be converted to an AND/OR diagram. The symbol of the gate in the last level is changed to an invert-OR. Looking for alternate levels, we find one more gate requiring a change of symbol as shown in Fig. 5-18(b). Any two circles along the same line are removed. Circles that go to external inputs are also removed, provided the corresponding input variable is complemented. The required AND/OR logic diagram is drawn in Fig. 5-18(c).

5-6 NOR LOGIC

The NOR function is the dual of the NAND function. For this reason, all procedures and rules for NOR logic form a dual of the corresponding procedures and rules developed for NAND logic. This section enumerates various methods for NOR logic implementation and analysis by following the same list of topics used for NAND logic. However, less detailed explanation is included so as to avoid excessive repetition of the material in Sec. 5-5.

Universal Gate

The NOR gate is universal because any Boolean function can be implemented with it, including a flip-flop circuit as shown in Sec. 6-2. The conversion of AND, OR, and NOT to NOR is shown in Fig. 5-19. The NOT operation is obtained from a one-input NOR gate, yet another symbol for an inverter circuit. The OR operation requires two NOR gates. The first produces the inverted OR and the second acts as an inverter to obtain the normal output. The AND operation is achieved through a NOR gate with additional inverters in each input.

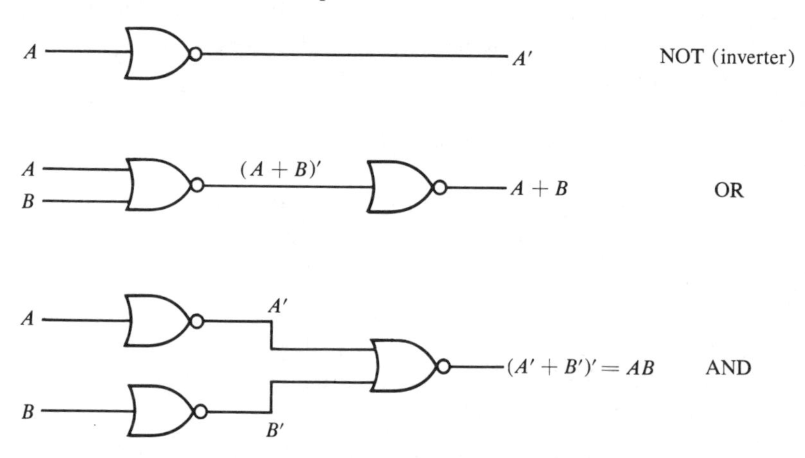

Figure 5-19 Implementation of NOT, OR and AND by NOR gates

Boolean Function Implementation—Block Diagram Method

The block diagram procedure for implementing Boolean functions with NOR gates is similar to the procedure outlined in the previous section for NAND gates.

1. Draw the AND/OR/NOT logic diagram from the given algebraic expression. Assume that both the normal and complement inputs are available.

2. Draw a second logic diagram with equivalent NOR logic as given in Fig. 5-19 substituted for each AND, OR, and NOT gate.

3. Remove pairs of cascaded inverters from the diagram. Remove inverters connected to single external inputs and complement the corresponding input variable.

The procedure is illustrated in Fig. 5-20 for the function:

$$F = A\,(B + CD) + BC'$$

The AND/OR implementation of the function is drawn in the logic diagram of Fig. 5-20(a). For each OR gate we substitute a NOR gate followed by an inverter. For each AND gate we substitute input inverters followed by a NOR gate. The pair of cascaded inverters from the OR box to the AND box is removed. The four inverters connected to external inputs are removed and the input variables complemented. The result is the NOR logic diagram shown in Fig. 5-20(c). The number of NOR gates in this example equals the number of AND/OR gates plus an additional inverter in the output (NOR gate number 6). In general, the number of NOR gates required to implement a Boolean function equals the number of AND/OR gates, except for an occasional inverter. This is true provided both normal and complement inputs are available because the conversion forces certain input variables to be complemented.

Two-Level Implementation—Product of Sums

The rule for obtaining the NOR logic diagram directly from a Boolean function expressed in product of sums is the same as that given for NAND logic for sum of products. This follows directly from the principle of duality:

1. Draw a NOR gate for each OR term of the function that has at least two literals. The inputs to each NOR gate are the literals of the term. This constitutes a group of first-level gates.

2. Draw a single NOR gate in the second level with inputs coming from outputs of first-level gates.

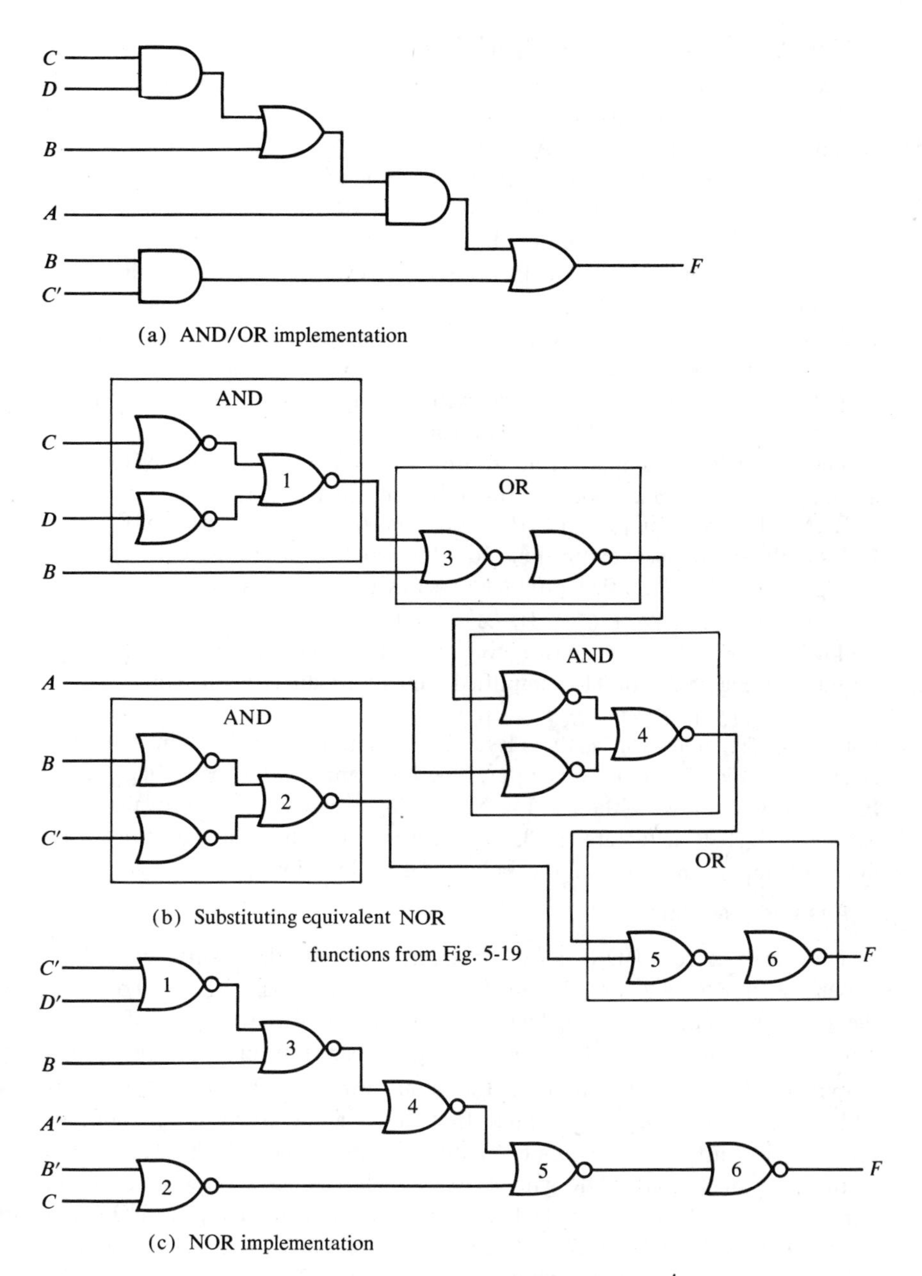

(a) AND/OR implementation

(b) Substituting equivalent NOR functions from Fig. 5-19

(c) NOR implementation

Figure 5-20 Implementation of $F = A(B + CD) + BC'$ with NOR gates

3. Any literal that appears as a term by itself is complemented and applied as input to the second-level gate.

Three-Level Implementation—Sum of Products

Again following the principle of duality, we find that a sum of products implementation with NOR gates follows a similar procedure as the product of sums implementation for NAND gates.

1. Draw a NOR gate for each AND term of the function that has at least two literals. The inputs to each NOR gate are the *complements* of the literals in the term. This constitutes a group of first-level gates.
2. Draw a single NOR gate in the second level with inputs coming from outputs of first-level gates.
3. Any literal that appears as a term by itself is applied as input to the second-level NOR gate.
4. Draw a third-level NOR gate with a single input connected to the output of the second-level NOR gate.

The following example demonstrates the derivation of the NOR logic diagrams from the same Boolean function used in Ex. 5-1.

EXAMPLE 5-2. Implement the function of Ex. 5-1 with NOR gates. The simplified Boolean expressions for this function are derived in the maps of Fig. 5-13. The simplified product of sums expression is:

$$F = (B' + C)(A + C' + D)(A' + B + C' + D')$$

Following the rules for a two-level implementation, we obtain the logic diagram of Fig. 5-21(a). The simplified sum of products expression is:

$$F = B'C' + A'CD + BCD + ACD'$$

Following the rules for a three-level implementation, we obtain the logic diagram of Fig. 5-21(b). It is interesting to compare the logic diagrams of Fig. 5-21 with those obtained for NAND logic in Fig. 5-14. Provided both NAND and NOR gates are available, the four logic diagrams constitute four different implementations of the same Boolean function.

Analysis Procedure

The analysis of NOR logic diagrams follows the same procedures presented in Sec. 4-7 for the analysis of combinational circuits. To derive the Boolean function from a logic diagram, we mark the outputs of various gates with arbitrary symbols. By repetitive substitutions we obtain the output variable as a function of the input variables. To obtain the truth table from a logic diagram without first deriving the Boolean function, we form a table listing the n input variables with 2^n rows of 1's and 0's. The truth table of various NOR gate outputs is derived in succession until the output truth table is obtained. The output function of a typical NOR gate is of the form $T = (A + B' + C)'$, so the truth table for T is marked with a 0 for those combinations where $A = 1$ or $B = 0$ or $C = 1$. The rest of the rows are filled with 1's.

Block Diagram Transformation

To convert a NOR logic diagram to its equivalent AND/OR/NOT logic diagram, we utilize the two symbols for NOR gates as shown in Fig. 5-22.

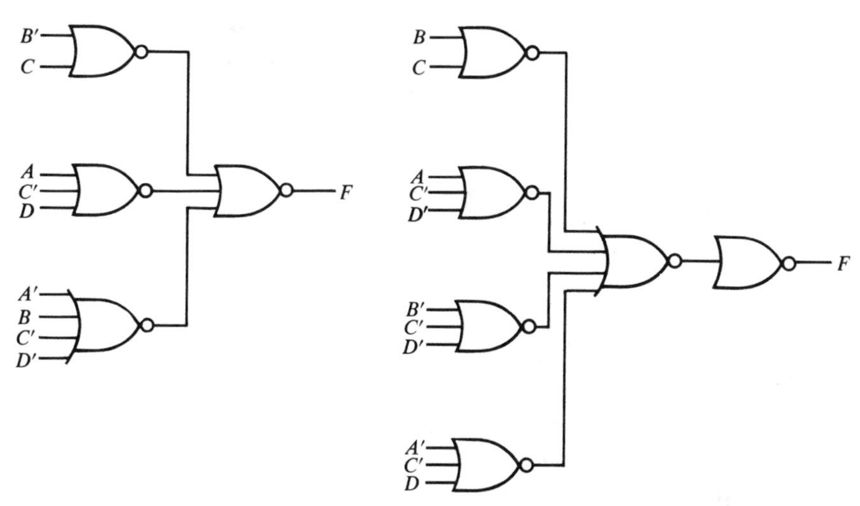

(a) Two-level from product of sums

$F = (B' + C)(A + C' + D)(A' + B + C' + D')$

(b) Three-level from sum of products

$F = B'C' + A'CD + BCD + ACD'$

Figure 5-21 NOR implementation of the Boolean function of Example 5-2

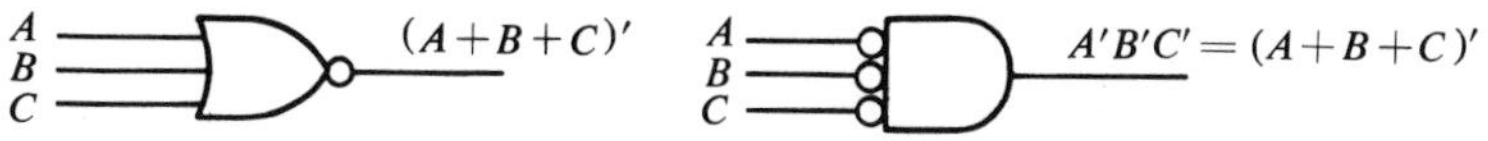

A B C $(A+B+C)'$

(a) OR-invert

A B C $A'B'C' = (A+B+C)'$

(b) invert-AND

Figure 5-22 Two symbols for NOR gate

The OR-invert is the normal symbol for a NOR gate and the invert-AND is a convenient alternative that utilizes De Morgan's theorem and the convention that small circles in the inputs denote complementation.

The conversion of a NOR logic diagram to an AND/OR/NOT diagram is achieved through a change in symbols from OR-invert to invert-AND starting from the last level and in alternate levels. Pairs of small circles along the same line are removed. One input AND or OR gate is removed, but if it has a small circle in the input or output, it is converted to an inverter.

This procedure is demonstrated in Fig. 5-23, where the NOR logic diagram in (a) is converted to an AND/OR/NOT diagram. The symbol of the gate in the last level (5) is changed to an invert-AND. Looking for alternate levels, we find one gate in level 3 and two in level 1. These three gates undergo a symbol change as shown in (b). Any two circles along the same line are removed. Circles that go to external inputs are also removed provided the corresponding input variable is complemented. The gate in level 5 becomes a one-input AND gate and is removed. The required AND/OR logic diagram is drawn in Fig. 5-23(c).

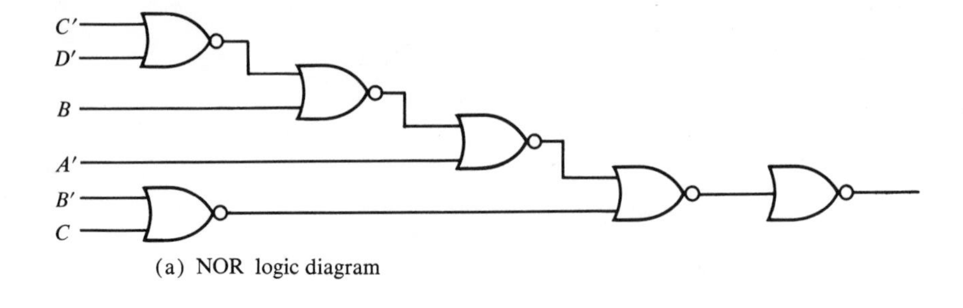

(a) NOR logic diagram

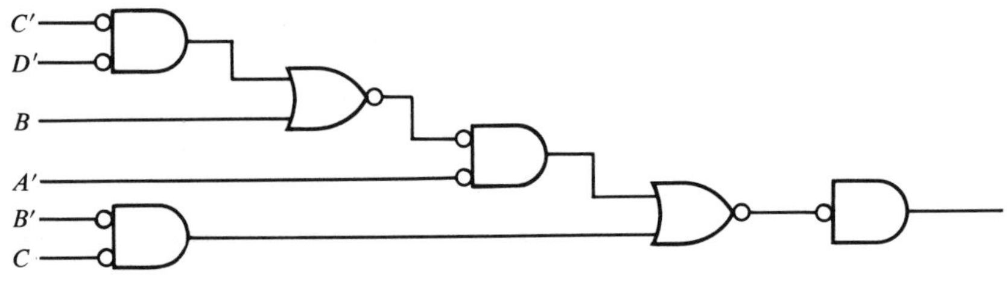

(b) Substitution of invert-AND symbols in alternate levels

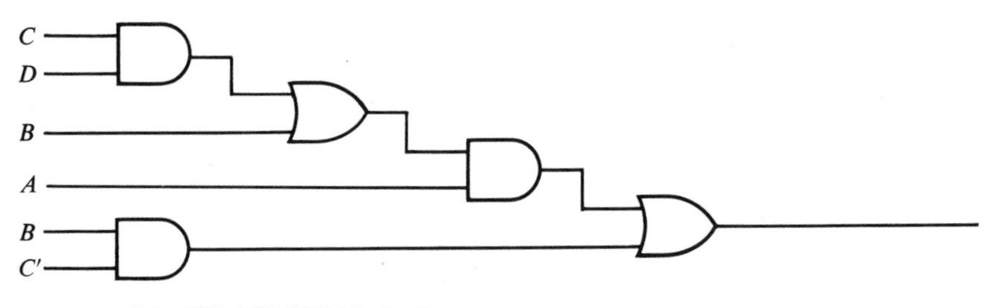

(c) AND/OR/NOT logic diagram

Figure 5-23 Conversion of NOR logic diagram to AND/OR/NOT

5-7 OTHER TWO-LEVEL IMPLEMENTATIONS

A general form of a two-level gate implementation is shown in Fig. 5-24. Each box labeled c_1 represents a first-level gate. The box labeled c_2 represents the second-level gate. For example, a Boolean function expressed in sum of products can be implemented in a two-level form with each c_1 box representing an AND gate and the c_2 box representing an OR gate. For convenience we use the notation c_1/c_2 to designate a two-level implementation and refer to it as an AND/OR two-level form.

In previous sections we have considered four types of gates: AND, OR, NAND, and NOR. If we assign one type of gate for c_1 and one type for c_2, we shall obtain 16 possible combinations of two-level forms. Eight of

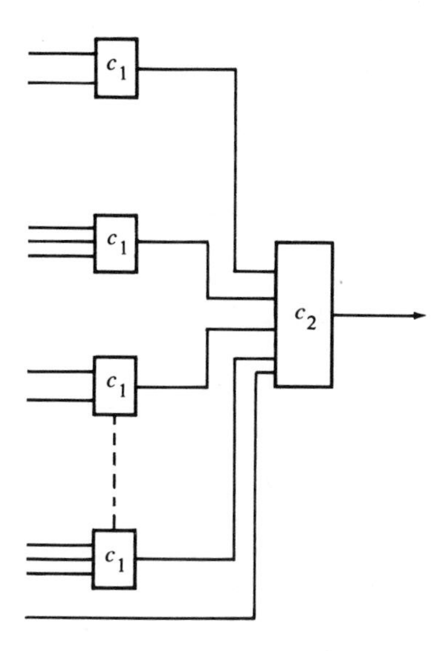

Figure 5-24 General two-level gate implementation

these combinations are said to be *degenerate forms*; they degenerate to a single operation (see Prob. 5-27). The other eight *nondegenerate forms* produce an implementation in sum of products or product of sums provided both normal and complement inputs are available. The eight nondegenerate forms are:

1. AND/OR	2. OR/AND
3. NAND/NAND	4. NOR/NOR
5. NOR/OR	6. NAND/AND
7. OR/NAND	8. AND/NOR

The four forms listed in the left column implement a function expressed in sum of products, while the four forms on the right implement a function expressed in product of sums. Note that any two forms listed in the same row are the dual of each other.

The AND/OR and OR/AND forms are basic two-level forms and have been discussed extensively in previous chapters. The NAND/NAND form was introduced and discussed in Sec. 5-5 and the NOR/NOR form in Sec. 5-6. The remaining two-level forms are investigated in this section.

One must realize that it is sometimes convenient, from various hardware considerations, to construct a two-level form as a nonseparable unit. For example, an AND/NOR two-level form may be available in an IC package with only the inputs to first-level gates and the output of the second-level gate connected to external pins. Another hardware facility that produces equivalent two-level forms is the *wired-OR* or *wired-AND* property of certain logic gates. *Wired* is a word used to explain the ability to produce a second-level operation (usually OR or AND but not both) by connecting (shorting) together the outputs of all first-level gates. For example, a NOR/OR form may be available from NOR gates that possess the property of producing a wired-OR function when two or more outputs are shorted.

Figure 5-25 may be conveniently used to determine the logic function of the various two-level forms. Each logic diagram has two first-level gates with inputs designated by A and B for one gate and C and D for the other gate. A fifth input designated by E goes directly to the second-level gate. The logic function performed by each two-level form can be determined algebraically and from it one can deduce the rule for implementing any Boolean function with the given two-level form.

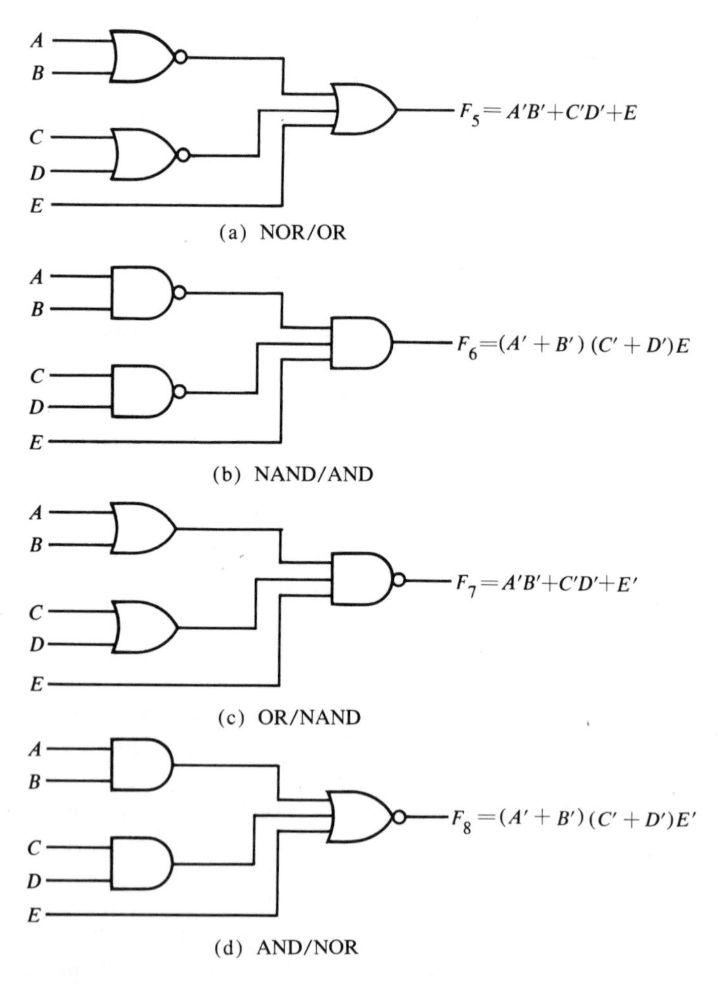

Figure 5-25 Two-level implementations of four non-degenerate forms

NOR/OR Form

The logic diagram for this two-level form is shown in Fig. 5-25(a). The output function for the circuit is:

$$F_5 = (A + B)' + (C + D)' + E = A'B' + C'D' + E$$

which shows that a Boolean function expressed in sum of products can be implemented with a NOR/OR form if the input literals are complemented except for a single literal forming an OR-term.

NAND/AND Form

The logic diagram for this two-level form is shown in Fig. 5-25(b). The output function for the circuit is:

$$F_6 = (AB)' \ (CD)' \ E = (A' + B') \ (C' + D') \ E$$

which shows that a Boolean function expressed in product of sums can be implemented with a NAND/AND form if the input literals are complemented (except for a single literal forming an AND-term).

OR/NAND Form

The logic diagram for this form shown in Fig. 5-25(c) produces the following output function:

$$F_7 = [(A + B) \ (C + D) \ E]' = A'B' + C'D' + E'$$

This result shows that a Boolean function expressed in sum of products can be implemented with an OR/NAND form provided all input literals are complemented.

Another way of looking at the OR/NAND form is to consider it as an OR/AND/INVERT function with the inversion done by the small circle of the second-level NAND gate. An OR/AND implementation requires a product of sums form. The inversion complements the function. Therefore, if we derive the *complement* of the function in product of sums, we can implement it with the OR/AND function and when it passes through the always present output inversion, it will generate the normal output function. This procedure produces the same implementation as the one outlined above.

AND/NOR Form

The logic diagram for this form produces the following output function:

$$F_8 = (AB + CD + E)' = (A' + B') \ (C' + D') \ E'$$

The result shows that a Boolean function expressed in product of sums can be implemented with an AND/NOR form provided all input literals are complemented.

Again, an AND/NOR form may be considered as an AND/OR/INVERT form, with the inversion done by the small circle of the second-level NOR gate. Therefore, if the *complement* of the function is expressed in sum of products, it can be implemented with the AND/OR function. When the complement passes through the output inversion, it generates the normal function. The implementation achieved by this method is the same as that outlined above.

The implementation of a Boolean function with any one of the eight nondegenerate two-level forms is summarized in Table 5-4. For completeness, we include the first four forms, which have been discussed in previous chapters.

EXAMPLE 5-3. Implement the Boolean function represented in the map of Fig. 5-26 with the four two-level forms NOR/OR, NAND/AND, OR/NAND, and AND/NOR.

The simplified Boolean function in sum of products is obtained by combining the 1's in the map:

$$F = AB + A'C + AC'D' \tag{5-1}$$

The complement of the function in sum of products is obtained by combining the 0's in the map:

$$F' = A'C' + AB'C + AB'D \tag{5-2}$$

Table 5-4 Method of Implementation of a Boolean Function With Non-degenerate Two-level Forms

Two-level form	*Standard form of function to be used*	*Inputs Literals to be complemented*
AND/OR	sum of products	none
OR/AND	product of sums	none
NAND/NAND	sum of products	single literal forming a term
NOR/NOR	product of sums	single literal forming a term
NOR/OR	sum of products	all, except for a single literal forming a term
NAND/AND	product of sums	all, except for a single literal forming a term
OR/NAND	sum of products	all
AND/NOR	product of sums	all

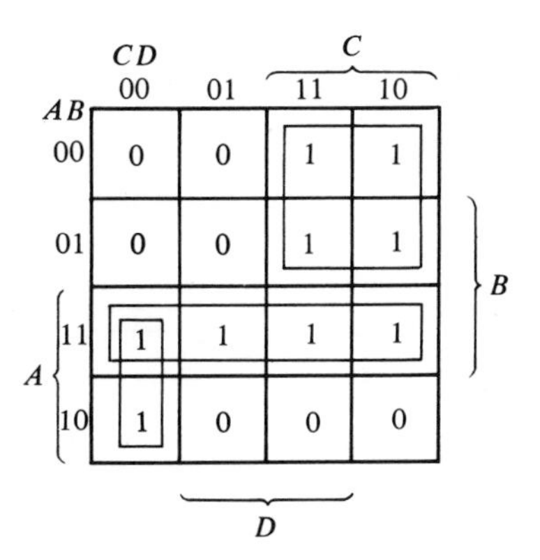

Figure 5-26 Map for Example 5-3

The product of sums expression is obtained from the complement of F' as expressed in Eq. 5-2:

$$F = (A + C)(A' + B + C')(A' + B + D') \tag{5-3}$$

The complement of the function in product of sums is obtained from the complement of F as expressed in Eq. 5-1:

$$F' = (A' + B')(A + C')(A' + C + D) \tag{5-4}$$

The NOR/OR implementation is shown in Fig. 5-27(a). The expression employed is the sum of products from Eq. 5-1, with all input literals complemented. (There is no term with a single literal in this expression.)

The NAND/AND implementation is shown in Fig. 5-27(b). The expression employed is the product of sums from Eq. 5-3, with all input literals complemented (again, there is no single-literal term in this expression).

The OR/NAND implementation is shown in Fig. 5-27(c). The expression used is the sum of products from Eq. 5-1, with all input literals complemented. The same result is obtained if we use the complement function F' expressed in product of sums from Eq. 5-4 to generate an OR/AND function. The small circle in the output complements F' to produce the normal function F.

The AND/NOR implementation is shown in Fig. 5-27(d). The expression employed is the product of sums from Eq. 5-3, with all input literals complemented. Again, the same result is obtained when F' is used from Eq. 5-2 to generate an AND/OR function. The small circle in the output complements F' to produce the normal function F.

It was shown in Secs. 5-5 and 5-6 that a Boolean function can be implemented with three levels of NAND or NOR gates. Similarly, other two-level forms can be extended to three levels, with the third level being

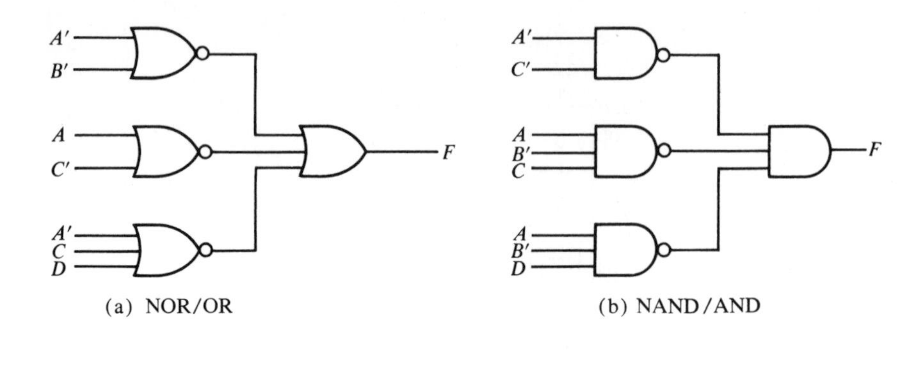

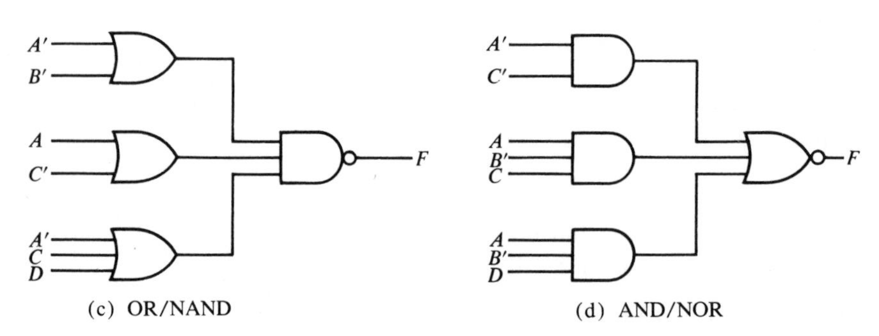

Figure 5-27 Two-level implementations for Example 5-3

either an inverter, a one-input NOR (when NOR gates are used), or a one-input NAND (when NAND gates are used). The third level acts as an inverter to complement what has been generated by the first two levels. The various methods for three-level implementation of a Boolean function can be listed in a table similar to Table 5-4 with the following changes: (a) each two-level form is changed to a three-level form (for example AND/OR is changed to AND/OR/NOT), (b) the standard form listed in Table 5-4 is interchanged–sum of products is changed to product of sums and vice versa, (c) the literals to be complemented are those listed *not* to be complemented in Table 5-4 (see Prob. 5-30). These rules are a direct consequence of the fact that a complemented function produces a dual function with all literals complemented.

5-8 EXCLUSIVE-OR AND EQUIVALENCE FUNCTIONS

Exclusive-or and equivalence, denoted by $\oplus$ and $\odot$ respectively, are binary operations that perform the following Boolean functions:

$$x \oplus y = xy' + x'y$$

$$x \odot y = xy + x'y'$$

The two operations are the complement of each other. Each is commutative and associative. Because of these two properties, a function of three or more variables can be expressed without parentheses as follows:

$$(A \oplus B) \oplus C = A \oplus (B \oplus C) = A \oplus B \oplus C$$

This would imply the possibility of using exclusive-or (or equivalence) gates with three or more inputs. However, multiple-input exclusive-or gates are very uneconomical from a hardware standpoint. In fact, even a two-input function is usually constructed with other types of gates. For example, Fig. 5-28(a) shows the implementation of a two-input exclusive-or function with AND, OR, and NOT gates. Figure 5-28(b) shows it with NAND gates. It is sometimes convenient to employ the gate symbols introduced in Fig. 5-2 instead of drawing the internal constructions in detail.

Only a limited number of Boolean functions can be expressed exclusively in terms of exclusive-or or equivalence operations. Nevertheless, these functions emerge quite often during the design of digital systems. The two functions are particularly useful in arithmetic operations and error detection and correction. In fact, we have already encountered these functions in two

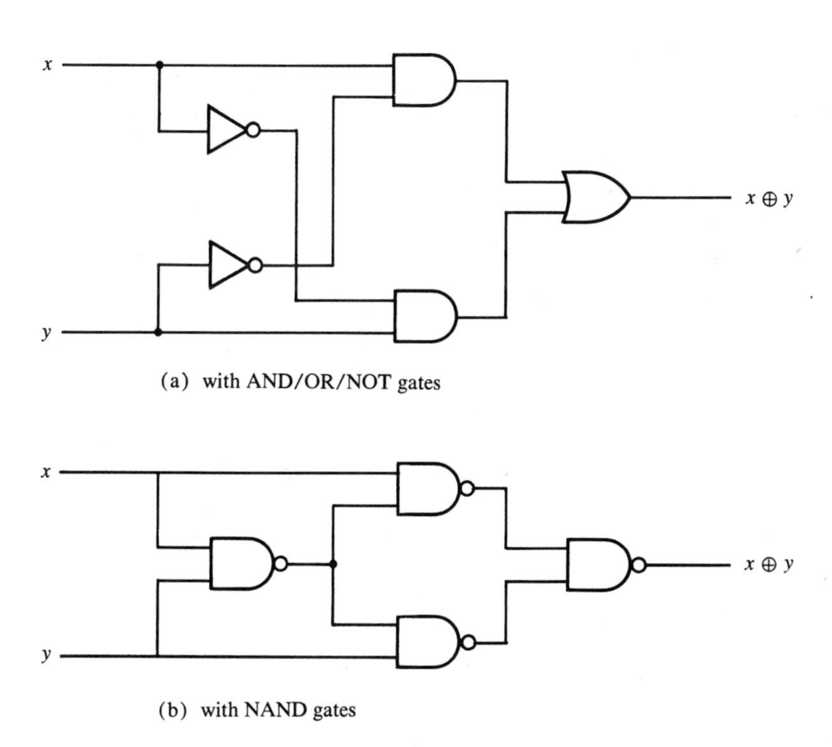

Figure 5-28 Exclusive-OR implementations

combinational circuits in Ch. 4: the S output of the half-adder in Fig. 4-2 is an exclusive-or function and the part of the comparator circuit of Fig. 4-10 used to generate the equality of each pair of binary digits comprises three equivalence functions.

An n-variable exclusive-or expression is equal to the Boolean function with $2^n/2$ minterms whose equivalent binary numbers have an odd number of 1's. This is demonstrated in the map of Fig. 5-29(a) for the four-variable case. There are 16 minterms for four variables. Half the minterms have a numerical value with an odd number of 1's; the other half have a numerical value with an even number of 1's. The numerical value of a minterm is determined from the row and column numbers of the square that represents the minterm. The map of Fig. 5-29(a) has 1's in the squares whose minterm numbers have an odd number of 1's. The function can be expressed in terms of the exclusive-or operations on the four variables. This is justified by the following algebraic manipulation:

$$\begin{aligned} A \oplus B \oplus C \oplus D &= (AB' + A'B) \oplus (CD' + C'D) \\ &= (AB' + A'B)(CD + C'D') \\ &\quad + (AB + A'B')(CD' + C'D) \\ &= \Sigma(1, 2, 4, 7, 8, 11, 13, 14) \end{aligned}$$

An n-variable equivalence expression is equal to the Boolean function with $2^n/2$ minterms, whose equivalent binary numbers have an even number of 0's. This fact is demonstrated in the map of Fig. 5-29(b) for the four-variable case. The squares with 1's represent the eight minterms with even number of 0's and the function can be expressed in terms of the equivalence operations on the four variables.

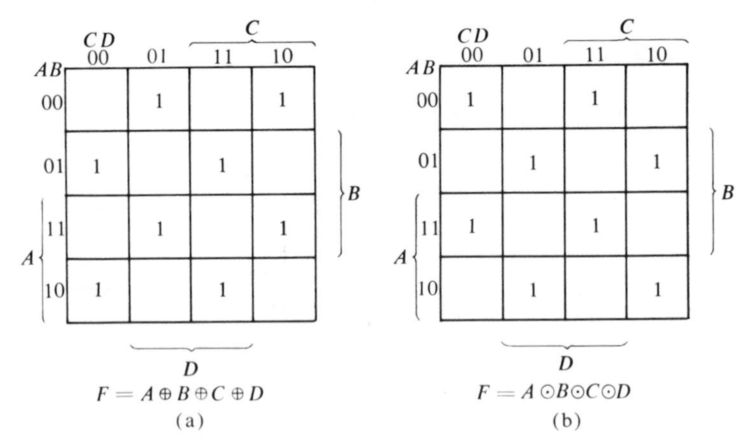

Figure 5-29 Map for a 4-variable (a) exclusive-OR function, (b) equivalence function

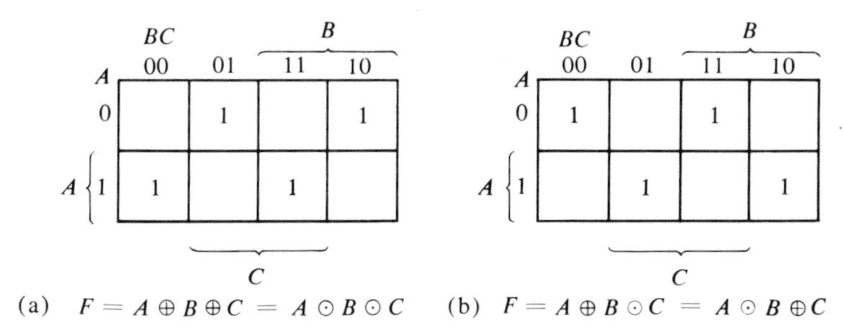

(a) $F = A \oplus B \oplus C = A \odot B \odot C$ (b) $F = A \oplus B \odot C = A \odot B \oplus C$

Figure 5-30 Map for 3-variable functions

When the number of variables in a function is odd, the minterms with an even number of 0's are the same as the minterms with an odd number of 1's. This fact is demonstrated in the three-variable map of Fig. 5-30(a). Therefore, an exclusive-or expression is equal to an equivalence expression when both have the same odd number of variables. However, they form the complement of each other when the number of variables is even, as is demonstrated in the two maps of Fig. 5-29(a) and (b).

When the minterms of a function with an odd number of variables have an even number of 1's (or equivalently, an odd number of 0's), the function can be expressed as the complement of either an exclusive-or or an equivalence expression. For example, the three-variable function shown in the map of Fig. 5-30(b) can be expressed as follows:

$$(A \oplus B \oplus C)' = A \oplus B \odot C$$

or $$(A \odot B \odot C)' = A \odot B \oplus C$$

The S output of a full-adder and the D output of a full-subtractor (Sec. 4-3) can be implemented with exclusive-or functions because each function consists of four minterms with numerical values having an odd number of 1's. The exclusive-or function is extensively used in the implementation of digital arithmetic operations because the latter are usually implemented through procedures that require a repetitive addition or subtraction operation.

Exclusive-or and equivalence functions are very useful in systems requiring error detection and correction codes. As discussed in Sec. 1-6, a parity bit is a scheme for detecting errors during transmission of binary information. A parity bit is an extra bit included with a binary message to make the number of 1's either odd or even. The message, including the parity bit, is transmitted and then checked in the receiving end for erros. An error is detected if the checked parity does not correspond to the one transmitted. The circuit that generates the parity bit in the transmitter is called a *parity generator*; the circuit that checks the parity in the receiver is called a *parity checker.*

As an example, consider a three-bit message to be transmitted with an odd parity bit. Table 5-5 shows the truth table for the parity generator. The three

Table 5-5 Odd Parity Generation

3-bit message			Parity-bit generated
x	y	z	P
0	0	0	1
0	0	1	0
0	1	0	0
0	1	1	1
1	0	0	0
1	0	1	1
1	1	0	1
1	1	1	0

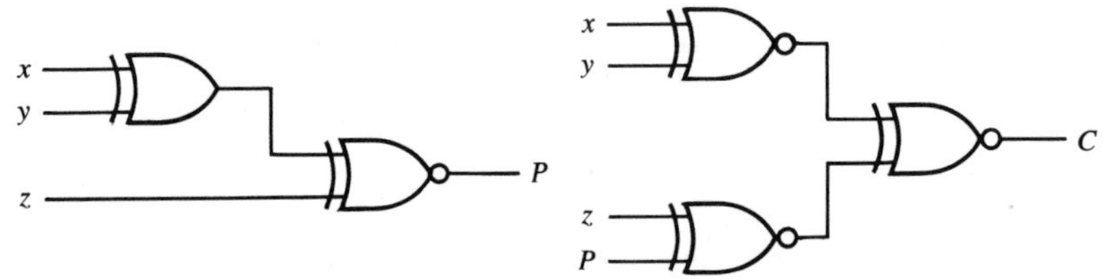

(a) 3-bit odd parity generator (b) 4-bit odd parity checker

Figure 5-31 Logic diagrams for parity generation and checking

bits x, y, and z constitute the message and are the inputs to the circuit. The parity bit P is the output. For odd parity, the bit P is generated so as to make the total number of 1's odd (including P). From the truth table we see that $P = 1$ when the number of 1's in x, y, and z is even. This corresponds to the map of Fig. 5-30(b), so that the function for P can be expressed as follows:

$$P = x \oplus y \odot z$$

The logic diagram for the parity generator is shown in Fig. 5-31(a). It consists of one two-input exclusive-or gate and one two-input equivalence gate. The two gates can be interchanged and still produce the same function since P is also equal to

$$P = x \odot y \oplus z$$

The three-bit message and the parity bit are transmitted to their destination, where they are applied to a parity checker circuit. An error occurs during transmission if the parity of the four bits received is even, since the binary information transmitted was originally odd. The output C of the parity checker should be a 1 when an error occurs; i.e., when the number of 1's in the four inputs is even. Table 5-6 is the truth table for the odd parity checker circuit. From it we see that the function for C consists of the eight minterms with numerical values having an even number of 0's. This corresponds to the map of

Table 5-6 Odd Parity Check

4-bits received				*Parity error check*
x	y	z	P	C
0	0	0	0	1
0	0	0	1	0
0	0	1	0	0
0	0	1	1	1
0	1	0	0	0
0	1	0	1	1
0	1	1	0	1
0	1	1	1	0
1	0	0	0	0
1	0	0	1	1
1	0	1	0	1
1	0	1	1	0
1	1	0	0	1
1	1	0	1	0
1	1	1	0	0
1	1	1	1	1

Fig. 5-29(b) so that the function can be expressed with equivalence operators as follows:

$$C = x \odot y \odot z \odot P$$

The logic diagram for the parity checker is shown in Fig. 5-31(b) and consists of three two-input equivalence gates.

It is worth noting that the parity generator can be implemented with the circuit of Fig. 5-31(b) if the input P is permanently held at logic-0 and the output is marked P, the advantage being that the same circuit can be used for both parity generation and checking.

It is obvious from the above example that parity generation and checking circuits always have an output function that includes half of the minterms whose numerical values have either an even or odd number of 1's. As a consequence, they can be implemented with equivalence and/or exclusive-or gates.

PROBLEMS

5-1. How many of the MSI functions introduced in Ch. 4 can be accommodated in one 14-pin IC package?

5-2. Discuss some of the advantages and difficulties that one may encounter when LSI devices are used in digital systems.

5-3. By substituting the AND/OR/NOT equivalent definition of the binary operations as defined in Sec. 2-6, show that:

(a) The inhibition and implication operators are neither commutative nor associative.

(b) The exclusive-or and equivalence operators are commutative and associative.

(c) The NAND operator is not associative.

(d) The NOR and NAND operators are not distributive.

5-4. A *majority* gate is a digital circuit whose output is logic-1 if the majority of the inputs are logic-1 and logic-0 otherwise. By means of truth tables, find the Boolean function implemented by the following majority (MAJ) gate circuit.

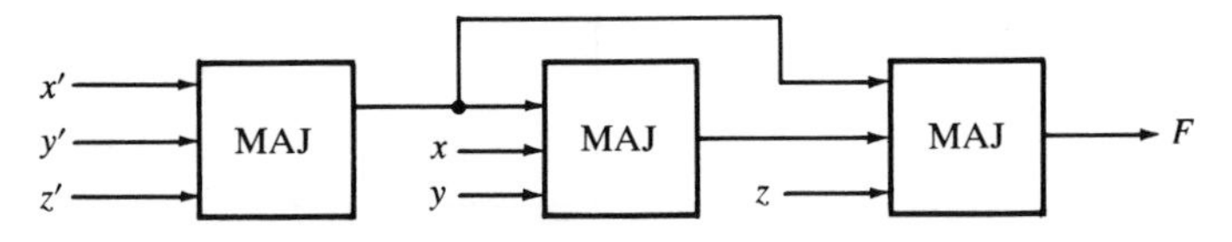

5-5. Repeat Prob. 5-4 for the following circuit. What is the function of the circuit?

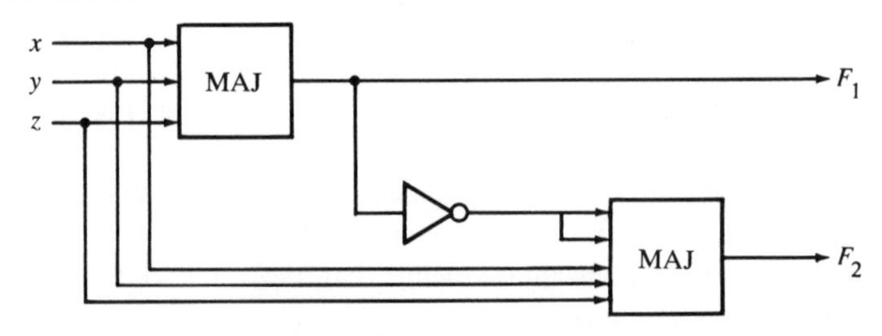

5-6. The gate shown below performs the *inhibition* operation $F = x'y$. Obtain the AND, OR, and NOT functions using inhibition gates only. Remember that 0 and 1 are valid values for inputs.

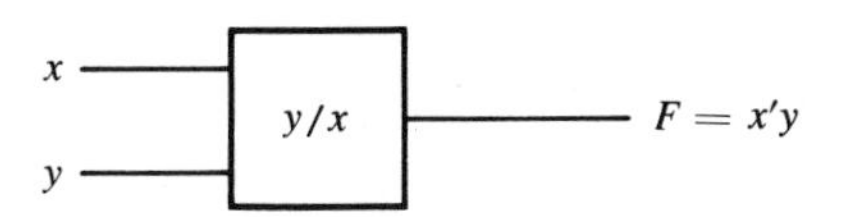

5-7. (a) Show that the NOR and NAND operations are the dual of each other.

(b) Show that one physical circuit can function as a positive logic NAND gate or as a negative logic NOR gate.

(c) Repeat part (b) for positive logic NOR and negative logic NAND.

5-8. The following example demonstrates the duality of positive and negative negative logic. (a) Draw the logic diagram for the function $F = AB + C$. (b) Obtain the truth table. (c) You are told that the gates operate with positive logic with logic-1 and logic-0 being +5 volt and -5 volt, respectively. Obtain a truth table showing voltage values instead of 1's and 0's. (d) Use the same physical gates as in (c) but change to a negative logic system; that is, let signal -5 volt represent logic-1 and +5 volt represent logic-0. Obtain the truth table in terms of 1's and 0's. (e) Show that the truth table obtained in (d) gives a Boolean function $F = (A + B)C$ which is the dual of the one in (a). (f) Now repeat steps (a) to (e) after changing step (c) to read: "You are told that the gates operate with negative logic with -5 volt and +5 volt representing 1 and 0 respectively." Show that the function obtained in step (e) is $AB + C$.

5-9. Count the number of logic levels in the circuits of Fig. P4-19 and P4-21. Obtain the equivalent two-level circuit for each.

5-10. What is the fan-in and fan-out of the AND gates and inverters specified in Prob. 4-27?

5-11. Obtain a two-level and a multilevel implementation AND and OR gates of the function

$$F(A, B, C, D) = \Sigma(2, 3, 5, 7, 10, 12, 13, 14)$$

with the restriction that inputs A and A' can provide only one load; i.e., each can be connected to one gate input only. Assume that both the normal and complement input variables are available.

5-12. Using the block diagram method, convert the following logic diagrams from AND/OR implementation to NAND implementation.

(a) BCD to excess-3 converter, Fig. 4-8.

(b) Three-bit comparator, Fig. 4-10.

(c) Full-adder, Fig. 4-11.

(d) Binary to octal decoder, Fig. 4-12.

(e) Octal to binary encoder, Fig. 4-13.

(f) Eight-input multiplexer, Fig. 4-15.

(g) Four-output demultiplexer, Fig. 4-17.

5-13. Repeat Prob. 5-12 for NOR implementation.

5-14. Obtain the NAND logic diagram of a full-adder from the Boolean functions:

$$C = xy + xz + yz$$

$$S = C'(x + y + z) + xyz$$

5-15. Simplify each of the following functions and implement them with NAND gates. Give two alternatives.

(a) $F_1 = AC' + ACE + ACE' + A'CD' + A'D'E'$

(b) $F_2 = (B' + D')(A' + C' + D)(A + B' + C' + D)$

$(A' + B + C' + D')$

5-16. Repeat Prob. 5-15 for NOR implementations.

5-17. Implement the following functions with NAND gates. Assume that both the normal and complement inputs are available.

(a) $BD + BCD + AB'C'D' + A'B'CD'$ with no more than six gates each having three inputs.

(b) $(AB + A'B')(CD' + C'D)$ with two-input gates.

5-18. Implement the following functions using the don't-care conditions. Assume that both the normal and complement inputs are available.

(a) $F = A'B'C' + AB'D + A'B'CD'$ with no more than two NOR gates.

$d = ABC + AB'D'$

(b) $F = (A + D)(A' + B)(A' + C')$ with no more than three NAND gates.

(c) $F = B'D + B'C + ABCD$ with NAND gates.

$d = A'BD + AB'C'D'$

5-19. Implement the following function with either NAND or NOR gates. Use only four gates. Only the normal inputs are available.

$$F = w'xz + w'yz + x'yz' + wxy'z$$

$$d = wyz$$

5-20. Implement the following functions with NOR gates. Assume that both the normal and complement inputs are available.

(a) $AB' + C'D' + A'CD' + DC'(AB + A'B') + DB(AC' + A'C)$

(b) $AB'CD' + A'BCD' + AB'C'D + A'BC'D$

5-21. Design the following combinational circuit using minimum number of NOR gates. The circuit has four inputs which represent a decimal digit in BCD and a single output, which is equal to 1 whenever the value of the input digit is less than 3 or more than 7.

5-22. Determine the Boolean function for the output F of the circuit in Fig. P5-22. Obtain an equivalent circuit with less NOR gates.

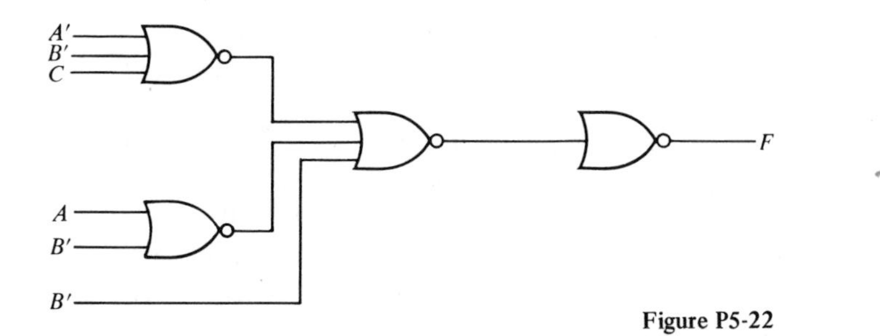

Figure P5-22

5-23. Determine the output Boolean functions of the circuits.

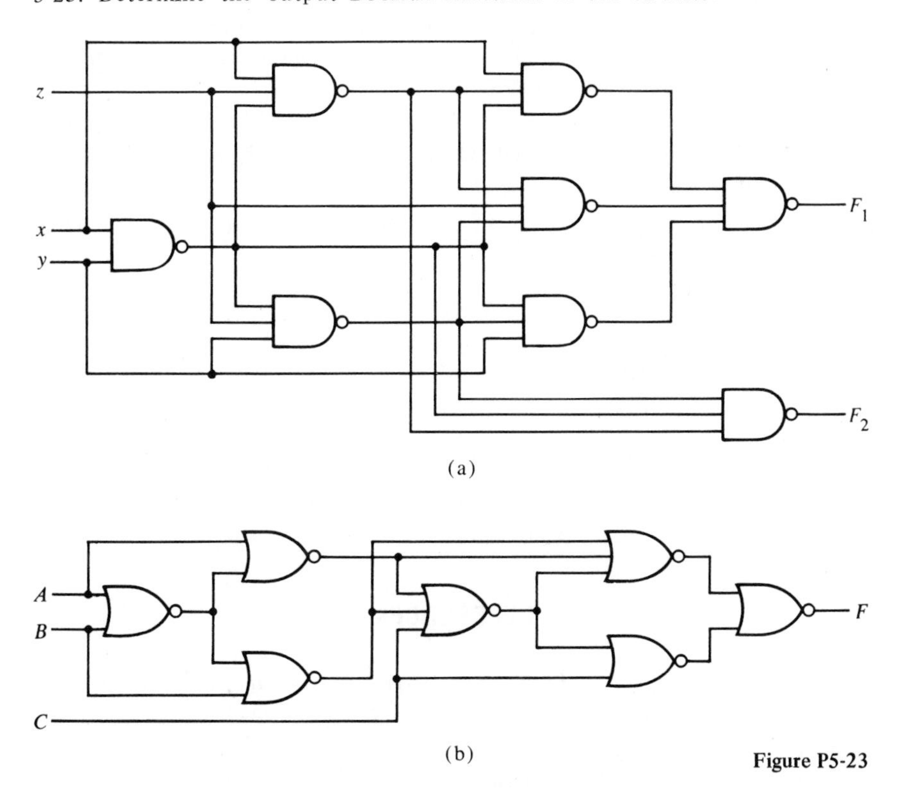

Figure P5-23

5-24. Obtain the truth table for the circuits of Fig. P5-23.

5-25. Obtain the equivalent AND/OR/NOT logic diagram of Fig. P5-23(a).

5-26. Obtain the equivalent AND/OR/NOT logic diagram of Fig. P5-23(b).

5-27. List the eight degenerate two-level forms and show that they reduce to a single operation. Explain how the degenerate two-level forms can be used to extend the fan-in of gates.

5-28. Implement the functions of Prob. 5-15 with the following two-level forms: NOR/OR, NAND/AND, OR/NAND, and AND/NOR.

5-29. Implement the functions of Prob. 5-15 with the three-level forms: NOR/OR/NOR, NAND/AND/NAND, OR/NAND/NAND, and AND/NOR/NOR.

5-30. List in a table (similar to Table 5-4) the method of implementation of Boolean functions with nondegenerate three-level forms. Implement the functions of Ex. 5-3 with all possible three-level forms.

5-31. Obtain the logic diagram of a two-input *equivalence* function using (a) AND, OR, NOT gates; (b) NOR gates; (c) NAND gates.

5-32. Show that the circuit of Fig. 5-28(b) is an exclusive-or.

5-33. Show that $A \odot B \odot C \odot D = \Sigma(0, 3, 5, 6, 9, 10, 12, 15)$.

5-34. Design a combinational circuit that converts a four-bit reflected code number (Table 1-4) to a four-bit binary number. Implement the circuit with exclusive-or gates.

5-35. Design a combinational circuit to *check* for even parity of four bits. A logic-1 output is required when the four bits do not constitute an even parity.

5-36. Implement the four Boolean functions listed using three half-adder circuits (Fig. 4-2).

(a) $D = A \oplus B \oplus C$

(b) $E = A'BC + AB'C$

(c) $F = ABC' + (A' + B')C$

(d) ABC

5-37. Implement the Boolean function:

$$F = AB'CD' + A'BCD' + AB'C'D + A'BC'D$$

with exclusive-or and AND gates.

REFERENCES

1. Maley, G. A., and J. Earle, *The Logic Design of Transistor Digital Computers.* Englewood Cliffs, N.J.: Prentice-Hall, 1963. Ch. 6.

2. Mano, M. M., "Converting to NOR and NAND Logic," *Electro-Technology,* Vol. 75, No. 4 (April 1965), 34-37.

3. Maley, G. A., *Manual of Logic Circuits,* Englewood Cliffs, N.J.: Prentice-Hall, 1970.

4. Garrett, L. S., "Integrated-Circuit Digital Logic Families," *IEEE Spectrum* (October, November, December 1970).

6 SEQUENTIAL LOGIC

6-1 INTRODUCTION

The digital circuits considered thus far have been combinational; that is, the outputs at any instant of time are entirely dependent upon the inputs present at that time. Although every digital system is likely to have combinational circuits, most systems encountered in practice also include memory elements, which require that the system be described in terms of *sequential logic.*

A block diagram of a sequential circuit is shown in Fig. 6-1. It consists of combinational logic gates which accept binary signals from external inputs and from outputs of memory elements and generate signals to external outputs and to inputs of memory elements. A memory element is a device capable of storing one bit of information. The binary information stored in memory elements can be changed by the outputs of the combinational circuit. The outputs of memory elements, in turn, go to the inputs of gates in the combinational circuit.

The combinational circuit, by itself, performs a specific information processing operation, part of which is used to determine the binary value to be stored in memory elements. The outputs of memory elements are applied to the combinational circuit and determine, in part, the circuit's outputs. This process clearly demonstrates that the external outputs of a sequential circuit are a function not only of external inputs but also of the present state of memory elements. (The *state* of a memory element or binary cell was defined in Sec. 1-7.) The next state of memory elements is a function of external inputs and the present state. Thus, a sequential circuit is specified by a time sequence of inputs, outputs, and internal states.

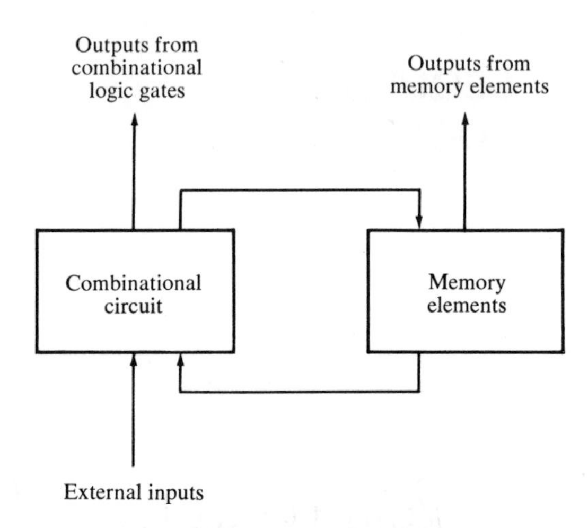

Figure 6-1 Block diagram of a sequential circuit

There are two main types of sequential circuits. Their classification depends on the timing of their signals. A *synchronous* sequential circuit is a system whose behavior can be defined from the knowledge of its signals at discrete instants of time. The behavior of an *asynchronous* sequential circuit depends upon the order in which its input signals change and can be affected at any instant of time. The memory elements commonly used in asynchronous sequential circuits are time-delay devices. The memory capability of a time-delay device is due to the fact that it takes a finite time for the signal to propagate through the device. In practice, the internal propagation delay of logic gates is of sufficient duration to produce the needed delay so that physical time-delay units may be unnecessary. In gate type asynchronous systems, the memory elements of Fig. 6-1 consist of logic gates whose propagation delays constitute the required memory. Thus, an asynchronous sequential circuit may be regarded as a combinational circuit with feedback. Because of the feedback among logic gates, an asynchronous sequential circuit may, at times, become unstable. The instability problem imposes many difficulties to the designer. Hence they are not as commonly used as synchronous systems.

A synchronous sequential logic system, by definition, must employ signals that affect the memory elements only at discrete instants of time. One way of achieving this goal is to use pulses of limited duration throughout the system so that one pulse amplitude represents logic-1 and another pulse amplitude (or the absence of a pulse) represents logic-0. The difficulty with a system of pulses is that any two pulses arriving from separate independent sources to the inputs of the same gate will exhibit unpredictable delays, separate the pulses slightly, and result in unreliable operation.

Practical synchronous sequential logic systems use fixed amplitudes such as voltage levels for the binary signals. Synchronization is achieved by a timing device called a master clock generator which generates a periodic train of *clock pulses*. The clock pulses are distributed throughout the system in such a way that memory elements are affected only with the arrival of the synchronization pulse. In practice, the clock pulses are applied into AND gates together with the signals that specify the required change in memory elements. The AND gate outputs can transmit signals only at instants which coincide with the arrival of clock pulses. Synchronous sequential circuits that use clock pulses in the inputs of memory elements are called *clocked sequential circuits*. Clocked sequential circuits are the type encountered most frequently. They do not manifest instability problems and their timing is easily broken down into independent discrete steps, each of which is considered separately. The sequential circuits discussed in this book are exclusively of the clocked type.

The memory elements used in clocked sequential circuits are called *flip-flops*. These circuits are binary cells capable of storing one bit of information. A flip-flop circuit has two outputs, one for the normal value and one for the complement value of the bit stored in it. Binary information can enter a flip-flop in a variety of ways, a fact which gives rise to different types of flip-flops. In the next section we shall examine the various types of flip-flops and define their logical properties.

6-2 FLIP-FLOPS

A flip-flop circuit can maintain a binary state indefinitely (as long as power is delivered to the circuit) until directed by an input signal to switch states. The major differences among various types of flip-flops are in the number of inputs they possess and in the manner in which the inputs affect the binary state. The most common types of flip-flops are discussed below.

Basic Flip-Flop Circuit

It was mentioned in Secs. 5-5 and 5-6 that a flip-flop circuit can be constructed from two NAND gates or two NOR gates. These constructions are shown in the logic diagrams of Figs. 6-2 and 6-3. Each circuit forms a basic flip-flop upon which other more complicated types can be built. The cross-coupled connection from the output of one gate to the input of the other gate constitutes a feedback path. For that reason, the circuits are classified as asynchronous sequential circuits. Each flip-flop has two outputs, Q and Q', and two inputs, *set* and *reset*. This type of flip-flop is sometimes called a *direct-coupled RS flip-flop*; the R and the S being the first letters of the two input names.

To analyze the operation of the circuit of Fig. 6-2, we must remember that the output of a NOR gate is 0 if any input is 1, and that the output is 1 only when all inputs are 0. As a starting point, assume that the set input is 1 and the reset input is 0. Since gate number 2 has an input of 1, its output Q' must be 0, which puts both inputs of gate number 1 at 0, so that output Q is 1. When the set input is returned to 0, the outputs remain the same. This is because output Q remains a 1, leaving one input of gate number 2 at 1. That causes output Q' to stay at 0, which leaves both inputs of gate number 1 at 0, so that output Q is a 1. In the same manner it is possible to show that a 1 in the reset input changes output Q to 0 and Q' to 1. When the reset input returns to 0, the outputs do not change.

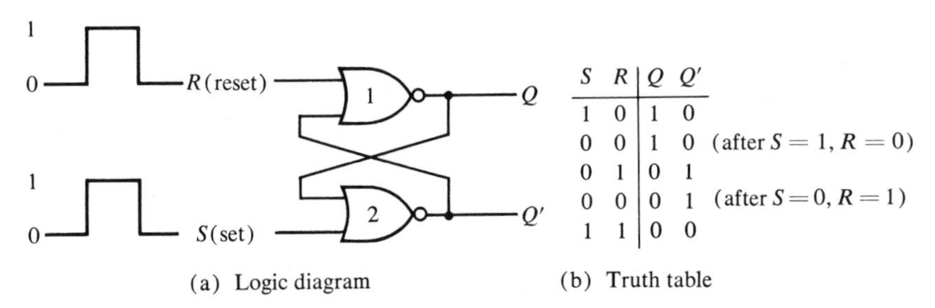

S	R	Q	Q'	
1	0	1	0	
0	0	1	0	(after $S = 1, R = 0$)
0	1	0	1	
0	0	0	1	(after $S = 0, R = 1$)
1	1	0	0	

(a) Logic diagram (b) Truth table

Figure 6-2 Basic flip-flop circuit with NOR gates

When a 1 is applied to both the set and reset inputs, both Q and Q' outputs go to 0. Strictly speaking, a flip-flop whose two outputs assume the same binary value while both the set and reset inputs have equal values is called a *latch*. In a true flip-flop circuit this condition is normally avoided.

A flip-flop has two useful states. When $Q = 1$ and $Q' = 0$ it is in the *set-state* (or 1-state). When $Q = 0$ and $Q' = 1$, it is in the *clear-state* (or 0-state). The outputs Q and Q' are complements of each other and are referred to as the normal and complement output, respectively. The binary state of the flip-flop is taken to be the value of the normal output.

Under normal operation, both inputs remain at 0 unless the state of the flip-flop has to be changed. The application of a momentary 1 to the set input causes the flip-flop to go to the set-state. The set input must go back to 0 before a 1 is applied to the reset input. A momentary 1 applied to the reset input causes the flip-flop to go the clear-state. When both inputs are initially 0, a 1 applied to the set input while the flip-flop is in the set state or a 1 applied to the reset input while the flip-flop is in the clear-state leaves the outputs unchanged. When a 1 is applied to both the set and reset inputs, both outputs go to 0. This state is undefined and is usually avoided. If both inputs now go to 0, the state of the flip-flop is indeterminate and depends on which input remains a 1 longer before the transition to 0.

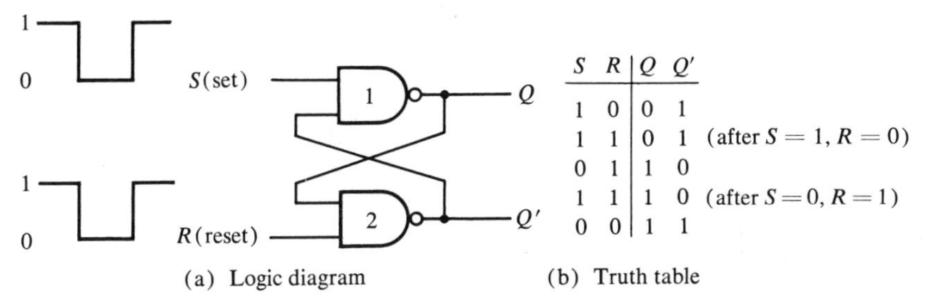

(a) Logic diagram

S	R	Q	Q'	
1	0	0	1	
1	1	0	1	(after $S = 1, R = 0$)
0	1	1	0	
1	1	1	0	(after $S = 0, R = 1$)
0	0	1	1	

(b) Truth table

Figure 6-3 Basic flip-flop circuit with NAND gates

The NAND basic flip-flop circuit of Fig. 6-3 operates with both inputs normally at 1 unless the state of the flip-flop has to be changed. The application of a momentary 0 to the set input causes output Q to go to 1 and Q' to go to 0, thus putting the flip-flop into the set-state. After the set input returns to 1, a momentary 0 to the reset input causes a transition to the clear-state. When both inputs go to 0, both outputs go to 1, a condition avoided in normal flip-flop operation.

Clocked *RS* Flip-Flop

The basic flip-flop as it stands is an asynchronous sequential circuit. By adding gates to the inputs of the basic circuit, the flip-flop can be made to respond to input levels during the occurrence of a clock pulse. The clocked *RS* flip-flop shown in Fig. 6-4(a) consists of a basic NOR flip-flop and two AND gates. The outputs of the two AND gates remain at 0 as long as the clock pulse (abbreviated *CP*) is 0, regardless of the S and R input values. When the clock pulse goes to 1, information from the S and R inputs is allowed to reach the basic flip-flop. The set-state is reached with $S = 1$, $R = 0$, and $CP = 1$. To change into the clear-state, the inputs must be $S = 0$, $R = 1$, and $CP = 1$. With both $S = 1$ and $R = 1$, the occurrence of a clock pulse causes both ouputs to momentarily go to 0. When the pulse is removed, the state of the flip-flop is indeterminate; i.e., either state may result, depending on whether the set or the reset input of the basic flip-flop remains a 1 longer before the transition to 0 at the end of the pulse.

Two symbols for the *RS* flip-flop are shown in Fig. 6-4(b). The AND gates with the clock-pulse input may be drawn external to the symbol, or a

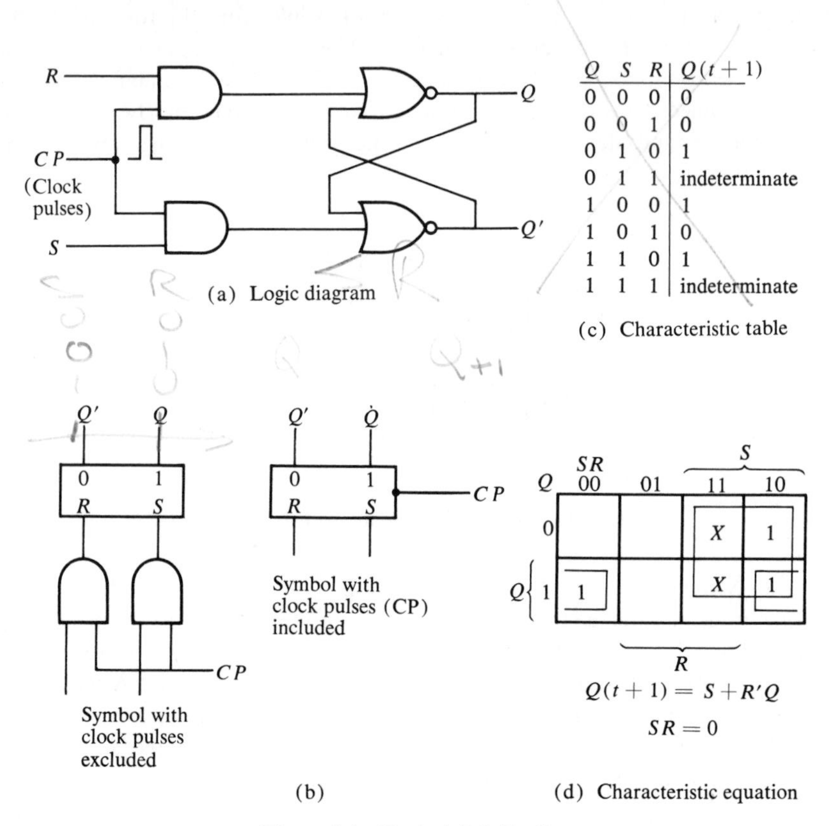

Q	S	R	Q(t + 1)
0	0	0	0
0	0	1	0
0	1	0	1
0	1	1	indeterminate
1	0	0	1
1	0	1	0
1	1	0	1
1	1	1	indeterminate

(c) Characteristic table

Figure 6-4 Clocked *RS* flip-flop

symbol with a *CP* label may be used to mean that flip-flop outputs are not affected unless a clock pulse occurs in the input marked *CP*.

The characteristic table for the flip-flop is shown in Fig. 6-4(c). This table summarizes the operation of the flip-flop in a tabular form. Q is the binary state of the flip-flop at a given time (referred to as *present state*), the S and R columns give the possible values of the inputs, and $Q(t + 1)$ is the state of the flip-flop after the occurrence of a clock pulse (referred to as *next state*).

The characteristic equation of the flip-flop is derived in the map of Fig. 6-4(d). This equation specifies the value of the next state as a function of the present state and the inputs. The characteristic equation is an algebraic expression for the binary information of the characteristic table. The two indeterminate states are marked by X's in the map since they may result in either a 1 or a 0. However, the relation $SR = 0$ must be included as part of the characteristic equation to specify that both S and R cannot equal 1 simultaneously.

D Flip-Flop

The D flip-flop shown in Fig. 6-5 is a modification of the clocked RS flip-flop. NAND gates 1 and 2 form a basic flip-flop and gates 3 and 4 modify it into a clocked RS flip-flop. The D input goes directly to the S input and its complement, through gate 5, is applied to the R input. As long as the clock pulse input is at 0, gates 3 and 4 have a 1 in their

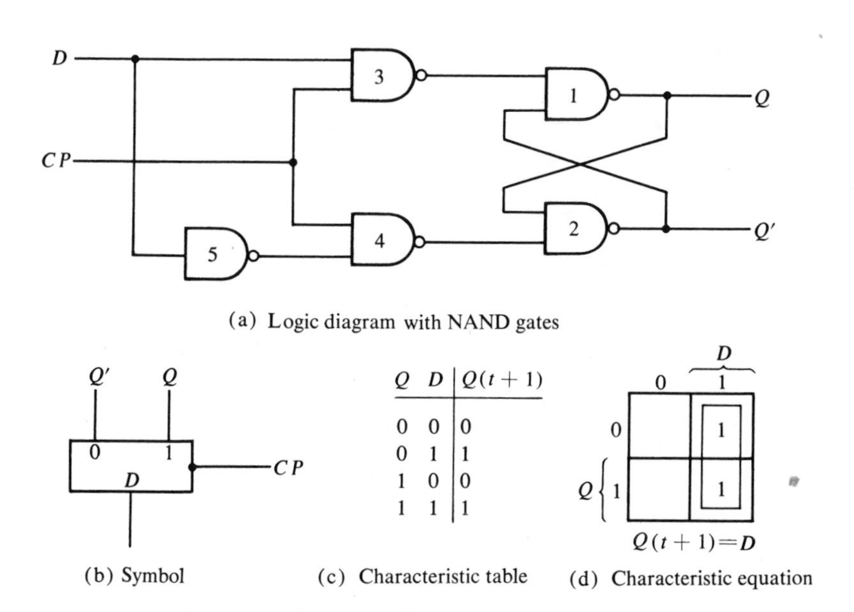

(a) Logic diagram with NAND gates

(b) Symbol

Q	D	$Q(t+1)$
0	0	0
0	1	1
1	0	0
1	1	1

(c) Characteristic table

(d) Characteristic equation

Figure 6-5 Clocked D flip-flop

outputs, regardless of the value of the other inputs. This conforms to the requirement that the two inputs of a basic NAND flip-flop (Fig. 6-3) remain initially at the 1 level. The D input is sampled during the occurrence of a clock pulse. If it is a 1, the output of gate 3 goes to 0, switching the flip-flop to the set-state (unless it was already set). If it is a 0, the output of gate 4 goes to 0, switching the flip-flop to the clear-state.

The D flip-flop receives the designation from its ability to transfer "data" into a flip-flop. It is basically an RS flip-flop with an inverter in the R input. The added inverter reduces the number of inputs from two to one.

The symbol for a clocked D flip-flop is shown in Fig. 6-5(b). The characteristic table is listed in part (c) and the characteristic equation is derived in part (d). The characteristic equation shows that the next state of the flip-flop is the same as the D input and is independent of the value of the present state.

JK Flip-Flop

A JK flip-flop is a refinement of the RS flip-flop in that the indeterminate state of the RS type is defined in the JK type. Inputs J and K behave like inputs S and R to set and clear the flip-flop (note that in a JK flip-flop, the first letter J is for *set* and the second letter K is for *clear*). When inputs are applied to both J and K simultaneously, the flip-flop switches to its complement state; that is, if $Q = 1$, it switches to $Q = 0$, and vice versa.

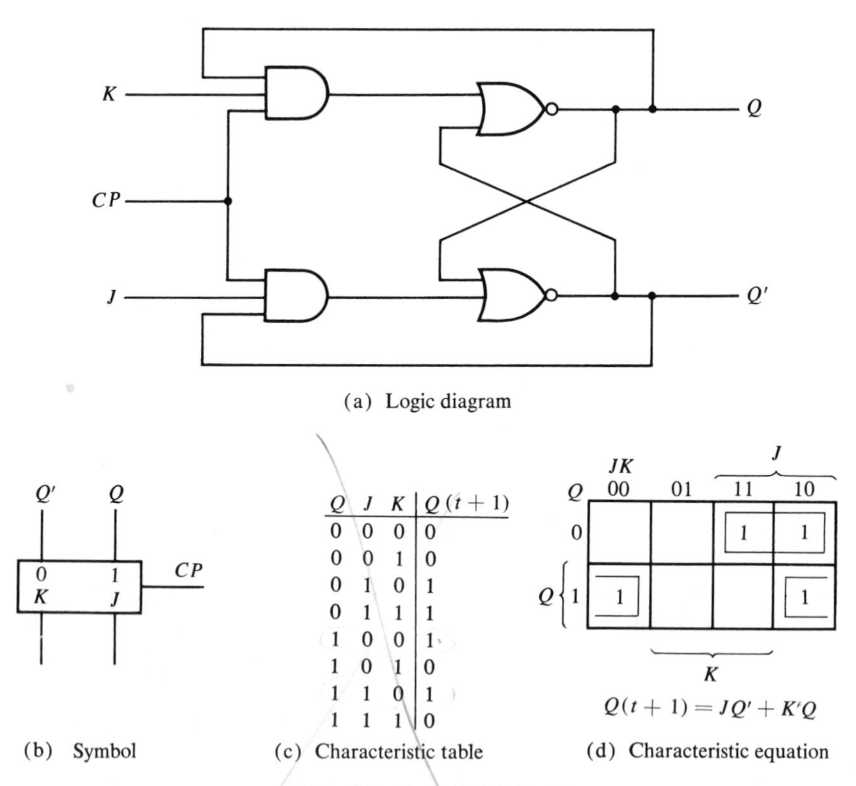

(a) Logic diagram

Q	J	K	$Q(t+1)$
0	0	0	0
0	0	1	0
0	1	0	1
0	1	1	1
1	0	0	1
1	0	1	0
1	1	0	1
1	1	1	0

(b) Symbol (c) Characteristic table (d) Characteristic equation

Figure 6-6 Clocked *JK* flip-flop

A clocked JK flip-flop is shown in Fig. 6-6(a). Output Q is ANDed with K and CP inputs so that the flip-flop is cleared during a clock pulse only if Q was previously 1. Similarly, output Q' is ANDed with J and CP inputs so that the flip-flop is set with a clock pulse only if Q' was previously 1.

When both J and K are 1, a clock pulse is transmitted through one AND gate; the one connected to the output equal to 1. Thus, if Q equals 1, the output of the upper AND gate becomes a 1 and the flip-flop is cleared. If Q' equals 1, the output of the lower AND gate becomes a 1 and the flip-flop is set. In either case, the state of the flip-flop is complemented. Note that if the *CP* signal remains a 1 after the outputs have been complemented, the flip-flop will go through a new transition. This timing problem is eliminated with a master-slave *JK* flip-flop.

The symbol, characteristic table, and characteristic equation of the *JK* flip-flop are shown in Fig. 6-6, parts (b), (c), and (d), respectively.

T Flip-Flop

The T flip-flop is a single-input version of the *JK* flip-flop. As shown in Fig. 6-7(a), the T flip-flop is obtained from a *JK* type if both inputs are

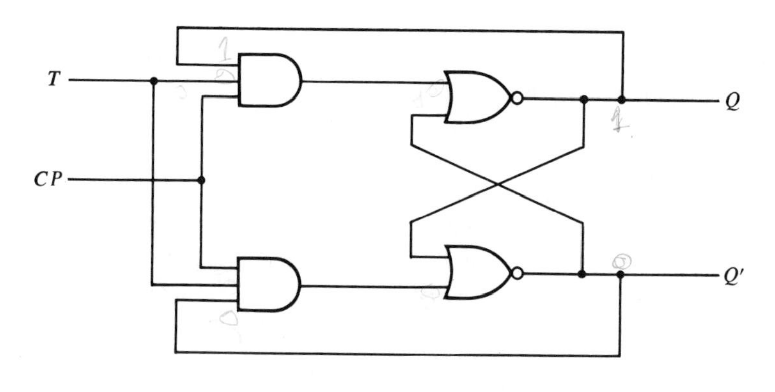

(a) Logic diagram

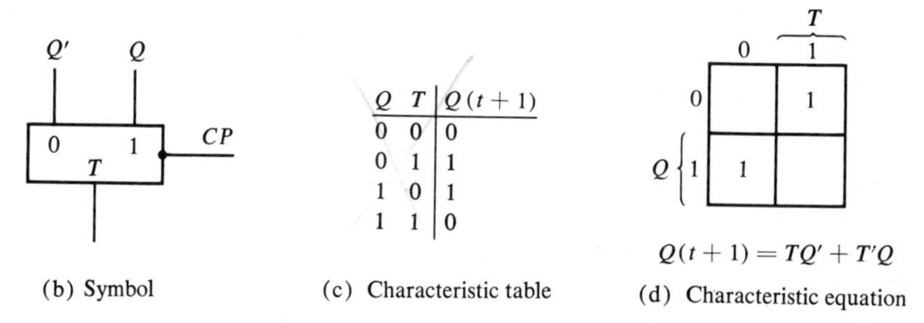

Q	T	$Q(t+1)$
0	0	0
0	1	1
1	0	1
1	1	0

$$Q(t+1) = TQ' + T'Q$$

(b) Symbol (c) Characteristic table (d) Characteristic equation

Figure 6-7 Clocked T flip-flop

tied together. The designation "T" comes from the ability of the flip-flop to "toggle," or change state. Regardless of the present state of the flip-flop, it assumes the complement state when the clock pulse occurs while input T is logic-1.

The symbol, characteristic table, and characteristic equation of the T flip-flop are shown in Fig. 6-7, parts (b), (c), and (d), respectively.

Other Considerations

The four types of flip-flops introduced in this section may be available in an unclocked version. Flip-flops without a clock input are useful for asynchronous operations. Unclocked flip-flop can be converted to clocked flip-flop by ANDing the clock pulse and the input information prior to their application to each input of the flip-flop.

The flip-flops introduced in this section are the most common types available commercially. The analysis and design procedures developed in this and the next chapter are applicable for any clocked flip-flop once its characteristic table is defined.

6-3 TRIGGERING OF FLIP-FLOPS

The state of a flip-flop is switched by a momentary change in the input signal. This momentary change is called a *trigger* and the transition which it causes is said to trigger the flip-flop. Asynchronous flip-flops, such as the basic circuits of Figs. 6-2 and 6-3, require an input trigger defined by a change of signal *level*. This level must be returned to its initial value (0 in the NOR and 1 in the NAND flip-flop) before applying a second trigger. Clocked flip-flops are triggered by *pulses*. A pulse starts from an initial value of 0, goes momentarily to 1, and after a short time, returns to its initial 0 value. The time interval from the application of the pulse until the output transition occurs is a critical factor that needs further investigation.

As seen from the block diagram of Fig. 6-1, a sequential circuit has a feedback path between the combinational circuit and the memory elements. This path can produce instability if the outputs of memory elements (flip-flops) are changing while the outputs of the combinational circuit that go to flip-flop inputs are being sampled by the clock pulse. This timing problem can be prevented if the outputs of flip-flops do not start changing until the pulse input has returned to 0. To insure such an operation, a flip-flop must have a signal propagation delay from input to output in excess of the pulse duration. This delay is usually very difficult to control if the designer depends entirely on the propagation delay of logic gates. One way of insuring the proper delay is to include within the flip-flop circuit a physical delay unit having a delay equal to or greater than the pulse duration. More commonly, the flip-flop is triggered during the trailing edge of the pulse as discussed below.

A clock pulse may be either positive or negative, depending on whether positive or negative logic is employed. A clock source remains at logic-0 between pulses and goes to logic-1 during the occurrence of a pulse. A pulse goes through two signal transitions: from 0 to 1 and the return from 1 back to 0. As shown in Fig. 6-8, the initial transition is defined as the *leading edge* and the return transition as the *trailing edge*. The two edges may be either rising or falling depending on whether we have positive or negative pulses.

The clocked flip-flops introduced in Sec. 6-2 are triggered during the leading edge of the pulse; that is, the state transition starts as soon as the pulse reaches the logic-1 level. The new state of the flip-flop may appear at the output terminals while the input pulse is still a 1. When this happens, the output of one flip-flop cannot be used as a condition for triggering another flip-flop when both respond to the same clock pulse. When flip-flops are triggered during the trailing edge of the pulse, the new state of the flip-flop appears at the output terminals after the pulse signal has gone back to 0. This is in effect a way of achieving the needed delay mentioned previously.

The time dependence of sequential circuits is best illustrated in a *timing diagram*. A timing diagram shows how the signals at various terminals vary over time. Each terminal in a timing diagram is assigned two coordinates: a horizontal coordinate to represent time and a vertical coordinate to represent signal amplitude. An example of a timing diagram that demonstrates the difficulty encountered when flip-flops are triggered on the leading edge of a clock pulse is shown in Fig. 6-9(a). The diagram shows the signals of a

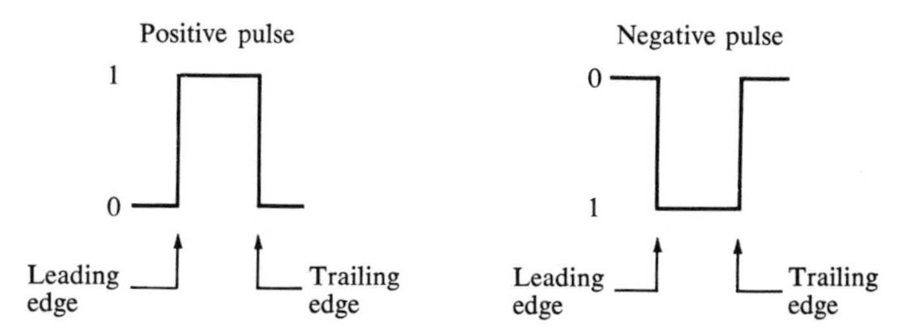

Figure 6-8 Definition of leading and trailing edge of a pulse

clock pulse (*CP*) terminal, a *J* input terminal and a *Q* output terminal of a clocked *JK* flip-flop. It is assumed that the signal at *J* comes from the output of some other flip-flop triggered by the same clock pulses. The signal at J starts its transition from 0 to 1 during the leading edge of clock pulse t_1. Output Q may or may not switch to 1 during clock pulse t_1. This is because input J is in a state of transition during the pulse interval and may sometimes be detected as a 0 signal and sometimes as a 1 signal. The best way to avoid this ambiguity is to insure that the flip-flop is not triggered while input J is in transition. It should be triggered at time t_2, when input J is maintained at logic-1, which requires additional control signals not indicated in the timing diagram. Under this condition, the signal at J must remain at logic-1 during pulse interval t_2 and can return to logic-0 during clock pulse t_3 or later.

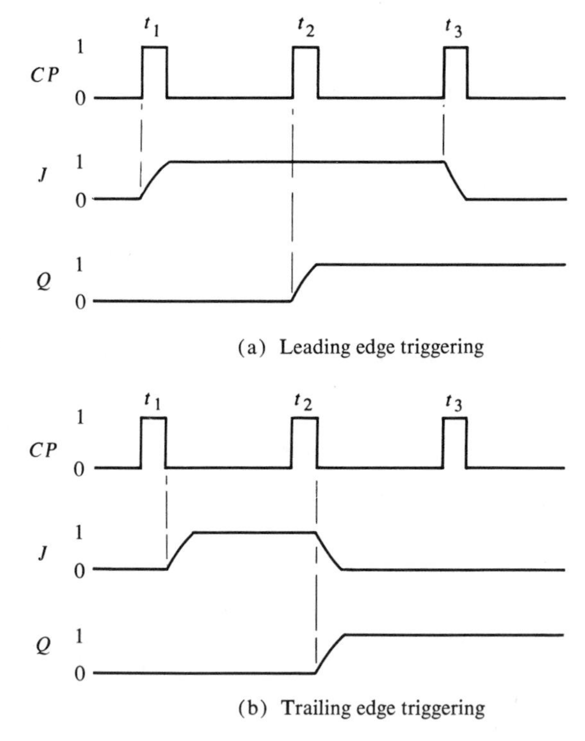

Figure 6-9 Timing diagram for a *JK* flip-flop

Now consider the same timing sequence with flip-flops being triggered on the trailing edge of the pulse as shown in Fig. 6-9(b). The signal at J becomes a 1 after the termination of pulse interval t_1. Output Q remains a 0 after t_1 because J is a 0 *during* the pulse interval. During pulse interval t_2, input J of the flip-flop is a 1, so output Q starts switching to 1 at the trailing edge of t_2. The same t_2 pulse is used to trigger the other flip-flop that returns the signal at J to 0. This is possible because the signal at J

remains a 1 until the end of the clock pulse. Thus, with trailing-edge triggering, it is possible to switch the output of a flip-flop and its input information with the same clock pulse. The trailing-edge trigger allows the application of the same clock pulse to all flip-flops. The state of all flip-flops can be changed simultaneously since the new state appears at the output terminals only after the clock pulse has returned to 0.

One way to make a flip-flop respond to the trailing edge of a pulse is to use AC coupling; i.e., insert a capacitor in the input. The capacitor forms a circuit capable of generating a spike as a response to a momentary change of input signal amplitude. A positive pulse emerges from such a circuit with a positive spike from the leading edge and a negative spike from the trailing edge. Trailing-edge operation is achieved by designing the flip-flop to neglect a positive spike and trigger on the occurrence of a negative spike.

Another method for achieving state transition during the trailing edge of a clock pulse is to employ two flip-flop circuits; one to hold the output state until the trailing edge and the other to sample the input information on the leading edge. Such a combination is called a *master-slave* flip-flop.

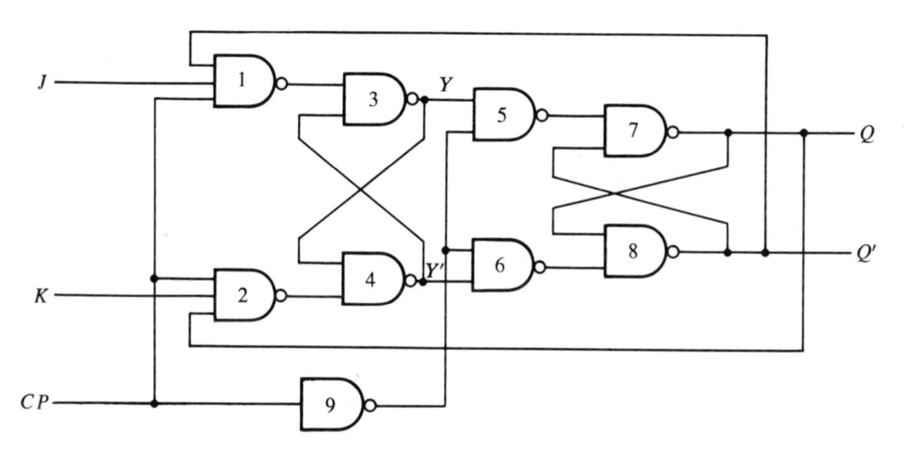

Figure 6-10 Clocked master-slave *JK* flip-flop

An example of a master-slave *JK* flip-flop constructed with NAND gates is shown in Fig. 6-10. It consists of two flip-flops; gates 1 through 4 form the master flip-flop and gates 5 through 8 form the slave flip-flop. The information present at the *J* and *K* inputs is transmitted to the master flip-flop on the leading edge of a clock pulse and held there until the trailing edge of the clock pulse occurs, after which it is allowed to pass through to the slave flip-flop. The clock input is normally 0, which keeps the outputs of gates 1 and 2 at the logic-1 level. This prevents the *J* and *K* inputs from affecting the master flip-flop. The slave flip-flop is a clocked *RS* type with the master flip-flop supplying the inputs and the clock input being inverted by gate number 9. When the clock is 0, the output of gate 9 is 1, so that output Q is equal to Y and Q' is equal to Y'. When the leading edge of a clock pulse occurs, the master flip-flop is affected and may switch states. The slave flip-flop is isolated as long as the clock is at logic-1 level, because the output of gate 9, being at 0, provides a 1 to both inputs of the NAND basic flip-flop of gates 7 and 8. When clock input returns to 0, the master flip-flop is isolated from the J and K inputs, and the slave flip-flop goes to the same state as the master flip-flop.

The master-slave combination can be constructed for any type of flip-flop. A leading-edge flip-flop is converted to a master-slave by the addition of a clocked *RS* flip-flop with an inverted clock input to form the slave flip-flop. We shall assume throughout the rest of the book that all flip-flop types change state after the trailing edge of the clock pulse.

Among the different types of flip-flops, the *JK* master-slave is the one used most in the design of digital systems. Some *JK* flip-flops available in IC packages provide additional inputs for setting and clearing the flip-flop asynchronously. These inputs are usually called "direct set" and "direct clear." They affect the flip-flop on a positive (or negative) swing of a signal amplitude without the need of clock pulses. These inputs are useful for bringing the flip-flops to an initial state prior to its clocked operation.

6-4 ANALYSIS OF CLOCKED SEQUENTIAL CIRCUITS

The behavior of a sequential circuit is determined from the inputs, the outputs, and the state of its flip-flops. Both the outputs and next state are a function of the inputs and the present state. The analysis of sequential circuits consists of obtaining a table or a diagram for the time sequence of inputs, outputs, and internal states. It is also possible to write Boolean expressions that describe the behavior of sequential circuits. However, these expressions must include the necessary time sequence either directly or indirectly.

A logic diagram is recognized as the circuit of a sequential circuit if it includes flip-flops. The flip-flops may be of any type and the logic diagram may or may not include combinational gates. In this section we shall first introduce a specific example of a clocked sequential circuit and then present various methods for describing the behavior of sequential circuits. The specific example will be used throughout the discussion to illustrate the various methods.

An Example of a Sequential Circuit

An example of a clocked sequential circuit is shown in Fig. 6-11. It has one input variable x, one output variable y, and two clocked RS flip-flops labeled A and B. The cross connections from outputs of flip-flops to inputs of gates are not shown by line drawings so as to facilitate the tracing of the circuit. Instead, the connections are recognized from the letter symbol marked in each input. For example, the input marked x' in gate number 1 designates an input from the complement of x. The second input marked A designates a connection to the normal output of flip-flop A.

We shall assume trailing-edge triggering in both flip-flops and in the source that produces the external input x. Therefore, the signals for a given present state are available during the time from the termination of a clock pulse to the termination of the next clock pulse, at which time the circuit goes to the next state.

State Table

The time sequence of inputs, outputs, and flip-flop states may be enumerated in a *state table**. The state table for the circuit of Fig. 6-11 is shown in Table 6-1. It consists of three sections labeled present state,

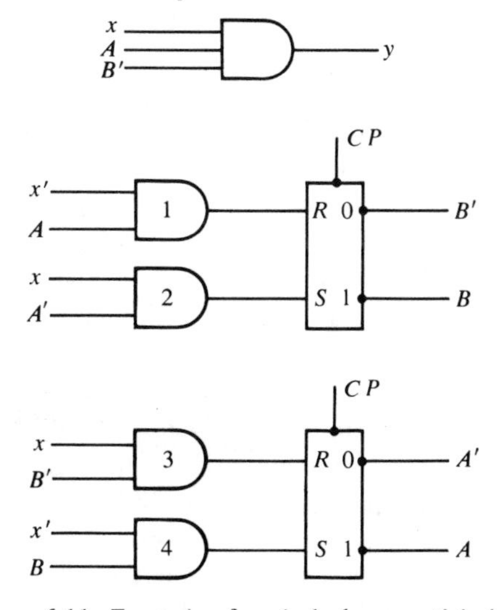

Figure 6-11 Example of a clocked sequential circuit

*Switching circuit theory books call this table a *transition table.* They reserve the name *state table* for a table with internal states represented by arbitrary symbols.

Table 6-1 State Table for Circuit of Fig. 6-11

Present State	Next state		Output	
	x = 0	*x = 1*	*x = 0*	*x = 1*
AB	*AB*	*AB*	*y*	*y*
00	00	01	0	0
01	11	01	0	0
10	10	00	0	1
11	10	11	0	0

next state, and output. The present state designates the state of flip-flops before the occurrence of a clock pulse. The next state shows the state of flip-flops after the application of a clock pulse, and the output section lists the value of the output variable during the present state. Both the next state and output sections have two columns, one for $x = 0$ and the other for $x = 1$.

The derivation of the state table starts from an assumed initial state. The initial state of most practical sequential circuits is defined to be the state with 0's in all flip-flops. Some sequential circuits have a different initial state and some have none at all. In either case, the analysis can always start from any arbitrary state. In this example, we start deriving the state table from the initial state 00.

When the present state is 00, $A = 0$ and $B = 0$. From the logic diagram we see that with both flip-flops cleared and $x = 0$, none of the AND gates produce a logic-1 signal. Therefore, the next state remains unchanged. With $AB = 00$ and $x = 1$, gate number 2 produces a logic-1 signal at the S input of flip-flop B and gate number 3 produces a logic-1 signal at the R input of flip-flop A. When a clock pulse triggers the flip-flops, A is cleared and B is set, making the next state 01. This information is listed in the first row of the state table.

In a similar manner, we can derive the next state starting from the other three possible present states. In general, the next state is a function of the inputs, the present state, and the type of flip-flop used. With *RS* flip-flops, for example, we must remember that a 1 in input S sets the flip-flop and a 1 in input R clears the flip-flop, irrespective of its previous state. A 0 in both the S and R inputs leaves the flip-flop unchanged, while a 1 in both the S and R inputs shows a bad design and an indeterminate state table.

The entries for the output section are easier to derive. In this example, output y is equal to 1 only when $x = 1$, $A = 1$, and $B = 0$. Therefore, the output columns are marked with 0's except when the present state is 10 and input $x = 1$, for which y is marked with a 1.

The state table of any sequential circuit is obtained by the same procedure used in the example. In general, a sequential circuit with m flip-flops and n input variables will have 2^m rows, one for each state. The next state and output sections each will have 2^n columns, one for each input combination.

The external outputs of a sequential circuit may come from logic gates or from memory elements. The output section in the state table is necessary only if there are outputs from logic gates. Any external output taken directly from a flip-flop is already listed in the present state column of the state table. Therefore, the output section of the state table can be excluded if there are no external outputs from logic gates. A sequential circuit whose state table includes an output section is sometimes called a *Mealy machine*. A circuit whose state table does not list the outputs explicitly is called a *Moore machine*. Mealy and Moore are the names of the original investigators of the respective sequential circuits.

State Diagram

The information available in a state table may be represented graphically in a *state diagram*. In this diagram, a state is represented by a circle and the transition between states is indicated by directed lines connecting the circles. The state diagram of the sequential circuit of Fig. 6-11 is shown in Fig. 6-12. The binary number inside each circle identifies the state the circle represents. The directed lines are labeled with two binary numbers separated by a /. The input value that causes the state transition is labeled first; the number after the symbol / gives the value of the output during the present state. For example, the directed line from state 00 to 01 is labeled 1/0, meaning that the sequential circuit is in a present state 00 while $x = 1$ and $y = 0$ and that on the termination of the next clock

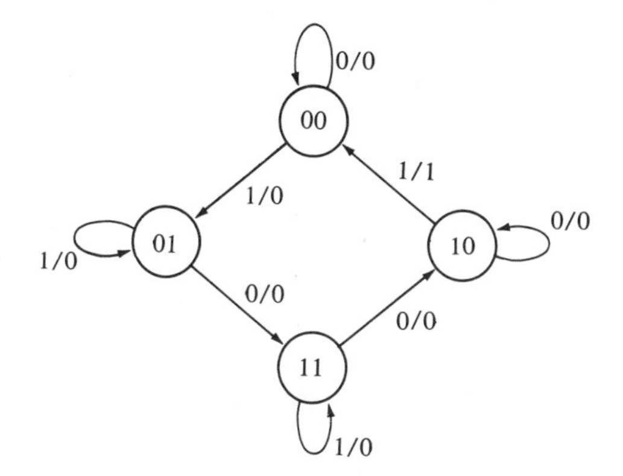

Figure 6-12 State diagram for circuit of Figure 6-11

pulse, the circuit goes to next state 01. A directed line connecting a circle with itself indicates that no change of state occurs. The state diagram provides the same information as the state table and is obtained directly from Table 6-1.

There is no difference between a state table and a state diagram except in the manner of representation. The state table is easier to derive from a given logic diagram and the state diagram follows directly from a state table. The state diagram gives a pictorial view of state transitions and is in a form suitable for human interpretation of the circuit operation. The state diagram is often used as the initial design specification of a sequential circuit.

State Equations

A *state equation* (also known as an *application equation*) is an algebraic expression that specifies the conditions for a flip-flop state transition. The left side of the equation denotes the next state of a flip-flop and the right side, a Boolean function that specifies the present state conditions that make the next state equal to 1. A state equation is similar in form to a flip-flop characteristic equation except that it specifies the next state conditions in terms of external input variables and other flip-flop values. The state equation is derived directly from a state table. For example, the state equation for flip-flop A is derived from inspection of Table 6-1. From the next state columns we note that flip-flop A goes to the 1-state four times: when $x = 0$ and $AB = 01$ or 10 or 11, or when $x = 1$ and $AB = 11$. This can be expressed algebraically in a state equation as follows:

$$A(t + 1) = (A'B + AB' + AB)x' + ABx$$

The right-hand side of the state equation is a Boolean function for a *present state.* When this function is equal to 1, the occurrence of a clock pulse will cause flip-flop A to have a next state of 1. When the function is equal to 0, the clock pulse will cause A to have a next state of 0. The left side of the equation identified the flip-flop by its letter symbol, followed by the time function designation $(t + 1)$ to emphasize that this value is to be reached by the flip-flop one pulse sequence later.

The state equation is a Boolean function with time included. It is applicable only in clock sequential circuits since $A(t + 1)$ is defined to change value with the occurrence of a clock pulse at discrete instants of time.

The state equation for flip-flop A is simplified by means of a map as shown in Fig. 6-13(a). With some algebraic manipulation the function can be expressed in the following form:

$$A(t + 1) = Bx' + (B'x)'A$$

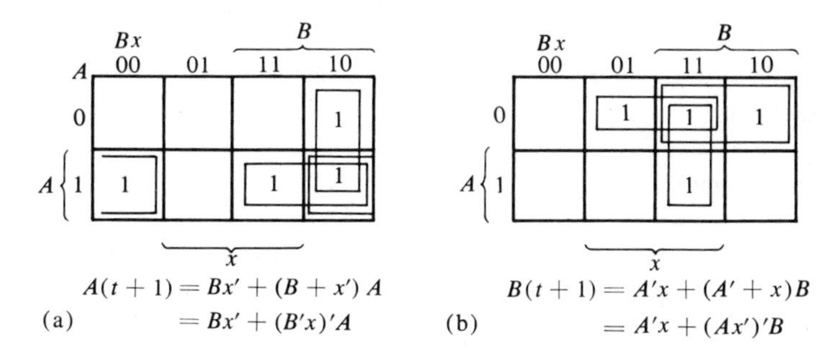

(a) $A(t+1) = Bx' + (B + x')A$
$= Bx' + (B'x)'A$

(b) $B(t+1) = A'x + (A' + x)B$
$= A'x + (Ax')'B$

Figure 6-13 State equations for flip-flops A and B

If we let $Bx' = S$ and $B'x = R$ we obtain the relation:

$$A(t + 1) = S + R'A$$

which is the characteristic equation of an *RS* flip-flop (Fig. 6-4(d)). This relation between the state equation and the flip-flop characteristic equation can be justified from inspection of the logic diagram of Fig. 6-11. In it we see that the S input of flip-flop A is equal to the Boolean function Bx' and the R input is equal to $B'x$. Substituting these functions into the flip-flop characteristic equation results in its state equation for this sequential circuit.

The state equation for a flip-flop in a sequential circuit may be derived from a state table or from a logic diagram. The derivation from the state table consists of obtaining the Boolean function specifying the conditions that make the next state of the flip-flop 1. The derivation from a logic diagram consists of obtaining the functions of the flip-flop inputs and substituting them into the flip-flop characteristic equation.

The derivation of the state equation for flip-flop B from the state table is shown in the map of Fig. 6-13(b). The 1's marked in the map are the present state and input combinations that cause the flip-flop to go to a next state of 1. These conditions are obtained directly from Table 6-1. The simplified form obtained in the map is manipulated algebraically and the state equation obtained is:

$$B(t + 1) = A'x + (Ax')'B$$

The state equation can be derived directly from the logic diagram. From Fig. 6-11 we see that the signal for input S of flip-flop B is generated by the function $A'x$ and the signal for input R by the function Ax'. Substituting $S = A'x$ and $R = Ax'$ into an *RS* flip-flop characteristic equation given by:

$$B(t + 1) = S + R'B$$

we obtain the state equation derived above.

The state equations of all flip-flops, together with the output functions, fully specify a sequential circuit. They represent, algebraically, the same information a state table represents in tabular form and a state diagram in graphical form.

Flip-Flop Input Functions

The logic diagram of a sequential circuit consists of memory elements and gates. The type of flip-flops and their characteristic table specify the logical properties of the memory elements. The interconnections among the gates form a combinational circuit and may be specified algebraically with Boolean functions. Thus, knowledge of the type of flip-flops and a list of the Boolean functions of the combinational circuit provide all the information needed to draw the logic diagram of a sequential circuit. The part of the combinational circuit that generates external outputs is described algebraically by the *circuit output functions*. The part of the circuit that generates the inputs to flip-flops are described algebraically by a set of Boolean functions called *flip-flop input functions*, or sometimes *input equations*.

We shall adopt the convention of using two letters to designate a flip-flop input variable; the first to designate the name of the input and the second the name of the flip-flop. As an example, consider the following flip-flop input functions:

$$JA = BC'x + B'Cx'$$

$$KA = B + y$$

JA and KA designate two Boolean variables. The first letter in each denotes the J and K input, respectively, of a JK flip-flop. The second letter A is the symbol name of the flip-flop. The right side of each equation is a Boolean function for the corresponding flip-flop input variable. The implementation of the two input functions is shown in the logic diagram of Fig. 6-14. The JK flip-flop has an output symbol A and two inputs labeled J and K. The combinational circuit drawn in the diagram is the implementation of the algebraic expression given by the input functions. The outputs of the combinational circuit are denoted by JA and KA in the input functions and go to the J and K inputs, respectively, of flip-flop A.

From this example we see that a flip-flop input function is an algebraic expression for a combinational circuit. The two-letter designation is a variable name for an *output* of the combinational circuit. This output is always connected to the *input* (designated by the first letter) of a flip-flop (designated by the second letter).

The sequential circuit of Fig. 6-11 has one input x, one output y, and two RS flip-flops denoted by A and B. The logic diagram can be expressed

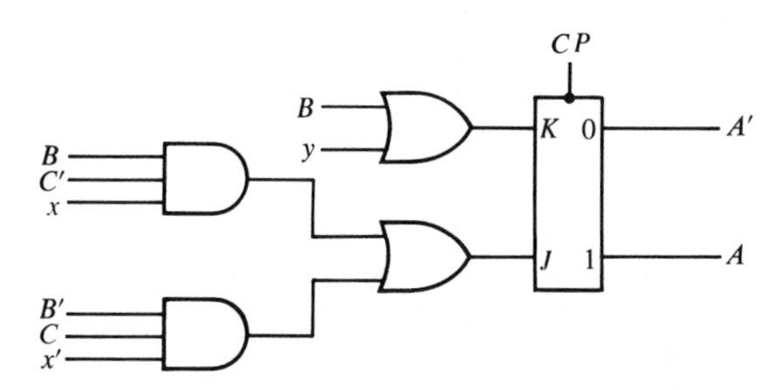

Figure 6-14 Implementation of the flip-flop input functions: $JA = BC'x + B'Cx'$ and $KA = B + y$

algebraically with four flip-flop input functions and one circuit output function as follows:

$$SA = Bx' \qquad RA = B'x$$

$$SB = A'x \qquad RB = Ax'$$

$$y = AB'x$$

This set of Boolean functions fully specifies the logic diagram. Variables SA and RA specify an RS flip-flop labeled A; variables SB and RB specify a second RS flip-flop labeled B. Variable y denotes the output. The Boolean expressions for the variables specify the combinational circuit part of the sequential circuit.

The flip-flop input functions constitute a convenient algebraic form for specifying a logic diagram of a sequential circuit. They imply the type of flip-flop from the first letter of the input variable and they fully specify the combinational circuit that drives the flip-flop. Time is not included explicitly in these equations but is implied from the clock-pulse operation. It is sometimes convenient to specify a sequential circuit algebraically with circuit output functions and flip-flop input functions instead of drawing the logic diagram.

6-5 COUNTERS

CONCENTRATE ON!

A sequential circuit that goes through a prescribed sequence of states upon the application of input pulses is called a counter. The input pulses, called *count pulses*, may be clock pulses or may originate from an external source and may occur at prescribed intervals of time or at random. The sequence of states in a counter may follow a binary count or any other sequence of states. Counters are found in almost all equipment containing digital logic. They are used for counting the number of occurrences of an event and are useful for generating timing sequences to control operations in a digital system.

Of the various sequences a counter may follow, the straight binary sequence is the simplest and most straightforward. A counter that follows the binary number sequence is called a *binary counter.* An n-bit binary counter consists of n flip-flops and can count in binary from 0 to $2^n - 1$. As an example, the state diagram of a four-bit binary counter is drawn in Fig. 6-15. As seen from the state sequence listed inside each circle, the flip-flop outputs repeat the binary count sequence with a return to 0000 after the count of 1111. The directed lines between circles are not marked with numbers because a counter is considered to be void of inputs and outputs. This is because input pulses are implied in clock sequential circuits and are not considered as input information. The outputs of a counter are the outputs of its flip-flops so that the output information is available from the present state. The present state of a counter as specified by the binary number inside a circle remains unchanged as long as an input pulse is absent. The directed line points to the next state reached after the occurrence of a count pulse.

Counters may have any sequence and may be operated synchronously or asynchronously. They are used extensively in the design of digital systems and are classified as standard items. A wide variety of counter circuits are available in MSI packages.

In this section we shall introduce a few simple counters and investigate their operation. A procedure for designing synchronous counters of any sequence is found in Sec. 7-5.

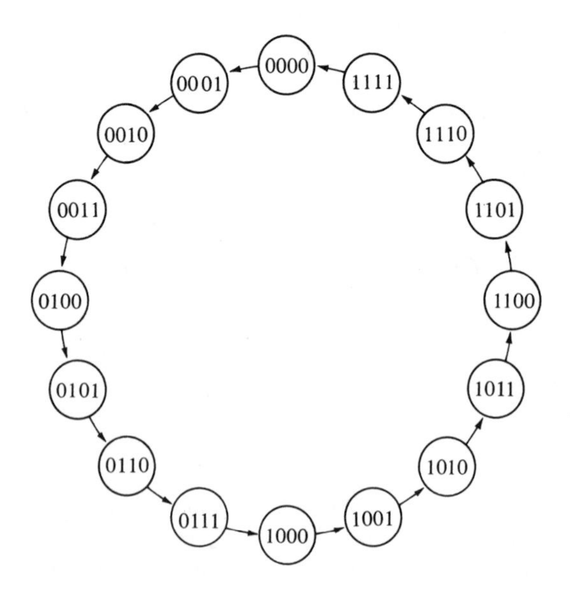

Figure 6-15 State diagram of a 4-bit binary counter

Binary Ripple Counter

A binary ripple counter is the most elementary digital circuit that performs the most elementary binary function. It consists of a series connection of T flip-flops without any logic gates. Each flip-flop is triggered by the output of its preceeding flip-flop. To understand the operation of the counter, it is necessary to go back to the state diagram of Fig. 6-15 and investigate the sequence of numbers inside circles. Going through all 16 numbers in sequence we note that the least significant bit is complemented with every state transition. The second significant bit is complemented during a transition from 1 to 0 of the least significant bit. Similarly, any significant bit in a binary count sequence is complemented during the transition from 1 to 0 of its preceeding significant bit. The binary ripple counter performs this binary count sequence with complementing T flip-flops as follows: the least significant flip-flop is complemented every count pulse and each succeeding flip-flop is complemented when its previous flip-flop goes from 1 to 0.

Two logic diagrams of a four-bit binary ripple counter are shown in Fig. 6-16. Each consists of four T flip-flops with the output of one flip-flop going into the input of the next on its left. The first flip-flop is complemented with each count pulse. The other flip-flops are complemented when the normal output of the preceeding flip-flop changes from 1 to 0. Each flip-flop serves as a source for triggering the next flip-flop and the

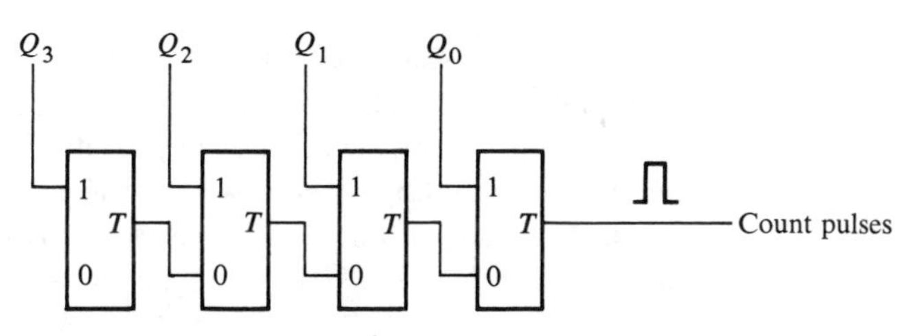

(a) With flip-flops triggered on positive-going edge

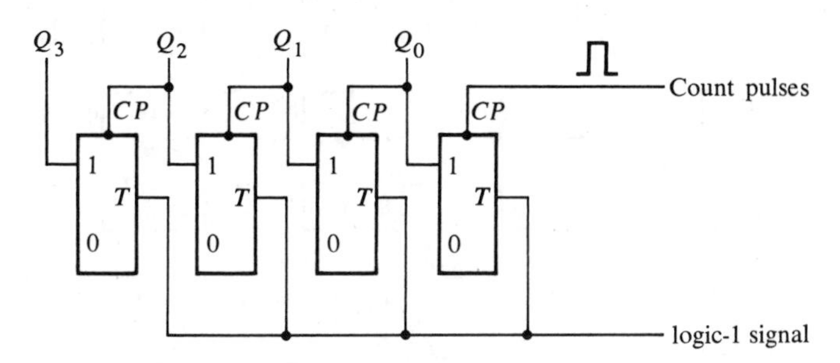

(b) With clocked flip-flops triggered on the trailing edge of the pulse

Figure 6-16 Asynchronous 4-bit binary counters

signal propagates through the counter in a "ripple" fashion; i.e., the flip-flops essentially change one at a time in rapid succession. This is in contrast to a synchronous counter where all flip-flops change state simultaneously with the incoming count pulse.

The binary ripple counter of Fig. 6-16(a) uses unclocked flip-flops triggered on the positive-going edge of a changing signal amplitude. In a positive logic system, this means that the flip-flop starts changing states when its input goes from 0 to 1. Therefore, we must connect the complement output of a flip-flop to the T input of the next flip-flop, since the complement output goes from 0 to 1 (the required signal) when the normal output goes from 1 to 0 (the required numerical condition).

The circuit of Fig. 6-16(b) uses clocked flip-flops with trailing-edge triggering. In a positive logic system, this means that the change of state occurs during a signal transition from 1 to 0. Therefore, the normal output of one flip-flop is connected to the CP input of the next. The T inputs are connected permanently to a logic-1 signal so that the state transition is controlled entirely by the CP input. The timing diagram for each flip-flop of Fig. 6-16(b) is shown in Fig. 6-17. Note that except for flip-flop Q_0, the state transition is not caused by the pulse input but by the signal transition from 1 to 0 of the previous flip-flop.

Synchronous Binary Counter

In a synchronous counter, all flip-flops are triggered simultaneously by the count pulse. As shown in Fig. 6-18, all CP terminals are connected to the count pulse, but the flip-flop is complemented only if its T input is equal to 1. The condition for state transition is now determined from the present state value of other flip-flops. These conditions are derived from inspection of the state diagram of Fig. 6-15. Going through the sequence of

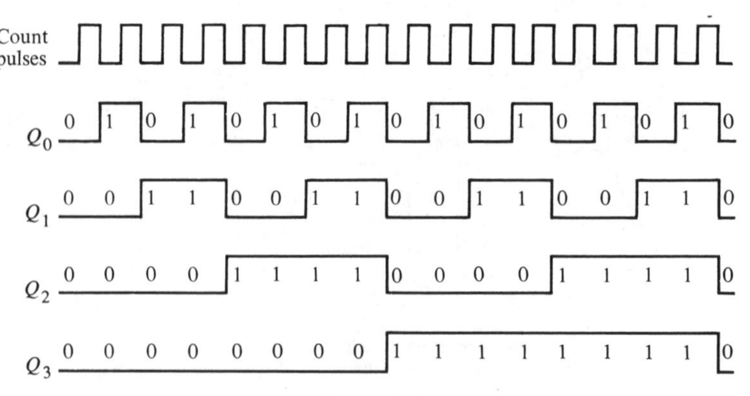

Figure 6-17 Timing diagram for counter of Figure 6-16 (b)

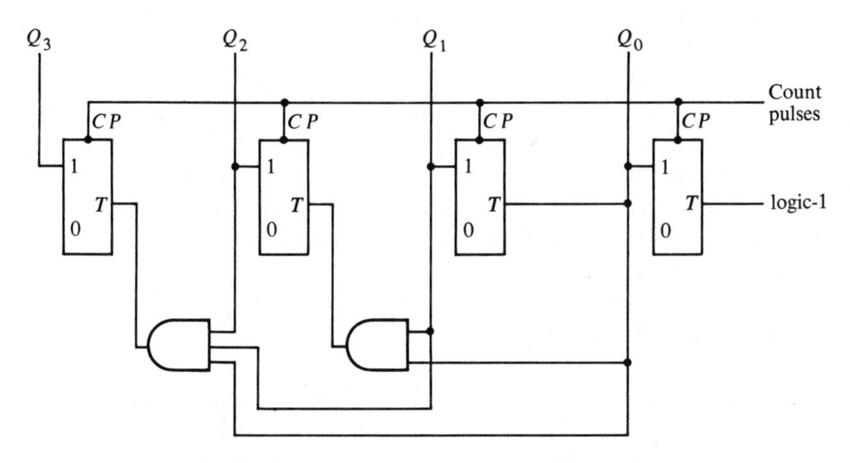

Figure 6-18 Synchronous 4-bit binary counter

binary numbers, we note that a bit is complemented if all previous least significant bits in the present state equal 1. Therefore, a flip-flop in a synchronous counter is allowed to be complemented by a count pulse if all previous flip-flops on its right have normal outputs equal to 1. This is shown in the logic diagram and also expressed algebraically by the following flip-flop input functions

$$TQ_0 = 1$$

$$TQ_1 = Q_0$$

$$TQ_2 = Q_0Q_1$$

$$TQ_3 = Q_0Q_1Q_2$$

An n-bit synchronous binary counter consists of n T flip-flops. The flip-flop input functions can be written in a concise form as follows:

$$TQ_0 = 1$$

$$TQ_i = \prod_{j=0}^{i-1} Q_j \quad \text{for } i = 1, 2, 3, \ldots, n$$

where Π is a product sign designating the AND operation. The T input of a flip-flop Q_i receives the output of an AND gate whose inputs come from flip-flops Q_j from $j = 0$ to $j = i - 1$. In addition the CP terminal in each flip-flop receives the input count pulses.

The advantage of a synchronous counter over a ripple counter is in its higher speed of operation. When the input count pulse arrives, all flip-flops are triggered and may change states simultaneously. The delay from the time the input is applied until all flip-flops reach the next state is equal to the signal transition time of a single flip-flop. The signal propagates through the AND gates during the time between pulses so that the signals in the *T* inputs of flip-flops are settled to a steady-state value waiting for the next count pulse. In a ripple counter, the signal propagates, or ripples, through all flip-flops and when all the flip-flops change state, as for example from 1111 to 0000, the total delay is equal to four flip-flop transitions. Ripple counters have logic simplicity and therefore cost less. They are preferable if speed of operation is not critical.

Decimal BCD Ripple Counter

A decimal counter follows a sequence of 10 states and returns to 0 after the count of 9. Such a counter must have at least four flip-flops to represent each decimal digit, since a decimal digit is represented by a binary code with at least four bits. The sequence of states in a decimal counter is dictated by the binary code used to represent a decimal digit. If BCD is used (Table 1-2), the sequence of states is as shown in the state diagram of Fig. 6-19. This is similar to a binary counter except that the state after 1001 (code for decimal digit 9) is 0000 (code for decimal digit 0).

The design of synchronous decimal counters using any code is presented in Sec. 7-5. The design of a decimal ripple counter or of any ripple counter not following the binary sequence is not a straightforward procedure. The formal tools of logic design can serve only as a guide. A satisfactory end product requires the ingenuity and imagination of the designer.

The logic diagram of a BCD ripple counter is shown in Fig. 6-20. The four outputs are designated by the letter symbol Q with a numeric subscript equal to the binary weight of the corresponding bit in the BCD code. The flip-flops are assumed to trigger on the trailing edge; i.e., when the *CP* signal goes from 1 to 0. Note that the output of Q_1 is applied to the *CP* inputs of both Q_2 and Q_8 and the output of Q_2 is applied to the *CP* input of Q_4. The J and K inputs are connected either to a permanent logic-1 signal or to outputs of flip-flops, as shown in the diagram.

A ripple counter is an asynchronous sequential circuit and cannot be described by Boolean equations developed for describing clocked sequential circuits. Signals that affect the flip-flop transition depend on the order in which they change from 1 to 0. The operation of the counter can be explained by a list of conditions for flip-flop transitions. These conditions are derived from the logic diagram and from knowledge of how a *JK* flip-flop operates . Remember that when the *CP* input goes from 1 to 0, the flip-flop is set if $J = 1$, is cleared if $K = 1$, is complemented if both $J = K = 1$, and is left unchanged if both $J = K = 0$. The following are the conditions for each flip-flop state transition:

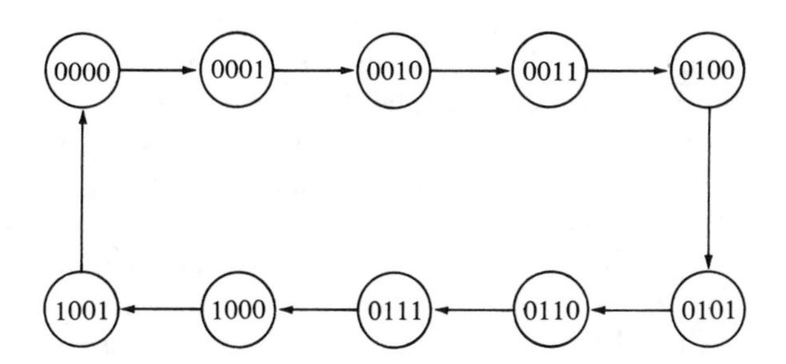

Figure 6-19 State diagram of decimal *BCD* counter

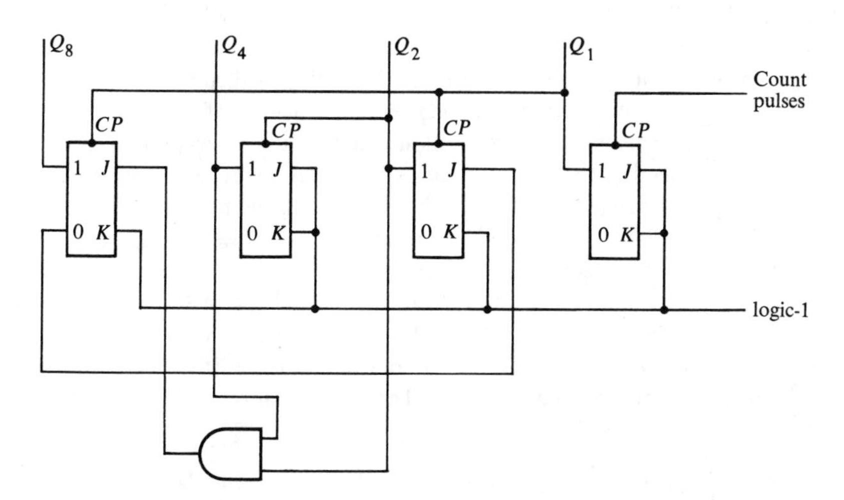

Figure 6-20 Logic diagram of a decade *BCD* ripple counter

Flip-flop Q_1 is complemented on the trailing edge of a count pulse.
Flip-flop Q_2 is complemented if $Q_8 = 0$ and Q_1 switches from 1 to 0. It is cleared if $Q_8 = 1$ and Q_1 switches from 1 to 0.
Flip-flop Q_4 is complemented when Q_2 switches from 1 to 0.
Flip-flop Q_8 is complemented if $Q_2 = 1$ and $Q_4 = 1$ and Q_1 switches from 1 to 0. It is cleared if $Q_2 = 0$ or $Q_4 = 0$ and Q_1 switches from 1 to 0.

To verify that these conditions result in the sequence required by a BCD counter, it is necessary to verify that the flip-flop transitions indeed follow a sequence of states as specified by the state diagram of Fig. 6-19. Another way to verify the operation of the counter is to derive the timing diagram for each flip-flop from the conditions listed above. This diagram is shown in Fig. 6-21, from which we see that the flip-flops follow the required sequence of states.

The decimal counter of Fig. 6-20 is a one decade counter; i.e., it counts from 0 to 9 only. To count higher than 9 we need decade counters, one for each decimal digit. For example, a three-decade decimal counter is shown in Fig. 6-22. Each box in the diagram represents a BCD decade counter similar to the one in Fig. 6-20 so that the total count ranges from 000 to 999. The inputs to the second and third decades come from Q_8 of the previous decade. When output Q_8 of one decade goes from 1 to 0, it triggers flip-flop Q_1 of the next decade, while its own decade goes from 1001 to 0000. Thus the carry from Q_8 in one decade is used to trigger the count in the next decade.

6-6 SHIFT-REGISTER

Binary information is stored in groups of flip-flops called *registers*. A group of n flip-flops, called an n-bit register, is capable of storing n-bits of information. In this section we shall investigate the operation of an often used register called *shift-register*. This register has many applications in the design of digital systems and is available in IC form as an MSI function.

The logic diagram of a four-bit shift-register is shown in Fig. 6-23. It is composed of four clocked flip-flops with D inputs. The register is used for temporary storage of a four-bit datum. Data can be transferred in and out of the register in three different modes dictated by control signals P, S_R, and S_L. The three modes are parallel transfer, serial shift-right transfer, and serial shift-left transfer, respectively. The three transfer modes operate as follows:

PARALLEL TRANSFER: with $P = 1$, $S_R = 0$, $S_L = 0$, the data available in inputs x_0 through x_3 are transferred into A_0 through A_3, respectively. This is called parallel transfer because all flip-flops receive new

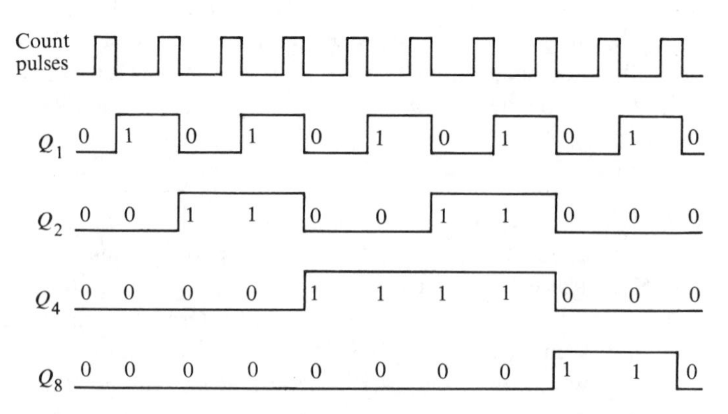

Figure 6-21 Timing diagram for decimal counter of Figure 6-20

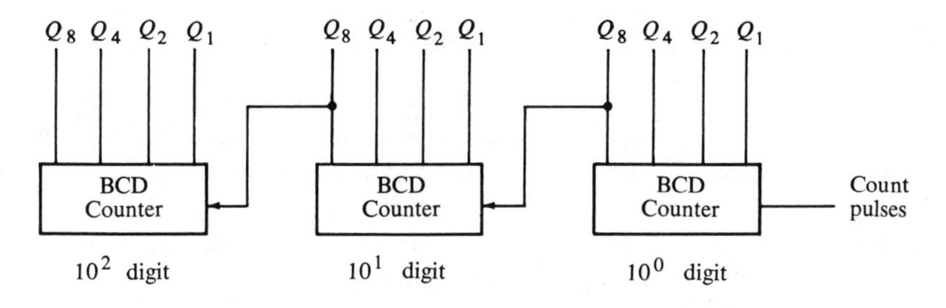

Figure 6-22 Block diagram of a 3-decade decimal *BCD* counter

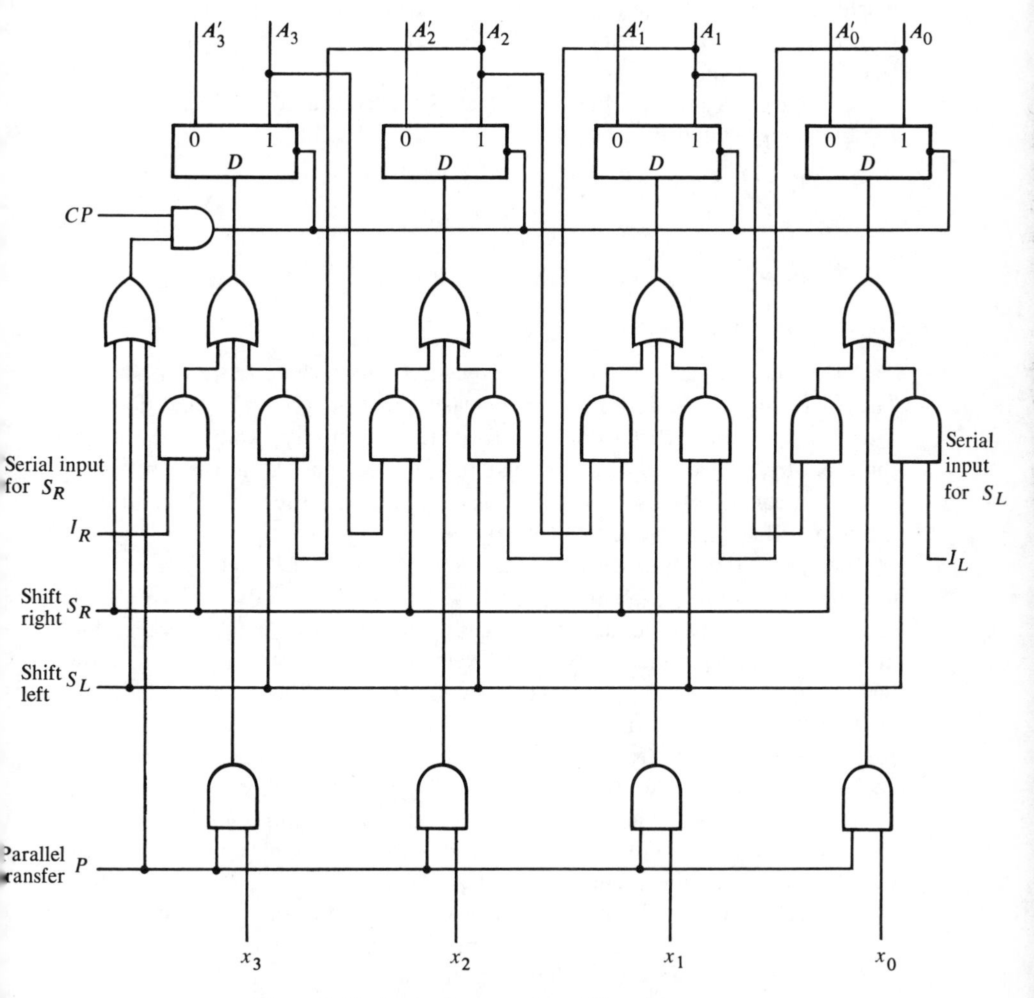

Figure 6-23 Logic diagram of a shift-register

data during one clock pulse. Data stored in the register is transferred out in parallel by sampling the outputs of the four flip-flops.

SHIFT-RIGHT: with $S_R = 1$, $P = 0$, $S_L = 0$, the data in the register is shifted to the right upon the occurrence of each clock pulse. Each flip-flop receives new data from its neighbor on the left and flip-flop A_3 receives its data from external input I_R. Data is transferred out by sampling the output of flip-flop A_0. This is called serial transfer because one bit is transferred in and out of the register during one clock pulse.

SHIFT-LEFT: with $S_L = 1$, $P = 0$, $S_R = 0$, the data in the register is transferred to the left. Each flip-flop receives its new data from the neighbor on its right and A_0 receives its data from external input I_L. Data is transferred out by sampling the output of A_3. This again is a serial transfer but its direction is toward the left.

The logic diagram of Fig. 6-23 may be specified algebraically by four flip-flop input functions as follows:

$$DA_0 = I_L S_L + x_0 P + A_1 S_R$$

$$DA_1 = A_0 S_L + x_1 P + A_2 S_R$$

$$DA_2 = A_1 S_L + x_2 P + A_3 S_R$$

$$DA_3 = A_2 S_L + x_3 P + I_R S_R$$

The D input of each flip-flop receives data from three different sources. The first, second, and third term in each function give the conditions for shift-left, parallel transfer, and shift-right, respectively. Only one of the three control signals should be equal to 1 at any time. Otherwise, if two or three control signals are equal to logic-1 simultaneously, the D input to a flip-flop will respond to the OR of the input data bits.

There are four flip-flops and nine inputs (plus the CP input) in the logic diagram of Fig. 6-23, so that a state table or diagram will consist of 16 states and 2^9 input combinations. Only some of the input combinations may occur during normal operation. Nevertheless, a table or diagram showing all allowable input combinations will be excessively complicated. The state table for this particular sequential circuit may be broken into three parts, one for each transfer mode, since the three modes are mutually exclusive. For example, the state table for the shift-left mode is listed in Table 6-2. In this mode, it is assumed that $S_L = 1$ and that the input to the circuit is I_L. An output section is not included in the table because the output is equal to the present state of A_3.

A binary number stored in a register is multiplied by 2 when it is shifted once to the left and divided by 2 when shifted once to the right. However, this is true only if the bit shifted out is not lost. To illustrate, consider a five-bit shift-register with the number 01100 stored in it. This

Table 6-2 State Table of a Shift-Left Register. (S_L = 1, S_R = 0, P = 0)

Present State				Next state							
				$I_L = 1$				$I_L = 0$			
A_3	A_2	A_1	A_0	A_3	A_2	A_1	A_0	A_3	A_2	A_1	A_0
0	0	0	0	0	0	0	1	0	0	0	0
0	0	0	1	0	0	1	1	0	0	1	0
0	0	1	0	0	1	0	1	0	1	0	0
0	0	1	1	0	1	1	1	0	1	1	0
0	1	0	0	1	0	0	1	1	0	0	0
0	1	0	1	1	0	1	1	1	0	1	0
0	1	1	0	1	1	0	1	1	1	0	0
0	1	1	1	1	1	1	1	1	1	1	0
1	0	0	0	0	0	0	1	0	0	0	0
1	0	0	1	0	0	1	1	0	0	1	0
1	0	1	0	0	1	0	1	0	1	0	0
1	0	1	1	0	1	1	1	0	1	1	0
1	1	0	0	1	0	0	1	1	0	0	0
1	1	0	1	1	0	1	1	1	0	1	0
1	1	1	0	1	1	0	1	1	1	0	0
1	1	1	1	1	1	1	1	1	1	1	0

binary number is equivalent to decimal 12. When shifted once to the left, the content of the register becomes 11000, the equivalent of decimal 24. Shifting the original number once to the right changes the content of the register to 00110, the equivalent of decimal 6.

A shift-register can temporarily store binary data and transfer this data in and out of the register. Strictly speaking, a shift-register transfers data serially a bit at a time. The transfer of data in parallel is an added feature included in most MSI packages so as to provide a more flexible register.

PROBLEMS

6-1. Show the logic diagram of a clocked *RS* flip-flop with four NAND gates.

6-2. Show the logic diagram of a clocked *D* flip-flop with AND/NOR gates.

6-3. Show that the clocked *D* flip-flop of Fig. 6-5(a) can be reduced by one gate.

6-4. Obtain the logic diagram of a master-slave *JK* flip-flop with AND and NOR gates. Include a provision for setting and clearing the flip-flop asynchronously (without a clock).

6-5. Consider a JK' flip-flop; i.e., a JK flip-flop with an inverter between external input K' and internal input K.

(a) Obtain the flip-flop characteristic table.

(b) Obtain the characteristic equation.

(c) Show that tying the two external inputs together forms a D flip-flop.

6-6. A set-dominate flip-flop has a set and a reset input. It differs from a conventional RS flip-flop in that an attempt to simultaneously set and reset results in setting the flip-flop.

(a) Obtain the characteristic table and characteristic equation for the set-dominate flip-flop.

(b) Obtain a logic diagram for an asynchronous set-dominate flip-flop.

6-7. The logic diagram of Fig. P6-7 has an input x and clock pulses (CP) as shown. Assume positive logic and that the flip-flop changes state at the trailing edge of the positive pulse. Draw the timing diagram of x, x', A, A', flip-flop inputs S and R, and output y. Start by assuming that the flip-flop is initially in the set-state and later justify your assumption. What is the function of this circuit?

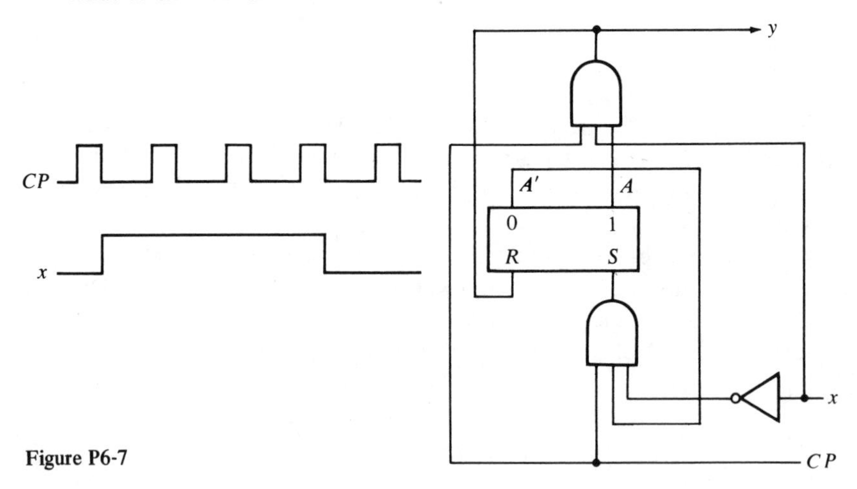

Figure P6-7

6-8. For the circuit of Fig. P6-7:

(a) Obtain the flip-flop input functions and the output Boolean function.

(b) Obtain the state table and state diagram.

(c) Derive the state equation from the state table and again directly from the logic diagram and show that both results are the same.

6-9. The full-adder of Fig. P6-9 receives two external inputs x and y, the third input z comes from the output of a D flip-flop. The carry output is transferred to the flip-flop every clock pulse. The external S output gives the sum of x, y, and z. Assume that x and y change after the trailing edge of the clock pulse. Obtain the state table and state diagram of the sequential circuit.

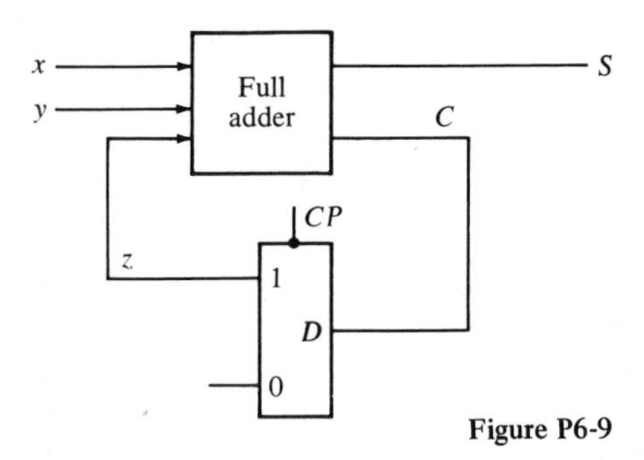

Figure P6-9

6-10. Derive the state table and state diagram of the sequential circuit shown. What is the function of the circuit?

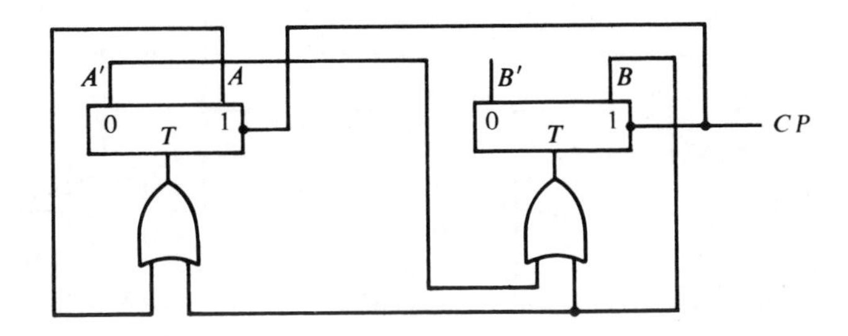

6-11. A sequential circuit has four flip-flops A, B, C, D and an input x. It is described by the following state equations:

$$A\ (t + 1) = (CD' + C'D)x + (CD + C'D')\ x'$$

$$B\ (t + 1) = A$$

$$C\ (t + 1) = B$$

$$D\ (t + 1) = C$$

Draw the logic diagram using D flip-flops and obtain the state diagram.

6-12. A sequential circuit has two flip-flops (A and B), two inputs (x and y), and an output (z). The flip-flop input functions and the circuit output function are as follows:

$$JA = xB + y'B' \qquad KA = xy'B'$$

$$JB = xA' \qquad KB = xy' + A$$

$$z = xyA + x'y'B$$

Obtain the logic diagram, state table, state diagram, and state equations.

6-13. Draw the timing diagram for the asynchronous binary counter of Fig. 6-16(a).

6-14. Draw the timing diagram for the decimal counter of Fig. 6-20 when negative logic signals are employed.

6-15. The asynchronous counter shown below uses clocked flip-flops that trigger at the trailing edge of the pulse. Determine the sequence of states starting from $ABC = 000$. What happens if the counter is initially in state 110 or 111?

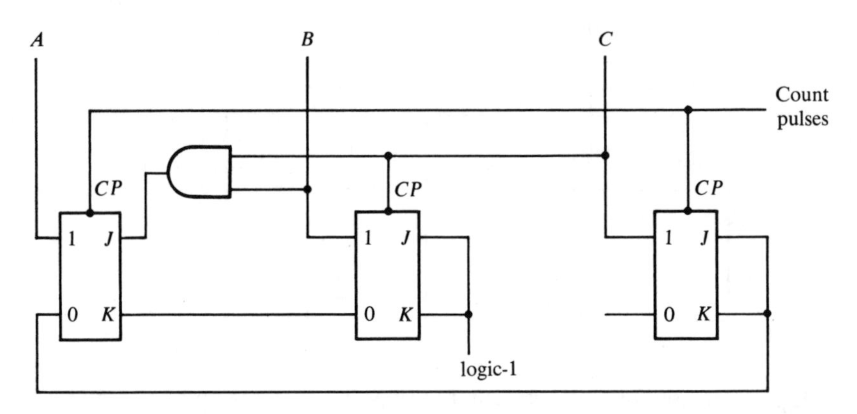

6-16. A count-down counter is a counter with reverse count; i.e., the binary count is reduced by one for every count pulse.

(a) Draw the state diagram of a four-bit count-down binary counter.

(b) Show that a flip-flop should be complemented when its previous flip-flop goes from 0 to 1. Draw the logic diagram of a four-bit count-down ripple binary counter.

(c) Draw the logic diagram of a synchronous four-bit count-down binary counter.

(d) Draw the logic diagram of a counter that counts either up or down. Use an external input x so that $x = 1$ for count up and $x = 0$ for count down.

6-17. The state equations of a 2-out-of-5 counter are as follows:

$$A(t + 1) = E \qquad D(t + 1) = A'C + AB$$

$$B(t + 1) = A \qquad E(t + 1) = D$$

$$C(t + 1) = A'B + AC$$

Determine the count sequence starting from the state $ABCDE = 01100$. Explain the reason for the name.

6-18. Given a six-bit shift-right register, the input to the left-most flip-flop is 101101. Assume that the register is initially cleared. Show the contents of the register after each clock pulse. Also draw a timing diagram for the output of each flip-flop.

6-19. A ring-counter is a shift-right register with the normal output of the right-most flip-flop connected to the input of the left-most flip-flop. Obtain the count sequence of a ring counter starting with the count of 1000.

6-20. A switch-tail ring-counter is a shift-right register with the complement output of the right-most flip-flop connected to the input of the left-most flip-flop. Obtain the count sequence of the counter starting with the count of 0000.

6-21. A feedback shift-register is a shift-register whose serial input is a function of the present state of the register. Obtain the state diagram of a feedback shift-right register whose serial input is $I_R = A_2A_0 + A_2'A_0$ (see Fig. 6-23).

7 DESIGN OF CLOCKED SEQUENTIAL CIRCUITS

7-1 INTRODUCTION

The design of a clocked sequential circuit starts from a set of specifications and culminates in a logic diagram or a list of Boolean functions from which the logic diagram can be obtained. In contrast to a combinational circuit, which is fully specified by a truth table, a sequential circuit requires a state table for its specification. The first step in the design of sequential circuits is to obtain a state table or an equivalent representation, such as a state diagram or state equations.

A synchronous sequential circuit is made up of flip-flops and combinational gates. The design of the circuit consists of choosing the flip-flops and then finding a combinational gate structure which, together with the flip-flops, produces a circuit that fulfils the stated specifications. The number of flip-flops is determined from the number of states needed in the circuit. The combinational circuit is derived from the state table by methods presented in this chapter. In fact, once the type and number of flip-flops are determined, the design process involves a transformation from the sequential circuit problem into a combinational circuit problem. In this way the techniques of combinational circuit design can be applied.

Any design process must consider the problem of minimizing the cost of the final circuit. The two most obvious cost reductions are reductions in the number of flip-flops and the number of gates. Because these two items seem the most obvious, they have been extensively studied and investigated. In fact, a large portion of the subject of switching theory is concerned with finding algorithms for minimizing the number of flip-flops and gates in sequential circuits. However, the reader is referred to Sec. 5-4 for a discussion of other criteria that must be considered in a practical situation.

The reduction of the number of flip-flops in a sequential circuit is referred to as the *state reduction* problem. State reduction algorithms are concerned with procedures for reducing the number of states in a state table while keeping the external input-output requirements unchanged. Since m flip-flops produce 2^m states, a reduction in the number of states may (or may not) result in a reduction in the number of flip-flops. An unpredictable effect in reducing the number of flip-flops is that sometimes the equivalent circuit (with less flip-flops) may require more combinational gates.

The cost of the combinational circuit part of a sequential circuit can be reduced by using the known simplification methods for combinational circuits. However, there is another factor that comes into play in minimizing the combinational gates, known as the *state assignment* problem. State assignment procedures are concerned with methods for choosing binary values to states in such a way as to reduce the cost of the combinational circuit that drives the flip-flops. This is particularly helpful when a sequential circuit is viewed from its external input-output terminals. Such a circuit may follow a sequence of internal states, but the binary values of the individual states may be of no consequence as long as the circuit produces the required sequence of outputs for any given sequence of inputs. This does not apply to circuits whose external outputs are taken directly from flip-flops with binary sequences fully specified, as in a binary counter.

Most digital systems considered in this book involve the use of registers with a prescribed number of flip-flops and binary value assignment. State reduction and state assignment procedures offer very little help in reducing the components in such circuits. However, state reduction and assignment play an important part in the design of certain sequential circuits and provide guidance to the problem of equipment minimization. For this reason we shall explain these two problems by an illustrative example in the next section. The enumeration of all possible algorithms and procedures for state reduction and state assignment is beyond the scope of this book.*

7-2 THE STATE REDUCTION AND ASSIGNMENT PROBLEMS

We shall illustrate the need for state reduction and state assignment with an example. We start with a sequential circuit whose specification is given in the state diagram of Fig. 7-1. In this example, only the input-output sequences are important; the internal states are used merely to provide the

*The interested reader will find these two topics covered in detail in any book on switching circuit theory (see references 1-7).

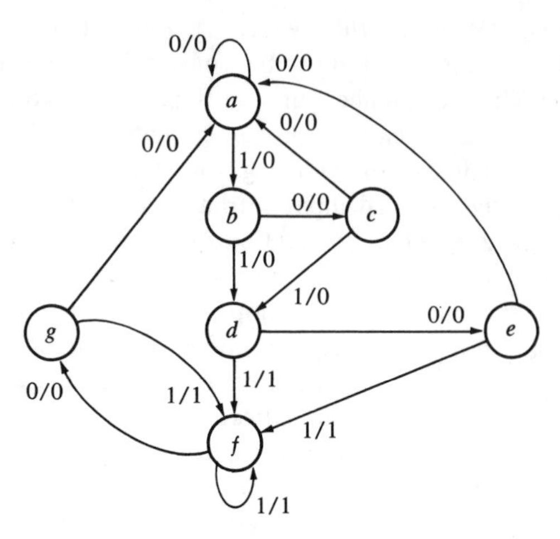

Figure 7-1 State diagram

required sequences. For this reason, the states marked inside the circles are denoted by letter symbols instead of their binary value. This is in contrast to a binary counter, where the binary value sequence of the states themselves are taken as the outputs.

There are an infinite number of input sequences that may be applied to the circuit; each will result in a unique output sequence. As an example, consider the input sequence 01010110100 starting from the initial state a. Each input of 0 or 1 will produce an output of 0 or 1 and cause the circuit to go to the next state. From the state diagram we obtain the output and state sequence for the given input sequence as follows: with the circuit in initial state a, and input of 0 produces an output of 0 and the circuit remains in state a. With present state a and input of 1, the output is 0 and the next state is b. With present state b and input of 0, the output is 0 and next state is c. Continuing this process, we find the complete sequence to be as follows:

state:	a	a	b	c	d	e	f	f	g	f	g	a
input:	0	1	0	1	0	1	1	0	1	0	0	
output:	0	0	0	0	0	1	1	0	1	0	0	

In each column we have the present state, input value, and output value. The next state is written on top of the next column. It is important to realize that in this circuit the states themselves are of secondary importance because we are interested only in output sequences caused by input sequences.

Now let us assume that we have found a sequential circuit whose state diagram has less than seven states and we wish to compare it with the circuit whose state diagram is given by Fig. 7-1. If identical input sequences are applied to the two circuits and identical outputs occur for all input sequences, then the two circuits are said to be equivalent (as far as the input-output is concerned) and one may be replaced by the other. The problem of state reduction is to find ways of reducing the number of states in a sequential circuit without altering the input-output relations.

We shall now proceed to reduce the number of states for this example. First, we need the state table; it is more convenient to apply procedures for state reduction here than in state diagrams. The state table of the circuit is listed in Table 7-1 and is obtained directly from the state diagram of Fig. 7-1.

An algorithm for the state reduction of a completely specified state table is given here without proof: "Two states are said to be equivalent if, for each member of the set of inputs, they give exactly the same output and send the circuit either to the same state or to an equivalent state. When two states are equivalent, one of them can be removed without altering the input-output relations."

We shall apply this algorithm to Table 7-1. Going through the state table we look for two present states that go to the same next state and have the same output for both input combinations. States g and e are two such states; they both go to states a and f and have outputs of 0 and 1 for $x = 0$ and $x = 1$, respectively. Therefore, states g and e are equivalent; one can be removed. The procedure of removing a state and replacing it by its equivalent is demonstrated in Table 7-2. The row with present state g is crossed-out and state g is replaced by state e everywhere it occurs in the next state columns.

Present state f now has next states e and f and outputs 0 and 1 for $x = 0$ and $x = 1$, respectively. The same next states and outputs appear in the row with present state d. Therefore, states f and d are equivalent; state f can be removed and replaced by d. The final reduced table is shown in Table 7-3. The state diagram for the reduced table consists of only five

Table 7-1 State Table

Present state	Next state		Output	
	$x = 0$	$x = 1$	$x = 0$	$x = 1$
a	a	b	0	0
b	c	d	0	0
c	a	d	0	0
d	e	f	0	1
e	a	f	0	1
f	g	f	0	1
g	a	f	0	1

Table 7-2 Reducing the State Table

Present state	Next state		Output	
	x = 0	*x = 1*	*x = 0*	*x = 1*
a	a	b	0	0
b	c	d	0	0
c	a	d	0	0
d	e	~~f~~ d	0	1
e	a	~~f~~ d	0	1
~~f~~	~~g~~ e	f	0	1
~~g~~	a	f	0	1

Table 7-3 Reduced State Table

Present state	Next state		Output	
	x = 0	*x = 1*	*x = 0*	*x = 1*
a	a	b	0	0
b	c	d	0	0
c	a	d	0	0
d	e	d	0	1
e	a	d	0	1

states and is shown in Fig. 7-2. This state diagram satisfies the original input-output specifications and will produce the required output sequence for any given input sequence. The following list derived from the state diagram of Fig. 7-2 is for the input sequence used previously. We note that the same output sequence results although the state sequence is different:

state:	*a*	*a*	*b*	*c*	*d*	*e*	*d*	*d*	*e*	*d*	*e*	*a*
input:	0	1	0	1	0	1	1	0	1	0	0	
output:	0	0	0	0	0	1	1	0	1	0	0	

In fact, this sequence is exactly the same as that obtained for Fig. 7-1 if we replace e by g and d by f.

It is worth noting that the reduction in the number of states of a sequential circuit is possible if one is interested only in external input-output relations. When external outputs are taken directly from flip-flops, the outputs must be independent of the number of states before applying state reduction algorithms.

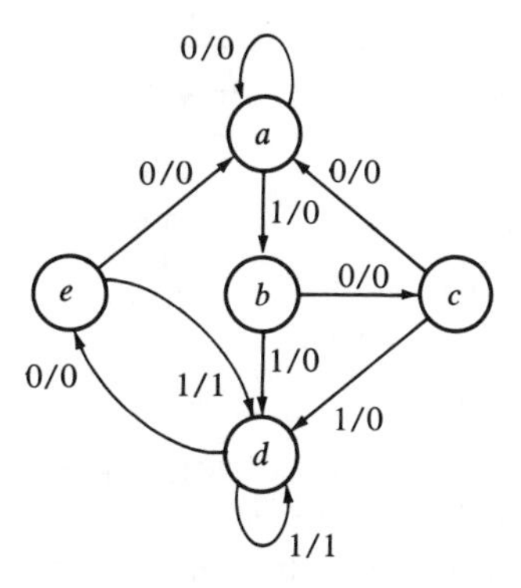

Figure 7-2 Reduced state diagram

The sequential circuit of this example was reduced from seven to five states. In either case, the representation of the states with physical components requires that we use three flip-flops because m flip-flops can represent up to 2^m distinct states. With three flip-flops, we can formulate up to eight binary states denoted by binary numbers 000 to 111 with each bit designating the state of one flip-flop. If the state table of Table 7-1 is used, we must assign binary values to seven states, the remaining state is unused. If the state table of Table 7-3 is used, only five states need binary assignment and we are left with three unused states. Unused states are treated as don't-care conditions during the design of the circuit. Since don't-care conditions usually help in obtaining a simpler Boolean function, it is more likely that the circuit with five states will require less combinational gates than the one with seven states. In any case, the reduction from seven to five states does not reduce the number of flip-flops. In general, reducing the number of states in a state table is likely to result in a circuit with less equipment. However, the fact that a state table has been reduced to less states does not guarantee a saving in the number of flip-flops or the number of gates.

It is now necessary to assign binary values to each state to replace the letter symbol used thus far. Remember that, in the present example, the binary value of the states is immaterial as long as their sequence maintains the proper input-output relations. For this reason, any binary number assignment is satisfactory as long as each state is assigned a unique number. Three examples of possible binary assignments are shown in Table 7-4 for the five states of the reduced table. Assignment 1 is a straight binary assignment for the sequence of states from a to e. The other two assignments are chosen arbitrarily. In fact, there are 140 different distinct assignments for this circuit (11).

Table 7-4 Three Possible Binary State Assignments

State	*Assignment 1*	*Assignment 2*	*Assignment 3*
a	001	000	000
b	010	010	100
c	011	011	010
d	100	101	101
e	101	111	011

Table 7-5 is the reduced state table with binary assignment 1 substituted for the letter symbols of the five states.* It is obvious that a different binary assignment will result in a state table with different binary values for the states, while the input-output relations remain the same. The binary form of the state table is used to derive the combinational circuit part of the sequential circuit. The complexity of the combinational circuit obtained depends on the binary state assignment chosen. In Sec. 7-3 we introduce a procedure for obtaining the combinational circuit from the state table. The effect of different binary assignments is demonstrated in Sec. 7-4 (Examples 7-1 and 7-2), where the design of the circuit is completed.

Various procedures have been suggested that lead to a particular binary assignment from the many available. The most common criteria is that the chosen assignment should result in a simple combinational circuit for the flip-flop inputs. However, to date, there are no state assignment procedures that guarantee a minimal cost combinational circuit. State assignment is one of the challenging problems of switching theory. The interested reader will find a rich and growing literature on this topic. Techniques for dealing with the state assignment problem are beyond the scope of this book.

7-3 FLIP-FLOP EXCITATION TABLES

The characteristic tables for the various flip-flops were presented in Sec. 6-2. A characteristic table defines the logical property of the flip-flop and completely characterizes its operation. Integrated circuit flip-flops are sometimes defined by a characteristic table tabulated somewhat differently. This second form of the characteristic tables for *RS*, *JK*, *D*, and *T* flip-flops is shown in Table 7-6. They represent the same information as the characteristic tables of Figs. 6-4(c) through 6-7(c).

Table 7-6 defines the state of each flip-flop as a function of its inputs and previous state. $Q(t)$ refers to the present state and $Q(t + 1)$ to the next state after the occurrence of a clock pulse. The characteristic table for the *RS* flip-flop shows that the next state is equal to the present state

*A state table with binary assignment is sometimes called a *transition table.*

Table 7-5 Reduced State Table with Binary Assignment 1

Present state	Next state		Output	
	$x = 0$	$x = 1$	$x = 0$	$x = 1$
001	001	010	0	0
010	011	100	0	0
011	001	100	0	0
100	101	100	0	1
101	001	100	0	1

Table 7-6 Flip-Flop Characteristic Tables

S	R	$Q(t+1)$
0	0	$Q(t)$
0	1	0
1	0	1
1	1	?

(a) *RS*

J	K	$Q(t+1)$
0	0	$Q(t)$
0	1	0
1	0	1
1	1	$Q'(t)$

(b) *JK*

D	$Q(t+1)$
0	0
1	1

(c) *D*

T	$Q(t+1)$
0	$Q(t)$
1	$Q'(t)$

(d) *T*

when both inputs S and R are 0. When the R input is equal to 1, the next clock pulse clears the flip-flop. When the S input is equal to 1, the next clock pulse sets the flip-flop. The question mark for the next state when S and R are both equal to 1 simultaneously designates an indeterminate next state.

The table for the *JK* flip-flop is the same as that for the *RS* when J and K are replaced by S and R, respectively, except for the indeterminate case. When both J and K are equal to 1, the next state is equal to the complement of the present state; i.e., $Q(t + 1) = Q'(t)$. The next state of the D flip-flop is completely dependent on the input D and independent of the present state. The next state of the T flip-flop is the same as the present state if $T = 0$ and complemented if $T = 1$.

The characteristic table is useful for analysis and for defining the operation of the flip-flop. It specifies the next state when the inputs and present state are known. During the design process we usually know the transition from present state to next state and wish to find the flip-flop input conditions that will cause the required transition. For this reason we need a table that lists the required inputs for a given change of state. Such a list is called an *excitation table.*

Table 7-7 lists the excitation tables for the four flip-flops. Each table consists of two columns $Q(t)$ and $Q(t+1)$ and a column for each input to show how the required transition is achieved. There are four possible transitions from present state to next state. The required input conditions for each of the four transitions are derived from the information available in the characteristic table. The symbol X in the tables represents a don't-care condition; that is, it does not matter whether the input is 1 or 0.

RS Flip-flop

The excitation table for the RS flip-flop is shown in Table 7-7(a). The first row shows the flip-flop in the 0-state at time t. It is desired to leave it in the 0-state after the occurrence of the pulse. From the characteristic table, we find that if S and R are both 0, the flip-flop will not change state. Therefore, both S and R inputs should be 0. However, it really doesn't matter if R is made a 1 when the pulse occurs since it results in leaving the flip-flop in the 0-state. Thus R can be 1 or 0 and the flip-flop will remain in the 0-state at $t+1$. Therefore, the entry under R is marked by the don't-care condition X.

If the flip-flop is in the 0-state and it is desired to have it go to the 1-state, then from the characteristic table, we find that the only way to make $Q(t+1)$ equal to 1 is to make $S = 1$ and $R = 0$. If the flip-flop is

Table 7-7 Flip-Flop Excitation Tables

$Q(t)$	$Q(t+1)$	S	R
0	0	0	X
0	1	1	0
1	0	0	1
1	1	X	0

(a) *RS*

$Q(t)$	$Q(t+1)$	J	K
0	0	0	X
0	1	1	X
1	0	X	1
1	1	X	0

(b) *JK*

$Q(t)$	$Q(t+1)$	D
0	0	0
0	1	1
1	0	0
1	1	1

(c) *D*

$Q(t)$	$Q(t+1)$	T
0	0	0
0	1	1
1	0	1
1	1	0

(d) *T*

to have a transition from the 1-state to the 0-state, we must have $S = 0$ and $R = 1$.

The last condition that may occur is for the flip-flop to be in the 1-state and remain in the 1-state. Certainly R must be 0; we do not want to clear the flip-flop. However, S may be either a 0 or a 1. If it is 0, the flip-flop does not change and remains in the 1-state; if it is a 1, it sets the flip-flop to the 1-state as desired. Therefore S is listed as a don't-care condition.

JK Flip-Flop

The excitation table for the *JK* flip-flop is shown in Table 7-7(b). When both present and next state are 0, the J input must remain at 0 and the K input can be either a 0 or a 1. Similarly, when both present and next state are 1, the K input must remain at 0 while the J input can be a 0 or a 1. If the flip-flop is to have a transition from the 0-state to the 1-state, J must be equal to 1 since the J input sets the flip-flop. However, input K may be either a 0 or a 1. If $K = 0$, the $J = 1$ condition sets the flip-flop as required; if $K = 1$ and $J = 1$, the flip-flop is complemented and goes from the 0-state to the 1-state as required. Therefore the K input is marked with a don't-care condition for the 0 to 1 transition. For a transition from the 1-state to the 0-state, we must have $K = 1$ since the K input clears the flip-flop. However, the J input may be either a 0 or a 1 since $J = 0$ has no effect and $J = 1$ together with $K = 1$ complements the flip-flop with a result transition from the 1-state to the 0-state.

The excitation table for the *JK* flip-flop illustrates the advantage of using this type when designing sequential circuits. The fact that it has so many don't-care conditions indicates that the combinational circuit for the input functions are likely to be simpler because don't-care terms usually simplify a function.

D Flip-flop

The excitation table for the D flip-flop is shown in Table 7-7(c). From the characteristic table, Table 7-6(c), we note that the next state is always equal to the D input and independent of the present state. Therefore, D must be 0 if $Q(t + 1)$ has to be 0 and 1 if $Q(t + 1)$ has to be 1, regardless of the value of $Q(t)$.

T Flip-Flop

The excitation table for the T flip-flop is shown in Table 7-7(d). From the characteristic table, Table 7-6(d), we find that when input $T = 1$, the state of the flip-flop is complemented; when $T = 0$, the state of the flip-flop remains unchanged. Therefore, when the state of the flip-flop must remain the same, the requirement is that $T = 0$. When the state of the flip-flop has to be complemented, T must equal 1.

Other Flip-Flops

The design procedure to be described in this chapter can be used with any flip-flop. It is necessary that the flip-flop characteristic table, from which it is possible to develop a new excitation table, be known. The excitation table is then used to determine the flip-flop input functions, as explained in the next section.

7-4 DESIGN PROCEDURE

In this section we present a procedure for the design of sequential circuits. Although intended to serve as a guide to the beginner, this procedure can be shortened with experience. The procedure is first summarized by a list of consecutive recommended steps as follows:

1. The word description of the circuit behavior is stated. This may be accompanied by a state diagram, a timing diagram, or other pertinent information.
2. From the given information about the circuit, obtain the state table.
3. The number of states may be reduced by state reduction methods if the sequential circuit can be characterized by input-output relations independent of the number of states.
4. Assign binary values to each state if the state table obtained in step 2 or 3 contains letter symbols.
5. Determine the number of flip-flops needed and assign a letter symbol to each.
6. Choose the type of flip-flop to be used.
7. From the state table, derive the circuit excitation and output tables.
8. Using the map or any other simplification method, derive the circuit output functions and the flip-flop input functions.
9. Draw the logic diagram.

The word specification of the circuit behavior usually assumes that the reader is familiar with digital logic terminology. It is necessary that the designer use his intuition and experience in order to arrive at the correct interpretation of the circuit specifications because word descriptions may be incomplete and inexact. However, once such a specification has been set down and the state table obtained, it is possible to make use of the formal procedure to design the circuit.

The state reduction and state assignment problems have been discussed in Sec. 7-2. The effect of state assignment on the complexity of the combinational circuit will be demonstrated subsequently in Exs. 7-1 and 7-2.

It has already been mentioned that the number of flip-flops is determined from the number of states. A circuit may have unused binary states if the total number of states is less than 2^m. The unused states are taken as don't-care conditions during the design of the combinational circuit part of the circuit.

The type of flip-flop to be used may be included in the design specifications or may depend on what is available to the designer. Many digital systems are constructed entirely with *JK* flip-flops because they are the most versatile available. When many types of flip-flops are available, it is advisable to use the *RS* or *D* flip-flop for those applications requiring transfer of data (such as shift-registers), the *T* type for applications involving complementation (such as binary counters), and the *JK* type for general applications.

The external output information is specified in the output section of the state table. From it we can derive the circuit output functions. The excitation table for the circuit is similar to that of the individual flip-flops, except that the input conditions are dictated by the information available in the present state and next state columns of the state table. The method of obtaining the excitation table and the simplified flip-flop input functions is best illustrated by an example.

We wish to design the clocked sequential circuit whose state diagram is given in Fig. 7-3. The type of flip-flops to be used are *JK*.

The state diagram consists of four states with binary values already assigned. Since the directed lines are marked with a single binary digit

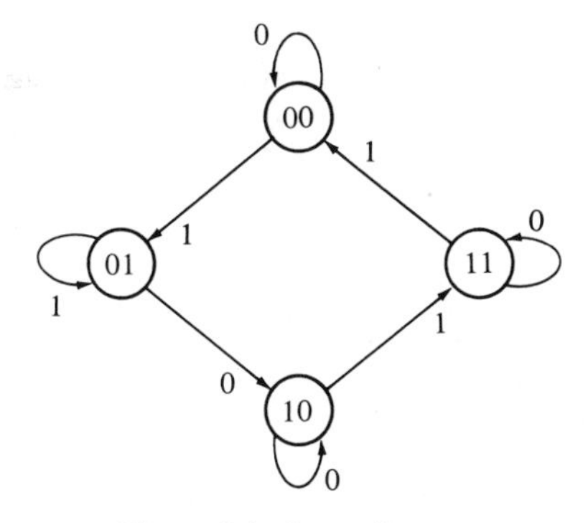

Figure 7-3 State diagram

without a (/), we conclude that there is one input variable and no output variables. (The state of the flip-flops may be considered the outputs of the circuit.) The two flip-flops needed to represent the four states are designated by A and B. The input variable is designated x.

The state table for this circuit, derived from the state diagram, is shown in Table 7-8. Note that there is no output section for this circuit. We shall now show the procedure for obtaining the excitation table and the combinational gate structure.

The derivation of the excitation table is facilitated if we arrange the state table in a different form. This form is shown in Table 7-9, where the present state and input variables are arranged in the form of a truth table. The next state value for each present state and input conditions is copied from Table 7-8. The excitation table of a circuit is a list of flip-flop input

Table 7-8 State Table

Present state		*Next state*			
		x = 0		*x = 1*	
A	*B*	*A*	*B*	*A*	*B*
0	0	0	0	0	1
0	1	1	0	0	1
1	0	1	0	1	1
1	1	1	1	0	0

Table 7-9 Excitation Table

Inputs of combinational circuit					*Outputs of combinational circuit*			
Present state		*Input*	*Next state*		*Flip-flop inputs*			
A	*B*	*x*	*A*	*B*	*JA*	*KA*	*JB*	*KB*
0	0	0	0	0	0	X	0	X
0	0	1	0	1	0	X	1	X
0	1	0	1	0	1	X	X	1
0	1	1	0	1	0	X	X	0
1	0	0	1	0	X	0	0	X
1	0	1	1	1	X	0	1	X
1	1	0	1	1	X	0	X	0
1	1	1	0	0	X	1	X	1

conditions that will cause the required state transitions and is a function of the type of flip-flop used. Since this example specified *JK* flip-flops, we need columns for the *J* and *K* inputs of flip-flops *A* (denoted by *JA* and *KA*) and *B* (denoted by *JB* and *KB*).

The excitation table for the *JK* flip-flop was derived in Table 7-7(b). This table is now used to derive the excitation table of the circuit. For example, in the first row of Table 7-9 we have a transition for flip-flop *A* from 0 in the present state to 0 in the next state. In Table 7-7(b) we find that a transition of states from 0 to 0 requires that input $J = 0$ and input $K = X$. So 0 and *X* are copied in the first row under *JA* and *KA*, respectively. Since the first row also shows a transition for flip-flop *B* from 0 in the present state to 0 in the next state, 0 and *X* are copied in the first row under *JB* and *KB*. The second row of Table 7-9 shows a transition for flip-flop *B* from 0 in the present state to 1 in the next state. From Table 7-7(b) we find that a transition from 0 to 1 requires that input $J = 1$ and input $K = X$. So 1 and *X* are copied in the second row under *JB* and *KB*, respectively. This process is continued for each row of the table and for each flip-flop; with the input conditions as specified in Table 7-7(b) being copied into the proper row of the particular flip-flop being considered.

Let us now pause and consider the information available in an excitation table such as Table 7-9. We know that a sequential circuit consists of a number of flip-flops and a combinational circuit. Figure 7-4 shows the two *JK* flip-flops needed for the circuit and a box to represent the combinational circuit. From the block diagram, it is clear that the outputs of the combinational circuit go to flip-flop inputs and external outputs (if specified). The inputs to the combinational circuit are the external inputs and the present state values of the flip-flops. Moreover, the Boolean functions that specify a combinational circuit are derived from a truth table that shows the input-output relations of the circuit. The truth table that describes the combinational circuit is available in the excitation table. The combinational circuit *inputs* are specified under the present state and input columns and the combinational circuit *outputs* are specified under the flip-flop input columns. Thus, an excitation table transforms a state diagram to the truth table needed for the design of the combinational circuit part of the sequential circuit.

The simplified Boolean functions for the combinational circuit can now be derived. The inputs are the variables *A*, *B*, and *x*; the outputs are the variables *JA*, *KA*, *JB*, and *KB*. The information from the truth table is transferred into the maps of Fig. 7-5, where the four simplified flip-flop input functions are derived:

$$JA = Bx' \qquad KA = Bx$$
$$JB = x \qquad KB = A \odot x$$

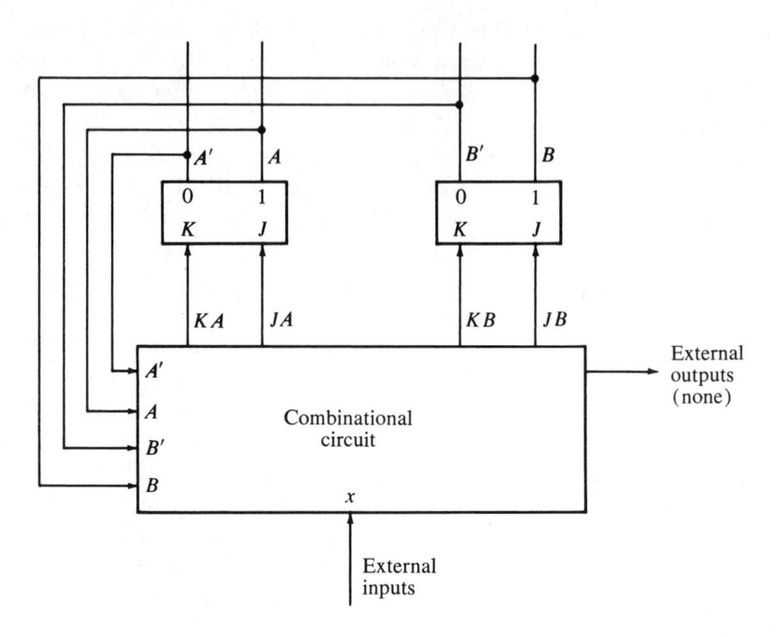

Figure 7-4 Block diagram of sequential circuit

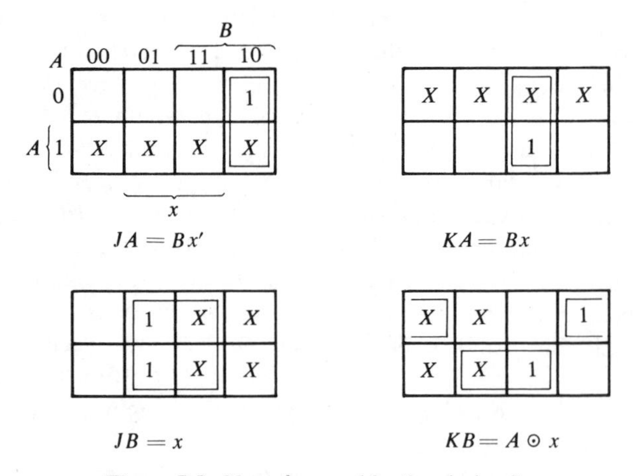

Figure 7-5 Maps for combinational circuit

The logic diagram is drawn in Fig. 7-6 and consists of two flip-flops, two AND gates, one equivalence gate, and one inverter.

With some experience, it is possible to reduce the amount of work involved in the design of the combinational circuit. For example, it is

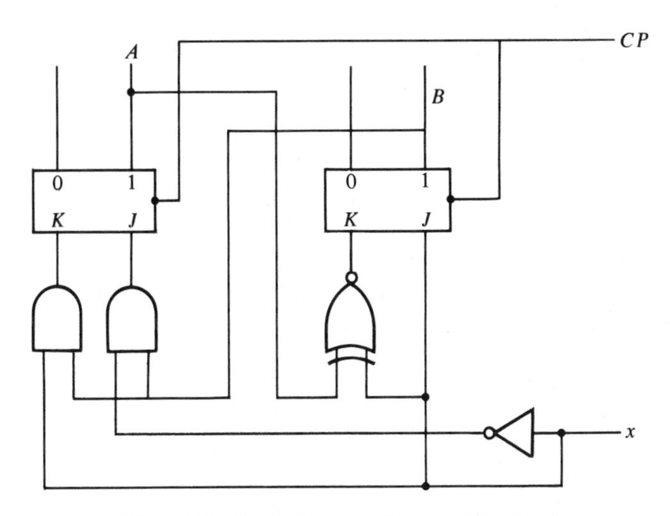

Figure 7-6 Logic diagram of sequential circuit

possible to obtain the information for the maps of Fig. 7-5 directly from Table 7-8, without having to derive Table 7-9. This is done by systematically going through each present state and input combination in Table 7-8 and comparing it with the binary values of the corresponding next state. The required input conditions as specified by the flip-flop excitation in Table 7-7 is then determined. Instead of inserting the 0, 1, or X thus obtained into the excitation table, it can be written down directly into the appropriate square of the appropriate map.

The excitation table of a sequential circuit with m flip-flops, k inputs per flip-flop, and n external inputs consists of $m + n$ columns for the present state and input variables and up to 2^{m+n} rows listed in some convenient binary count. The next state section has m columns, one for each flip-flop. The flip-flop input values are listed in mk columns, one for each input of each flip-flop. If the circuit contains j outputs, the table must include j columns. The truth table of the combinational circuit is taken from the excitation table by considering the $m + n$ present state and input columns as *inputs* and the $mk + j$ flip-flop input values and external outputs as *outputs.*

It is common practice to go from the Boolean functions expressed in algebraic form to a wiring list that gives the interconnections among the external terminals of flip-flops and gates in IC packages. In that case, the design need not go any further than the required simplified circuit output functions and flip-flop input functions. A logic diagram, however, may be helpful for visualizing the implementation of the circuit with flip-flops and gates.

7-5 DESIGN EXAMPLES

The design of clocked sequential circuits is illustrated in this section by means of seven examples. Examples 7-1 and 7-2 demonstrate the fact that different binary assignments produce different combinational circuits. They also show how unused binary states are treated as don't-care conditions. Example 7-3 shows how to design with *RS* flip-flops. Example 7-4 shows how to design with *T* and *D* flip-flops and how to treat unspecified next states. The design of a circuit initially specified by a timing diagram is demonstrated in Ex. 7-5. Circuits specified by state equations are designed in Exs. 7-6 and 7-7.

Design With Excitation Tables

EXAMPLE 7-1. Complete the design of the sequential circuit presented in Sec. 7-2. Use the reduced state table (Table 7-3), with binary assignment 1 of Table 7-4. Use *JK* flip-flops.

The state table with the required binary assignment is found in Table 7-5. From this table, we obtain the entries for the present state, input, and next state columns of Table 7-10. From the present and next state columns and from the excitation requirements of a *JK* flip-flop, we deduce the conditions necessary to excite the flip-flop inputs. The entries for the output column are obtained directly from Table 7-5. Table 7-10 is the excitation table for the circuit. From its information, we can derive the required combinational circuit.

There are three unused binary states in the state table: 000, 110, and 111. When an input of 0 or 1 is included with each unused state, six unused combinations result. These combinations have equivalent binary numbers 0, 1, 12, 13, 14, and 15 and are not listed in the table under the columns with labels "present state" and "input." These columns represent the input variables of the

Table 7-10 Excitation Table for Example 7-1

Present state			*Input*	*Next state*			*Flip-flop inputs*						*Output*
A	*B*	*C*	*x*	*A*	*B*	*C*	*JA*	*KA*	*JB*	*KB*	*JC*	*KC*	*y*
0	0	1	0	0	0	1	0	*X*	0	*X*	*X*	0	0
0	0	1	1	0	1	0	0	*X*	1	*X*	*X*	1	0
0	1	0	0	0	1	1	0	*X*	*X*	0	1	*X*	0
0	1	0	1	1	0	0	1	*X*	*X*	1	0	*X*	0
0	1	1	0	0	0	1	0	*X*	*X*	1	*X*	0	0
0	1	1	1	1	0	0	1	*X*	*X*	1	*X*	1	0
1	0	0	0	1	0	1	*X*	0	0	*X*	1	*X*	0
1	0	0	1	1	0	0	*X*	0	0	*X*	0	*X*	1
1	0	1	0	0	0	1	*X*	1	0	*X*	*X*	0	0
1	0	1	1	1	0	0	*X*	0	0	*X*	*X*	1	1

combinational circuit and therefore, the above numbers are treated as don't-care minterms.

The design of the combinational circuit consists of finding the six flip-flop input functions and the circuit output function. These functions are taken from the excitation table and simplified in the maps of Fig. 7-7. The four input variables in each map are the three flip-flop variables A, B, and C and the input variable x. Each map has X's in the six squares that belong to the don't-care minterms. The other don't-care terms in the maps are placed there because of the characteristic property of JK flip-flops. The abundance of don't-care terms for the inputs of JK flip-flops, together with the six don't-care terms from the

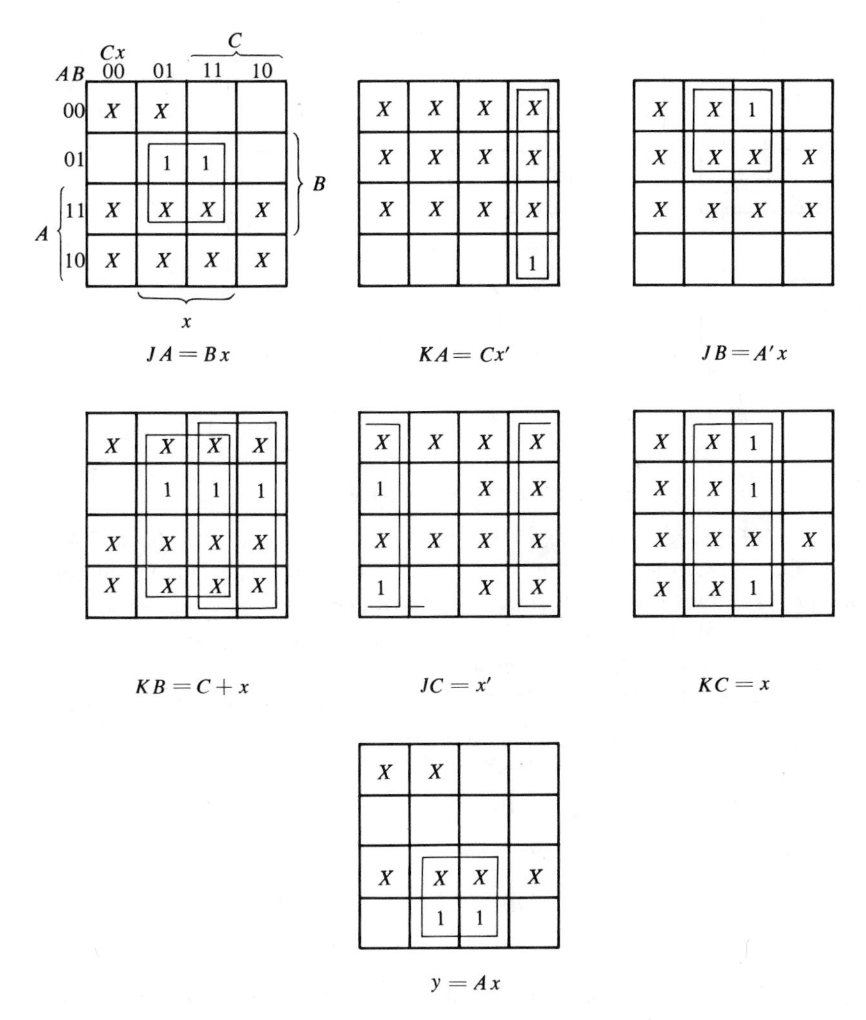

Figure 7-7 Maps for simplifying combinational circuit of Example 7-1

unused combinations, contribute a large amount of X's to the maps and are the main reason for the relatively simple expressions obtained:

$$JA = Bx \qquad KA = Cx'$$

$$JB = A'x \qquad KB = C + x$$

$$JC = x' \qquad KC = x$$

$$y = Ax$$

The logic diagram for the circuit is drawn in Fig. 7-8. It consists of three flip-flops, four AND gates, and one OR gate.

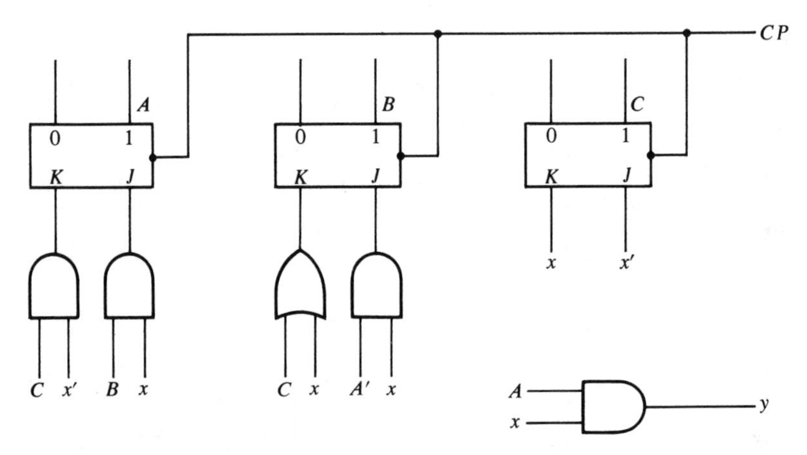

Figure 7-8 Logic diagram for Example 7-1

EXAMPLE 7-2. Repeat Ex. 7-1 with binary assignment 2 of Table 7-4.

Table 7-11 is the excitation table for this example. The binary values for the present state and next state are determined from the substitution of assignment 2 of Table 7-4 into Table 7-3. Again there are six unused input minterms. For this assignment, they are 2, 3, 8, 9, 12, and 13. The simplification of the six flip-flop input functions and the circuit output function should include these don't-care terms. The map simplifications are not shown, but the reader can easily verify the following results (see Prob. 7-16):

$$JA = Bx \qquad KA = Bx'$$

$$JB = Cx' + C'x \qquad KB = C + x$$

$$JC = B \qquad KC = Bx'$$

$$y = Ax$$

The logic diagram for the circuit is drawn in Fig. 7-9. It consists of three flip-flops, six AND gates and two OR gates.

Table 7-11 Excitation Table for Example 7-2

Present state			*Input*	*Next state*			*Flip-flop inputs*						*Output*
A	*B*	*C*	*x*	*A*	*B*	*C*	*JA*	*KA*	*JB*	*KB*	*JC*	*KC*	*y*
0	0	0	0	0	0	0	0	*X*	0	*X*	0	*X*	0
0	0	0	1	0	1	0	0	*X*	1	*X*	0	*X*	0
0	1	0	0	0	1	1	0	*X*	*X*	0	1	*X*	0
0	1	0	1	1	0	1	1	*X*	*X*	1	1	*X*	0
0	1	1	0	0	0	0	0	*X*	*X*	1	*X*	1	0
0	1	1	1	1	0	1	1	*X*	*X*	1	*X*	0	0
1	0	1	0	1	1	1	*X*	0	1	*X*	*X*	0	0
1	0	1	1	1	0	1	*X*	0	0	*X*	*X*	0	1
1	1	1	0	0	0	0	*X*	1	*X*	1	*X*	1	0
1	1	1	1	1	0	1	*X*	0	*X*	1	*X*	0	1

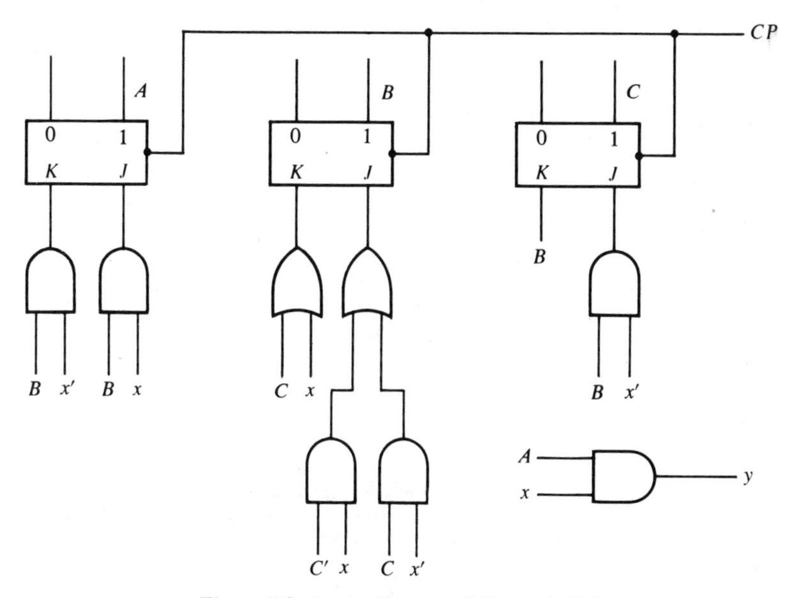

Figure 7-9 Logic diagram of Example 7-2

Examples 7-1 and 7-2 demonstrate how a different combinational circuit is obtained for the same circuit when a different binary assignment is made. The circuits of Figs. 7-8 and 7-9 are identical as far as the specifications given by Table 7-3 are concerned. The choice of binary states for the letter symbols dictates the complexity of the combinational circuit obtained.

EXAMPLE 7-3. Design a sequential circuit whose state diagram is the solution of Prob. 6-9 and is given by Fig. 7-10. Use *RS* flip-flops.

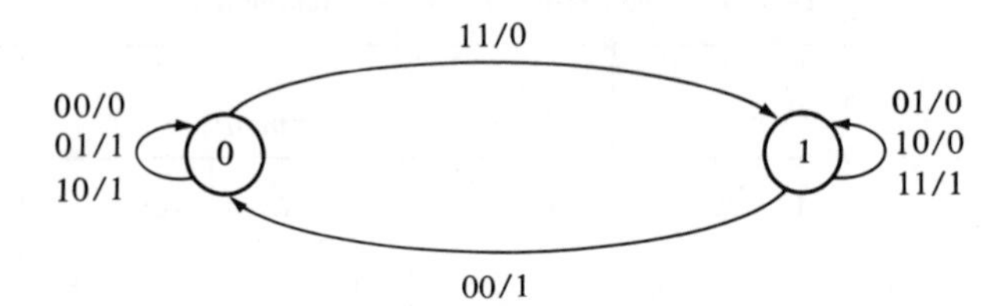

Figure 7-10 State diagram for Example 7-3

The state diagram consists of two states, with binary values already assigned. The directed lines are marked with two inputs and one output. Three of the input combinations leave the circuit in the same state. Only inputs 00 and 11 cause a transition of states.

The excitation table for the circuit is shown in Table 7-12. The present state, input, next state, and output columns are derived directly from the state diagram of Fig. 7-10. The flip-flop input conditions are obtained from the state transition and from the excitation requirements of an *RS* flip-flop as listed in Table 7-7(a). The flip-flop output is denoted by A, the inputs by x and y, and the output by z.

The simplified Boolean functions for the combinational circuit are derived from the maps of Fig. 7-11. The entries for the map are obtained from Table 7-12. The simplified flip-flop input functions and the output functions are:

$$SA = xy$$

$$RA = x'y'$$

$$z = x \oplus y \oplus A$$

Table 7-12 Excitation Table for Example 7-3

PS	*inputs*		*NS*	*Output*	*FF inputs*	
A	x	y	A	z	SA	RA
0	0	0	0	0	0	X
0	0	1	0	1	0	X
0	1	0	0	1	0	X
0	1	1	1	0	1	0
1	0	0	0	1	0	1
1	0	1	1	0	X	0
1	1	0	1	0	X	0
1	1	1	1	1	X	0

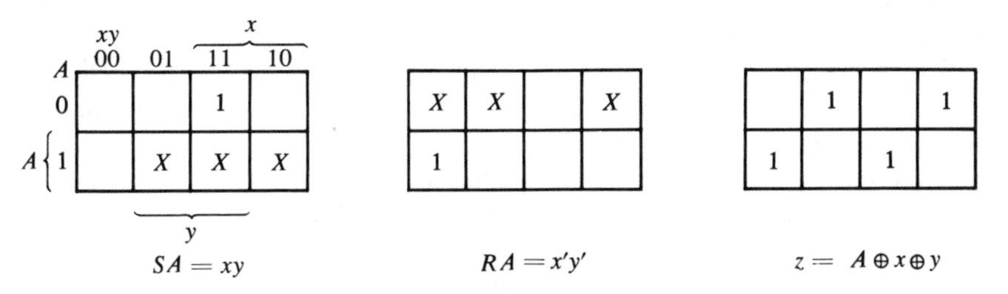

Figure 7-11 Maps for Example 7-3

The logic diagram of the circuit is shown in Fig. 7-12.

EXAMPLE 7-4. Design a sequential circuit with two flip-flops and one input. When the input is equal to 1, the flip-flop outputs repeat the sequence 00, 01, 10. When the input is equal to 0, they repeat the sequence 11, 10, 01. Design the circuit with: (a) T flip-flops and (b) D flip-flops.

The state diagram, derived from the word description, is shown in Fig. 7-13(a). Each sequence is repeated as long as the input remains the same. The statement of the problem does not specify the next state when the input changes while the circuit is in state 00 or 11. The state table of Fig. 7-13(b) is derived from the state diagram. The two dashes in the table designate the fact that the next states are not specified for these conditions. The question now arises, how do we treat these unspecified next states? In a practical situation, the designer would go back to the source of the statement of the problem and ask for further clarification. If this is impossible, the designer must decide for himself what to do. One possible alternative is to leave the circuit in state 00 or 11 when the input is 0 or 1, respectively. A second alternative is to force a transition to either state 01 or 10 so the proper sequence may continue. A third alternative is to

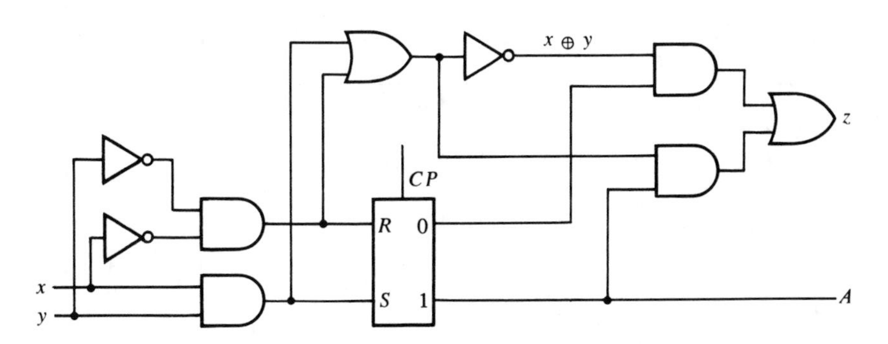

Figure 7-12 Logic diagram for Example 7-3

assume that it does not matter what the next state is going to be. We shall arbitrarily assume the third alternative.

The excitation table is shown in Table 7-13. The two unspecified next states are marked by X's and treated as don't-care conditions. The flip-flop input values are listed for the T type and the D type. The input of the T flip-flop must be a 1 if there is a change from 0 to 1 or from 1 to 0 during the state transition. It is equal to 0 if there is no change of state from the present state to the next state. The input for the D flip-flop is exactly the same as the entry in the next state column. The flip-flop excitation tables from Table 7-7 may be consulted to verify these conditions.

The maps of Fig. 7-14(a) are used if the circuit is to have T flip-flops. The simplified input functions are:

$$TA = A \oplus B$$

$$TB = A \oplus x$$

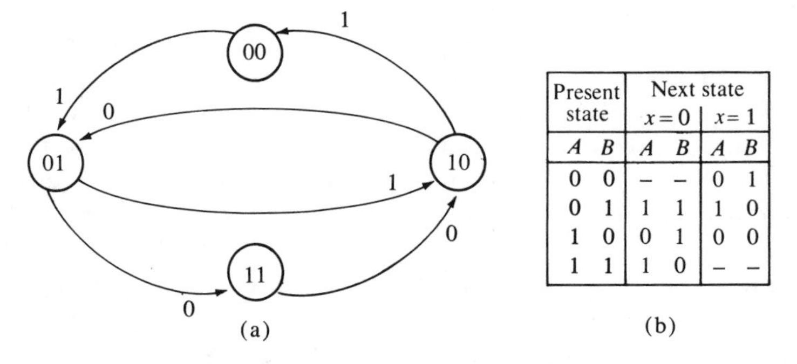

Present state		Next state x=0		Next state x=1	
A	B	A	B	A	B
0	0	–	–	0	1
0	1	1	1	1	0
1	0	0	1	0	0
1	1	1	0	–	–

Figure 7-13 State diagram and state table for Example 7-4

Table 7-13 Excitation Table for Example 7-4

Present State		*Input*	*Next State*		*Flip-flop Inputs*			
					T Type		*D Type*	
A	*B*	*x*	*A*	*B*	*TA*	*TB*	*DA*	*DB*
0	0	0	*X*	*X*	*X*	*X*	*X*	*X*
0	0	1	0	1	0	1	0	1
0	1	0	1	1	1	0	1	1
0	1	1	1	0	1	1	1	0
1	0	0	0	1	1	1	0	1
1	0	1	0	0	1	0	0	0
1	1	0	1	0	0	1	1	0
1	1	1	*X*	*X*	*X*	*X*	*X*	*X*

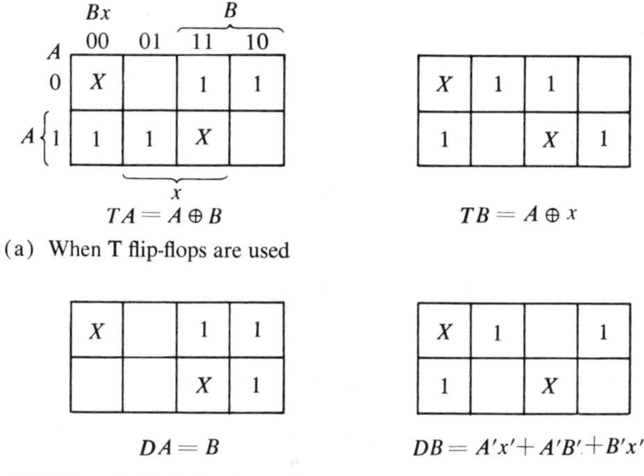

Figure 7-14 Maps for Example 7-4

The maps of Fig. 7-14(b) are used if the circuit is to have D flip-flops. The input functions obtained are:

$$DA = B$$

$$DB = A'x' + A'B' + B'x'$$

It is interesting to check the assignment given to the unspecified next states. From the maps of Fig. 7-14(a) we note that, for the circuit with T flip-flops, the X's are included with the 0's. This means that no change of state occurs; the circuit stays in state 00 or 11 when the input is 0 or 1, respectively. For the circuit with D flip-flops, we have chosen 01 to be the next state from present state 00 with $x = 0$ and 10 to be the next state from present state 11 with $x = 1$.

EXAMPLE 7-5. Design a circuit with one flip-flop and two inputs to conform with the timing diagram of Fig. 7-15. Flip-flop Q is set when $A = 1$ and $B = 0$; it is cleared when $A = 1$ and $B = 1$ and is left in the same state otherwise.

The reader should immediately realize that this is a simple sequential circuit with functions AB' and AB, respectively, needed to set and clear the flip-flop. The type appropriate for these functions is either an RS or JK flip-flop. The flip-flop input functions for an RS type are:

$$SQ = AB'$$
$$RQ = AB$$

Now, let us assume that the solution was not immediately apparent and proceed to determine the flip-flop input functions by means of an excitation

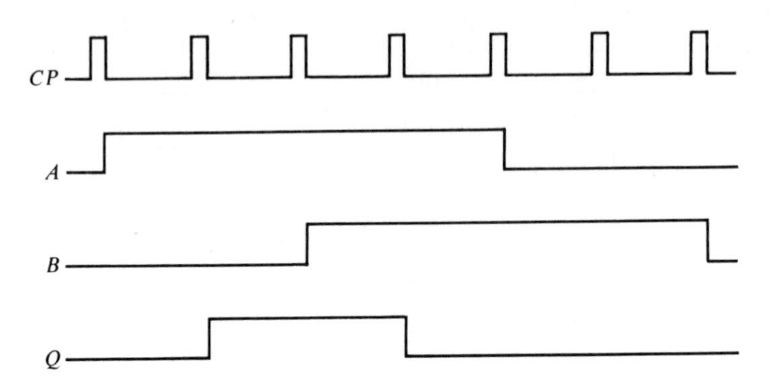

Figure 7-15 Timing diagram for Example 7-5

table. The next state of Q in Table 7-14 has to be a 1 when the inputs AB = 10 and a 0 when AB = 11, irrespective of the present state of Q. For the four remaining combinations, the next state of Q must remain the same as its present state. From the *RS* flip-flop input values, we obtain the maps of Fig. 7-16. The simplified input functions from the maps are as predicted.

Design With State Equations

EXAMPLE 7-6. Design a sequential circuit that behaves according to the following state equations:

$$A(t+1) = CD' + C'D$$
$$B(t+1) = A$$
$$C(t+1) = B$$
$$D(t+1) = C$$

From the four given state equations, we conclude that there are four flip-flops in the circuit denoted by A, B, C, and D and that there are no external inputs or outputs since none are specified. The specified circuit is a shift-register with the input into A being dependent on the present state of C and D. Such a circuit is called a *feedback shift-register*. In a feedback shift-register, each flip-flop shifts its content to the next flip-flop when a clock pulse occurs and, at the same time, the states of certain flip-flops determine the next state of the first flip-flop. The most convenient flip-flops to use are the D type.

As shown in Sec. 6-4, the state equations represent the same information as a state table. It is possible to obtain the state table and flip-flop excitation requirements for this circuit but, when D flip-flops are used, it will be a complete waste of time. This is because the entries for the flip-flop inputs are exactly the same as the next state and the conditions for the next state are already listed in the specifications. Therefore, the

Table 7-14 Excitation Table for Example 7-5

PS	Inputs		NS	FF Inputs	
Q	A	B	Q	SQ	RQ
0	0	0	0	0	X
0	0	1	0	0	X
0	1	0	1	1	0
0	1	1	0	0	X
1	0	0	1	X	0
1	0	1	1	X	0
1	1	0	1	X	0
1	1	1	0	0	1

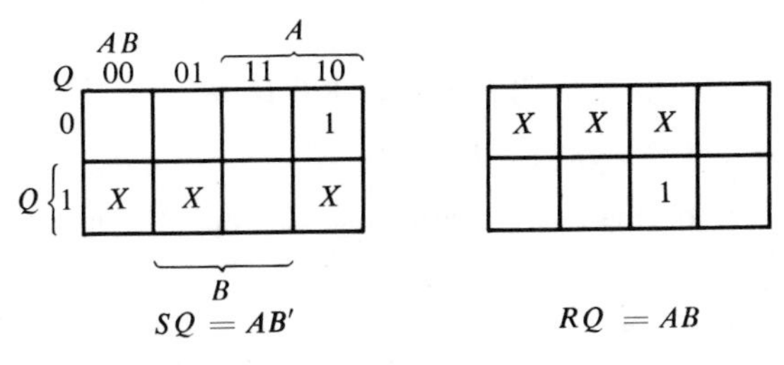

Figure 7-16 Maps for Example 7-5

flip-flop input functions for this circuit are taken directly from the specifications, with the next state symbol replaced by the flip-flop input variable as follows:

$$DA = CD' + C'D$$

$$DB = A$$

$$DC = B$$

$$DD = C$$

EXAMPLE 7-7. Design a sequential circuit with *JK* flip-flops to satisfy the following state equations:

$$A(t + 1) = A'B'CD + A'B'C + ACD + AC'D'$$

$$B(t + 1) = A'C + CD' + A'BC'$$

$$C(t + 1) = B$$

$$D(t + 1) = D'$$

The combinational circuit can be found directly from the state equations without having to draw the state table or excitation table. This is accomplished by means of a matching process between the state equation and the characteristic equation of the JK flip-flop. The characteristic equation of the JK flip-flop in terms of an output variable Q was derived in Fig. 6-6(d) and is repeated here:

$$Q(t + 1) = (J)Q' + (K')Q$$

Input variables J and K' are enclosed in parentheses so as not to confuse the AND terms in the characteristic equation with the two-letter convention which has been used to represent a flip-flop input variable.

The matching process consists of arranging and manipulating the state equation until it is in the form of the characteristic equation. Once this is done, the functions for J and K can be extracted and simplified.

The input functions for flip-flop A are derived by this method by arranging the state equation and matching it with the characteristic equation as follows:

$$A(t + 1) = (B'CD + B'C)A' + (CD + C'D')A$$
$$= (J)A' + (K')A$$

From the equality of the two equations, we deduce the input functions for flip-flop A to be:

$$J = B'CD + B'C = B'C$$
$$K = (CD + C'D')' = CD' + C'D$$

The state equation for flip-flop B can be arranged as follows:

$$B(t + 1) = (A'C + CD') + (A'C')B$$

However, this form is not suitable for matching with the characteristic equation because the variable B' is missing. If the first quantity in parentheses is ANDed with $(B' + B)$, the equation remains the same but with the variable B' included. Thus,

$$B(t + 1) = (A'C + CD')\,(B' + B) + (A'C')B$$
$$= (A'C + CD')B' + (A'C + CD' + A'C')B$$
$$= (J)B' + (K')B$$

From the equality of the two equations, we deduce the input functions for flip-flop B:

$$J = A'C + CD'$$
$$K = (A'C + CD' + A'C')' = AC' + AD$$

The state equation for flip-flop C can be manipulated as follows:

$$C(t+1) = B = B(C' + C) = BC' + BC$$
$$= (J)C' + (K')C$$

The input functions for flip-flop C are:

$$J = B$$
$$K = B'$$

Finally, the state equation for flip-flop D may be manipulated for the purpose of matching as follows:

$$D(t+1) = D' = 1.D' + 0.D$$
$$= (J)D' + (K')D$$

which gives the input function:

$$J = K = 1$$

The derived input functions can be accumulated and listed together. The two-letter convention to designate the flip-flop input variable, not used in the above derivation, is used below:

$$JA = B'C \qquad KA = CD' + C'D$$
$$JB = A'C + CD' \qquad KB = AC + AD'$$
$$JC = B \qquad KC = B'$$
$$JD = 1 \qquad KD = 1$$

The design procedure introduced in Exs. 7-6 and 7-7 is an alternative method for determining the flip-flop input functions of a sequential circuit when D or JK flip-flops are used. To use this procedure when a state diagram or a state table is initially specified, it is necessary that the state equations be derived by the procedure outlined in Sec. 6-4. The state equation method for determining flip-flop input functions must be modified somewhat if the circuit has unused binary states which are considered as don't-care conditions (see Prob. 7-10). The application of this procedure to circuits with RS and T flip-flops is possible but involves a considerable amount of algebraic manipulation (see Prob. 7-11).

7-6 DESIGN OF COUNTERS

State transitions in clocked sequential circuits occur during a clock pulse; the circuit is assumed to remain in its present state if no pulse occurs. For this reason, the clock pulse does not appear explicitly as an input variable in a state table. From this point of view, a counter may be regarded as a sequential circuit with no input variables, since the only input is the count pulse. The next state of a counter depends entirely on its present state and the state transition occurs every time a pulse occurs. Because of this property, a counter can be completely specified by a list of the *count sequence*; that is, the sequence of binary states that it undergoes.

Consider for example a BCD counter whose count sequence is listed in the first column of Table 7-15. The next number in the sequence represents the next state reached by the circuit upon the application of a count pulse. The count is assumed to repeat itself in such a way that state 0000 is the next state after 1001. The count sequence gives all the information needed to construct the state table. However, it is unnecessary to list the next states in a separate column because they can be read from the next number in the sequence. The design of the combinational circuit part of a counter follows the same procedure as that outlined in Sec. 7-4, except that the excitation table can be obtained directly from the count sequence. We shall illustrate this procedure by designing a BCD counter with T flip-flops.

Table 7-15 is the excitation table for the BCD counter. The entries for the flip-flop input conditions are determined from the characteristics of a T flip-flop and by inspecting the state transition from a given count to the next below it. For example, the row with count 0101 is compared with 0110, the next count below it. The bits for flip-flops Q_8 and Q_4 do not undergo a change, so TQ_8 and TQ_4 in row 0101 are marked with a 0. Flip-flop Q_2 changes from 0 to 1 and Q_1 changes from 1 to 0, and, since both have to be complemented to reach the next count, it is necessary that their T inputs be a 1. Therefore, TQ_2 and TQ_1 in row 0101 are marked with a 1. The last row with count 1001 is compared with count 0000 of the first row to obtain the entries for the flip-flop inputs in the last row.

Table 7-15 Excitation Table for BCD Counter

Count sequence				*Flip-flop inputs*			
Q_8	Q_4	Q_2	Q_1	TQ_8	TQ_4	TQ_2	TQ_1
0	0	0	0	0	0	0	1
0	0	0	1	0	0	1	1
0	0	1	0	0	0	0	1
0	0	1	1	0	1	1	1
0	1	0	0	0	0	0	1
0	1	0	1	0	0	1	1
0	1	1	0	0	0	0	1
0	1	1	1	1	1	1	1
1	0	0	0	0	0	0	1
1	0	0	1	1	0	0	1

The flip-flop input functions from the excitation table are simplified in the maps of Fig. 7-17. The six X's in each map represent the don't-care conditions from the six unused states. The simplified flip-flop input functions are:

$$TQ_1 = 1$$
$$TQ_2 = Q_8'Q_1$$
$$TQ_4 = Q_2Q_1$$
$$TQ_8 = Q_8Q_1 + Q_4Q_2Q_1$$

The circuit requires four T flip-flops, four AND gates, and one OR gate.

Counter-Decoder Circuits

Counters, together with decoders, are used to generate timing and sequencing signals that control the operations of digital systems. Figure 7-18 shows a simple configuration for generating timing signals. It consists of a counter with m flip-flops that goes through a sequence of up to 2^m states. The outputs of the flip-flops go into a decoder (Sec. 4-8) that has up to 2^m outputs. As the count sequence progresses from 1 to n, decoder outputs t_1 to t_n become logic-1 in consecutive order. These outputs can be used as command signals to perform a

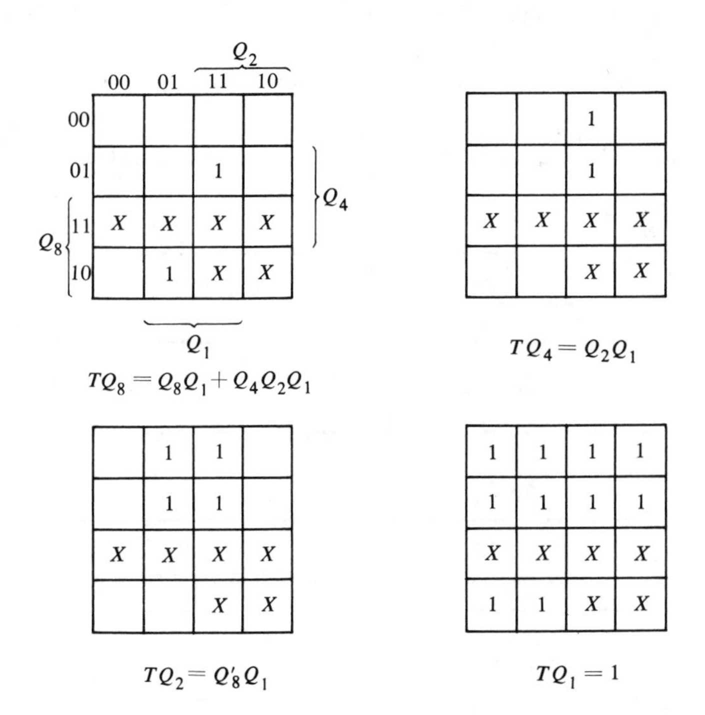

Figure 7-17 Maps for BCD counter

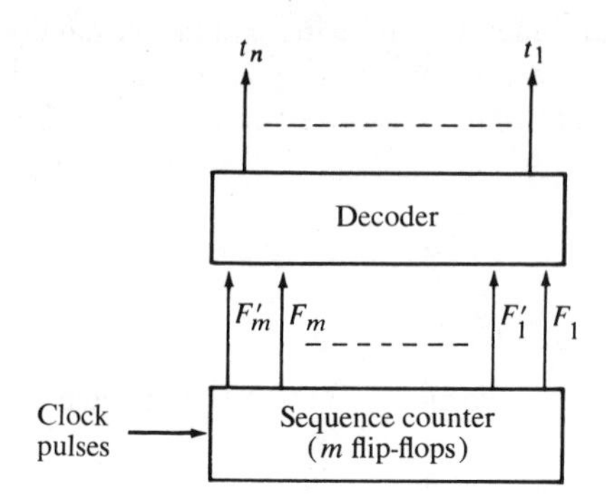

Figure 7-18 Block diagram of counter-decoder

sequence of operations in a digital system.

A counter-decoder circuit can be designed to give any desired number of repeated timing sequence. Consider for example the sequence of six timing signals shown in Fig. 7-19. To generate the six outputs t_1 to t_6, we need a counter that goes through a repeated sequence of six states and a decoder whose output t_k becomes logic-1 when the counter is in the corresponding state k for $k = 1, 2, \ldots, 6$. We shall now proceed to design two counter-decoder circuits capable of generating the six timing signals shown in Fig. 7-19.

Three flip-flops produce up to eight binary states. To generate six timing

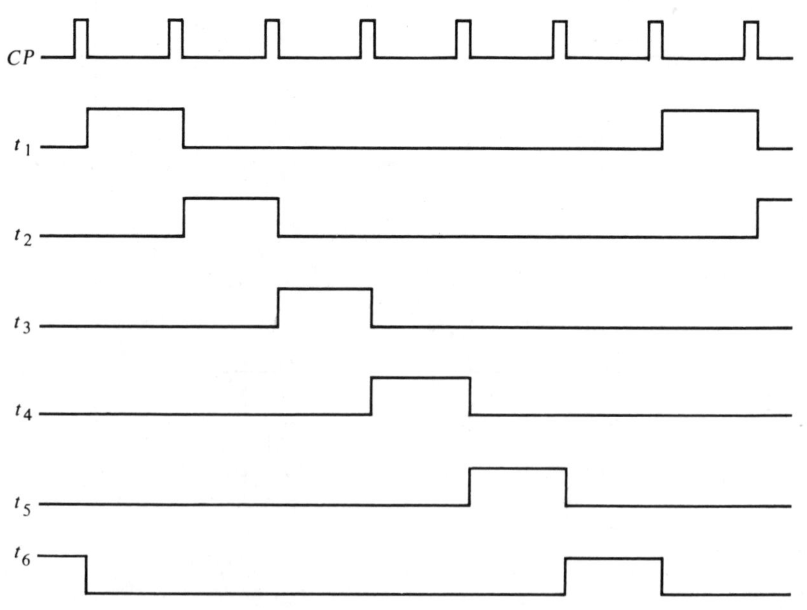

Figure 7-19 Sequence of 6 timing signals

signals, we need a counter that goes through a sequence of six states. It does not matter which six states are chosen or in what sequence they are ordered; the outputs of the circuit are taken from the decoder which can be designed to decode any combination of six states. We are confronted here with the state assignment problem; that is, we have to choose a sequence of six states in such a manner as to minimize the number of gates in the combinational circuit part of the counter and in the decoder. Two assignments will be chosen out of the many possible; the first is a straight binary assignment and the second is an assignment derived to produce a counter without the need of gates. We shall employ *JK* flip-flops since they are the most versatile for this application.

Table 7-16 is the excitation table for a count-by-6 counter with the count sequence chosen to be a straight binary count from 000 to 101. The two unused states, 110 and 111, are taken as don't-care terms. The simplified flip-flop input functions are easily derived to be:

$$JA = BC \qquad KA = C$$

$$JB = A'C \qquad KB = C$$

$$JC = 1 \qquad KC = 1$$

The Boolean functions for the decoder are simplified somewhat when the unused states are taken as don't-care terms:

$$t_1 = A'B'C' \qquad t_4 = BC$$

$$t_2 = A'B'C \qquad t_5 = AC'$$

$$t_3 = BC' \qquad t_6 = AC$$

The logic diagram for the circuit is drawn in Fig. 7-20. It requires three flip-flops and eight AND gates.

A second possible assignment of six states is shown in Table 7-17. In this assignment, the sequence for flip-flops *B* and *C* is a repetition of the binary count 00, 01, 10, while flip-flop *A* is chosen to alternate between 0

Table 7-16 Excitation Table for First Count-by-6 Counter

Count sequence			*Flip-flop inputs*					
A	*B*	*C*	*JA*	*KA*	*JB*	*KB*	*JC*	*KC*
0	0	0	0	*X*	0	*X*	1	*X*
0	0	1	0	*X*	1	*X*	*X*	1
0	1	0	0	*X*	*X*	0	1	*X*
0	1	1	1	*X*	*X*	1	*X*	1
1	0	0	*X*	0	0	*X*	1	*X*
1	0	1	*X*	1	0	*X*	*X*	1

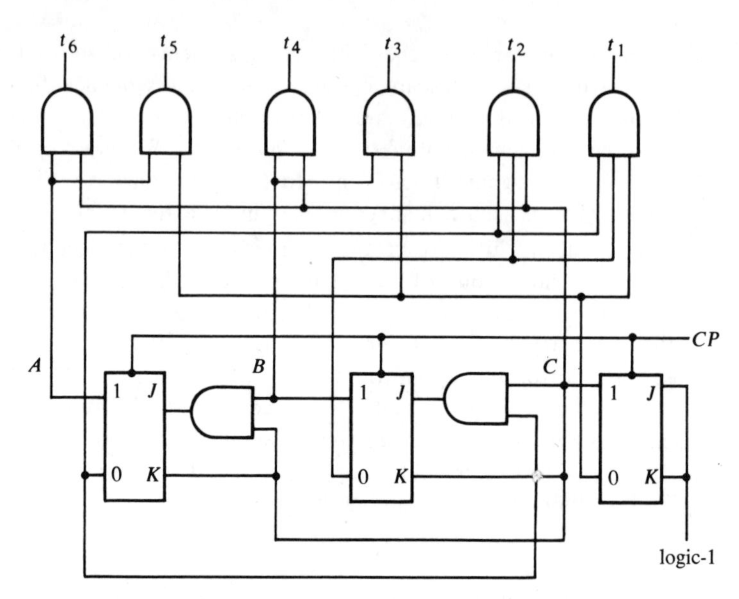

Figure 7-20 Count-by-6 circuit with straight binary count

Table 7-17 Excitation Table for Second Count-by-6 Counter

Count sequence			*Flip-flop inputs*					
A	*B*	*C*	*JA*	*KA*	*JB*	*KB*	*JC*	*KC*
0	0	0	0	*X*	0	*X*	1	*X*
0	0	1	0	*X*	1	*X*	*X*	1
0	1	0	1	*X*	*X*	1	0	*X*
1	0	0	*X*	0	0	*X*	1	*X*
1	0	1	*X*	0	1	*X*	*X*	1
1	1	0	*X*	1	*X*	1	0	*X*

and 1 every three counts. The unused states for this count are 011 and 111. The simplified functions for this circuit are:

$JA = B \qquad KA = B$

$JB = C \qquad KB = 1$

$JC = B' \qquad KC = 1$

$t_1 = A'B'C' \qquad t_4 = AB'C'$

$t_2 = A'C \qquad t_5 = AC$

$t_3 = A'B \qquad t_6 = AB$

The logic diagram is drawn in Fig. 7-21. The counter does not require any gates, the decoder requires six AND gates. The circuits of Figs. 7-20 and 7-21 perform identical operations and produce identical timing sequences, as specified in Fig. 7-19, yet the one in Fig. 7-21 requires two less gates.

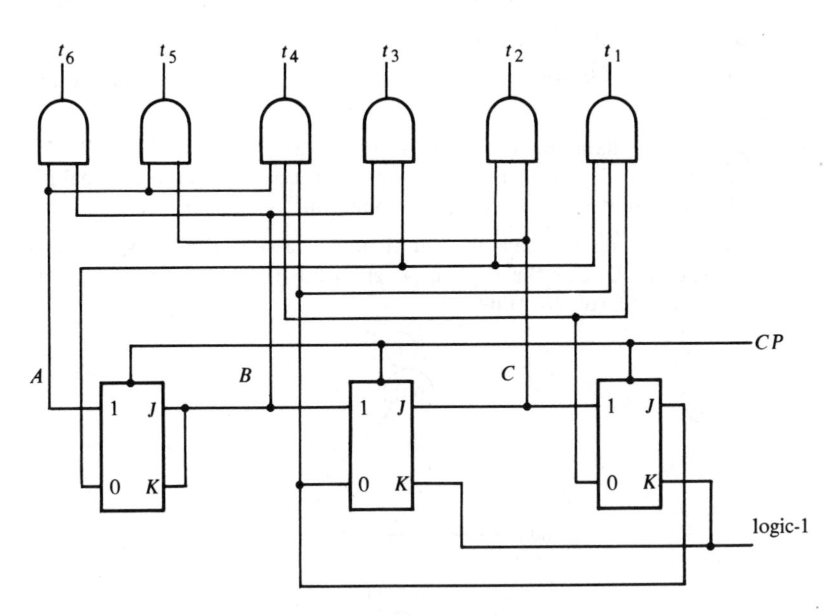

Figure 7-21 Second form of a count-by-6 counter

PROBLEMS

7-1. Reduce the number of states in the following state table and tabulate the reduced state table.

Present state	Next state		Output	
	x = 0	*x = 1*	*x = 0*	*x = 1*
a	*f*	*b*	0	0
b	*d*	*c*	0	0
c	*f*	*e*	0	0
d	*g*	*a*	1	0
e	*d*	*c*	0	0
f	*f*	*b*	1	1
g	*g*	*h*	0	1
h	*g*	*a*	1	0

7-2. Starting from state a of the state table in Prob. 7-1, find the output sequence generated with an input sequence 01110010011.

7-3. Repeat Prob. 7-2 using the reduced table of Prob. 7-1. Show that the same output sequence is obtained.

7-4. Substitute binary assignment 2 of Table 7-4 to the states in Table 7-3 and obtain the binary state table. Repeat with binary assignment 3.

7-5. Obtain the excitation table of the JK' flip-flop described in Prob. 6-5.

7-6. Obtain the excitation table of the set-dominate flip-flop described in Prob. 6-6.

7-7. Obtain the characteristic table and excitation table of an RST flip-flop. This is a three-input flip-flop with both RS and T capabilities. Only one input can be excited at one time.

7-8. A sequential circuit has one input and one output. The state diagram is as shown. Design the sequential circuit with (a) T flip-flops, (b) RS flip-flops, and (c) JK flip-flops.

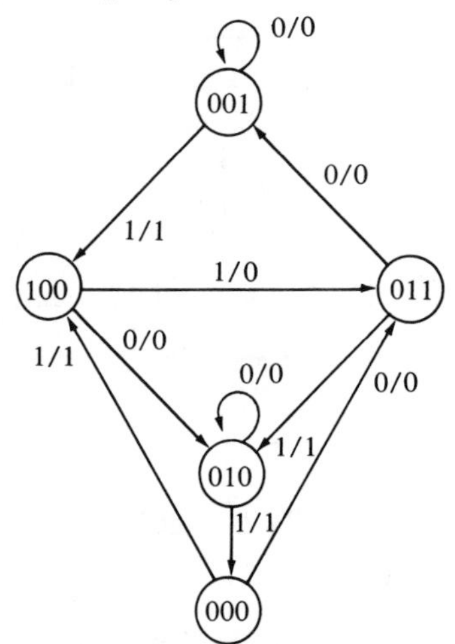

7-9. Design the circuit of a four-bit register that converts the binary number stored in the register to its 2's complement value when input $x = 1$. The flip-flops of the register are of the RST type. This flip-flop has three inputs: two inputs have RS capabilities and one has a T capability. The RS inputs are used to transfer the four-bit number when an input $y = 1$. Use the T input for the conversion.

7-10. (a) Derive the state equations for the sequential circuit specified by Table 7-5, Sec. 7-2. List the don't-care terms. (b) Derive the flip-flop input functions from the state equations (and don't-care terms) using the method outlined in Ex. 7-7. Check your answers with the solution given in Ex. 7-1.

7-11. The state equation method for determining flip-flop input functions as outlined in Exs. 7-6 and 7-7 can be extended to circuits with T or RS flip-flops. Chapter 5 of the book by Phister (8) shows that the following relations hold:

type of flip-flop	*characteristic equation*	*state equation*	*flip-flop function from g_1 and g_2*	*condition*
T	$T'Q + TQ'$	$g_1 Q + g_2 Q'$	$T = g_2$	if $g_2 = g_1'$
			$T = g_1' Q + g_2 Q'$	none
RS	$S + R'Q$	$g_1 Q + g_2 Q'$	$S = g_2,\ R = g_1'$	if $g_1' g_2 = 0$
			$S = g_2 Q', R = g_1' Q$	if $g_1' g_2 \neq 0$

Repeat Exs. 7-3 and 7-4 using the above table.

7-12. Using only T flip-flops, design a shift-register that shifts both right and left.

7-13. Derive the state diagram for the circuit of Ex. 7-6.

7-14. Repeat Ex. 7-1 with binary assignment 3 of Table 7-4.

7-15. Verify the circuit obtained in Ex. 7-7 by using the excitation table method.

7-16. Verify the flip-flop input functions obtained for Ex. 7-2.

7-17. Repeat Ex. 7-4 using RS flip-flops.

7-18. Repeat Ex. 7-7 using D flip-flops.

7-19. Design the sequential circuit described by the following state equations. Use JK flip-flops.

$$A(t + 1) = xAB + yA'C + xy$$

$$B(t + 1) = xAC + y'CB'$$

$$C(t + 1) = x'B + yAB'$$

7-20. Design a synchronous BCD counter with JK flip-flops.

7-21. Design a decimal counter using the 2,4,2,1 code and T flip-flops.

7-22. Design a counter-decoder circuit that recycles every 12 input pulses. The decoder has four outputs to detect the occurrence of the 3rd, 6th, 9th, and 12th input pulse, respectively. Use T flip-flops. Also draw the timing diagram for the four decoder outputs.

7-23. Design the following counter-decoders using JK flip-flops: (a) count-by-3; (b) count-by-5; (c) count-by-7; (d) count-by-8.

7-24. Design a counter with the following binary sequence: 0, 1, 3, 2, 6, 4, 5, 7 and repeat. Use RS flip-flops.

7-25. Design a counter with the following binary sequence: 0, 1, 3, 7, 6, 4 and repeat. Use T flip-flops.

7-26. Design a counter with the following binary sequence: 0, 4, 2, 1, 6 and repeat. Use JK flip-flops.

REFERENCES

1. Marcus, M. P., *Switching Circuits for Engineers*, 2nd ed. Englewood Cliffs, N.J.: Prentice-Hall, 1965.

2. McCluskey, E. J., *Introduction to the Theory of Switching Circuits*, New York: McGraw-Hill Book Co., 1965.

3. Miller, R. E., *Switching Theory*, two volumes. New York: John Wiley and Sons, 1965.

4. Krieger, M., *Basic Switching Circuit Theory*. New York: The Macmillan Col, 1967.

5. Hill, F. J., and G. R. Peterson, *Introduction to Switching Theory and Logical Design*. New York: John Wiley and Sons, 1968.

6. Givone, D. D., *Introduction to Switching Circuit Theory*. New York: McGraw-Hill Book Co., 1970.

7. Kohavi, Z., *Switching and Finite Automata Theory*. New York: McGraw-Hill Book Co., 1970.

8. Phister M., *The Logical Design of Digital Computers*. New York: John Wiley and Sons, 1958.

9. Paull, M. C., and S. H. Unger, "Minimizing the Number of States in Incompletely Specified Sequential Switching Functions," *IRE Trans. on Electronic Computers*, Vol. EC-8, No. 3 (September 1959), 356-66.

10. Hartmanis, J., "On the State Assignment Problem for Sequential Machines I," *IRE Trans. on Electronic Computers*, Vol. EC-10, No. 2 (June 1961), 157-65.

11. McCluskey, E. J., and S. H. Unger, "A Note on the Number of Internal Assignments for Sequential Circuits," *IRE Trans. on Electronic Computer*, Vol. EC-8, No. 4 (December 1959), 439-40.

8 OPERATIONAL AND STORAGE REGISTERS

8-1 REGISTER TRANSFER

A digital system is a sequential logic system which can be constructed with memory elements and combinational gates. As long as the number of flip-flops and inputs is small, the circuit can be designed by means of state tables and excitation tables. A large system, such as a digital computer, would be difficult, if not impossible, to define by a single state table. To overcome this difficulty, large systems are invariably designed using a modular approach. The system is partitioned into a number of standard modules, each of which performs some functional task. A computer processor unit for example, is designed using modules such as registers, decoders, arithmetic circuits, and control circuits. The modules are in turn constructed with submodules until one arrives at a circuit with a small number of elements which can be designed by means of excitation tables or equivalent procedures. With such an approach, each module is separately designed and tested. The various modules are then interconnected with common data-transfer and control-signal paths to form a processor unit.

The standard module used for storing binary information is the register. A register was defined in Sec. 1-7 as a group of binary cells. A group of flip-flops constitutes a register since a flip-flop is a binary cell capable of storing one bit of information. In addition to the flip-flops, a register may have combinational gates that perform certain data processing tasks. In its broadest definition, a register consists of a group of flip-flops and the gates that affect their transition. The flip-flops hold the binary information while the combinational gates process it. The shift-register is an example of a register whose data processing task is the transfer of information. A counter may be considered a register whose function is to count the number of incoming pulses.

Data transfer among modules and units is accomplished by means of *inter-register transfer* operations. These operations consist of a direct transfer of binary information from one register to another. The destination register that receives the information assumes the previous value of the source register. The value of the source register normally does not change because of the transfer.

Information transfer from one register to another can be performed in either series or parallel. In series transfer, both the source and destination registers are shift-registers. The information is transferred one bit at a time by shifting the bits out of the source register into the destination register. In order not to lose the information stored in the source register, it is necessary that the information shifted out of the source register be recirculated and shifted back at the same time. Parallel transfer consists of a simultaneous transfer of all the bits from the source register to the destination register. Parallel transfer is done during one clock pulse, while serial transfer needs a number of clock pulses equal to the number of bits transferred.

Figure 8-1 shows a parallel transfer of binary information from an n-bit source register to four destination registers. Each register is represented by a box and contains n D-type flip-flops. Each flip-flop in a destination register uses an AND gate with two inputs. One input contains the corresponding information bit and the other is a control signal. The four control signals determine which register receives the information. For example, when control line A is logic-1 (while control lines B, C, and D are logic-0) and a clock pulse occurs, register A receives the information from the source register. The information in the other three registers and the source register remains the same.

In a system with many registers, the transfer from each register to another requires that lines be connected from the output of each flip-flop in one register to the input of each flip-flop in all the other registers. Consider for example the requirement for transfer among three registers as shown in Fig. 8-2. There are six data paths between registers. If each register consists of n flip-flops, there is a need for $6n$ lines. As the number of registers increases, the number of lines increases considerably. However, if we restrict the transmission of data among registers to one at a time, the number of paths among all registers can be reduced to just one per flip-flop for a total of n lines. This is shown in Fig. 8-3, where the output and input of each flip-flop is connected to a common line through an electronic circuit that acts like a switch. All the switches are normally open until a transfer is required. For a transfer from F_1 to F_3, for example, switches S_1

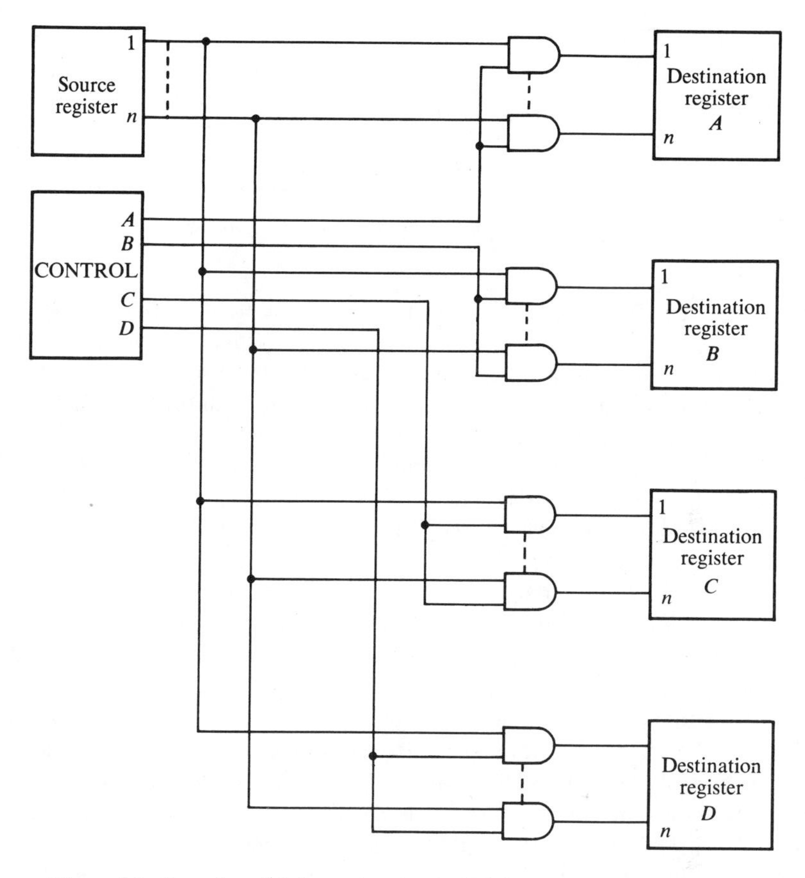

Figure 8-1 Transfer of information in parallel from one source to four destinations

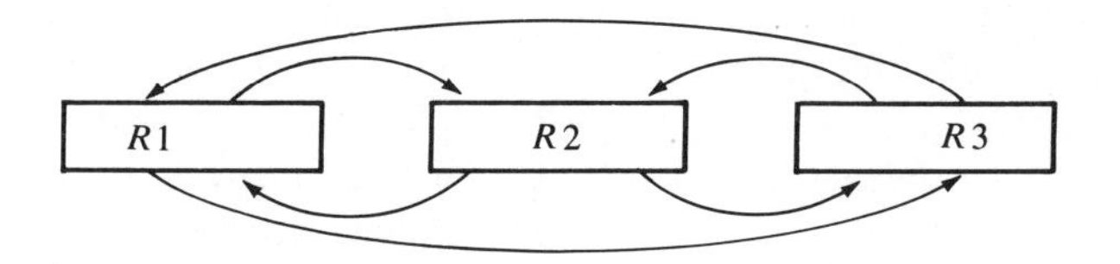

Figure 8-2 Transfer among three registers

and S_4 are closed to form the required path. This scheme can be extended to registers with n flip-flops and requires n common lines, since each flip-flop of the register must be connected to one common line.

A group of wires through which binary information is transferred one at a time among registers is called a *bus*. For a parallel transfer, the number

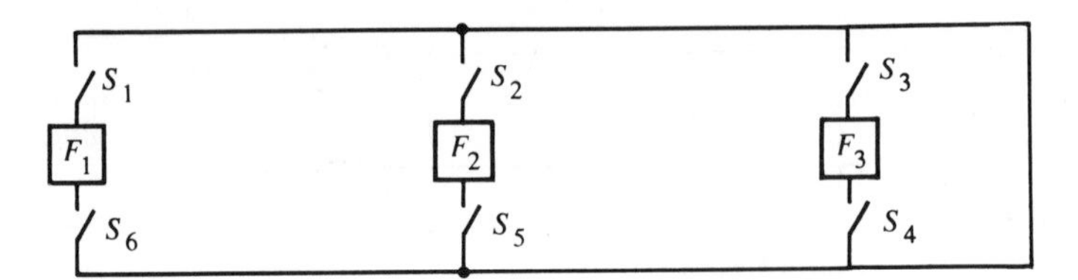

Figure 8-3 Transfer through one common line

of wires in the bus is equal to the number of flip-flops in the register. The idea of a bus transfer is analogous to a central transportation system used to bring commuters from one point to another. Instead of each commuter using his own private car to go from one location to another, a bus system is used with each commuter waiting in line until transportation is available.

Figure 8-4 shows one possible implementation of inter-register transfer through a bus. Three registers, each with n flip-flops, are connected to the bus lines through circuits labeled BD (Bus Driver). The output of each flip-flop goes to one input of a BD circuit and the other input of the BD is connected to a control signal. The two inputs in each BD produce a logic AND operation. The outputs of the BD have a wired-OR function (Sec. 5-7); that is, their connection to a common wire produces a logic OR operation. For a transfer from register R1 to register R3, for example, control signals S_1 and S_4 become logic-1. This connects the flip-flops from register R1 to the bus lines and at the same time enables the AND gates in R3 to accept the information from the bus lines. A clock pulse to all flip-flops results in the transfer of the binary information from R1 to R3, while the information in R1 and R2 remains the same.

Transfer through a bus is limited to one transmission at a time. If two transfers are required at the same time, two buses must be used. A large digital system will normally employ a number of buses with each of its registers connected to one or more buses to form the various paths needed for the transfer of information.

A bus system may be formed with multiplexer and demultiplexer circuits (Sec. 4-9). A multiplexer circuit selects data from many lines and directs it to a single output line. A demultiplexer circuit receives information from one line and distributes it over a large number of destinations. Thus, multiplexers can function as BD circuits and demultiplexers as the input gating circuits. The selection of the registers is done by the selection lines of the two circuits. The multiplexer selection lines determine the source register and the demultiplexer selection lines determine the destination register.

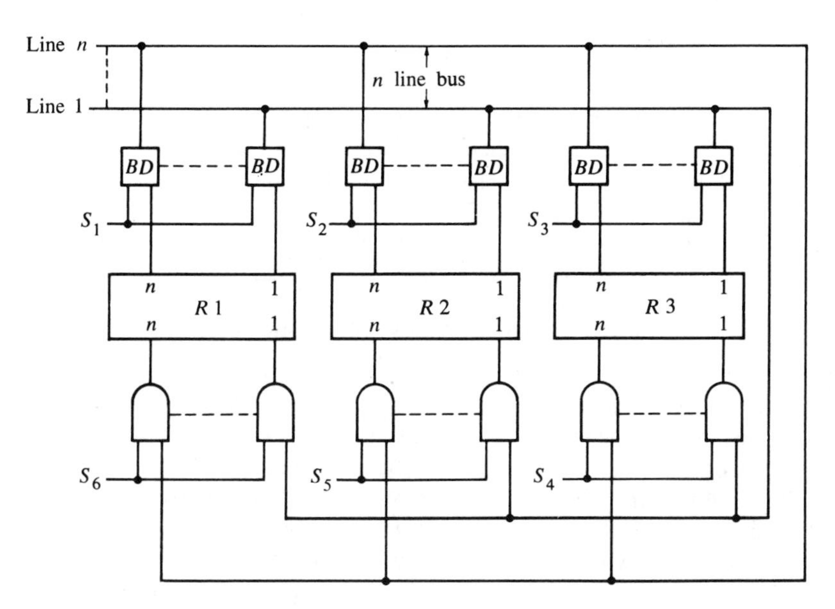

Figure 8-4 Transfer through a bus

8-2 REGISTER OPERATIONS

An *elementary operation* is an operation performed during one clock pulse on the information stored in a register. The result of the operation may replace the previous binary information of the register or may be transferred to another register. An inter-register transfer is an elementary operation. Other examples are shift, count, add, and clear. Since an elementary operation is completed during one clock pulse, it is sometimes called a *micro-operation.* It is the most basic operation that exists in a digital system.

A register capable of performing elementary operations is an *operational* register. As shown in Fig. 8-5, an operational register consists of a group of flip-flops and a combinational circuit that performs the required operations. All processor registers, including shift-registers and those involved with parallel transfers, are classified as operational registers. For this reason, when dealing with register operations we shall refer to an operational register as a register; that is, the word *register* will include the group of flip-flops and its associated combinational circuit.

The following chapters will show that all data processing performed in digital systems is implemented through a sequence of elementary operations

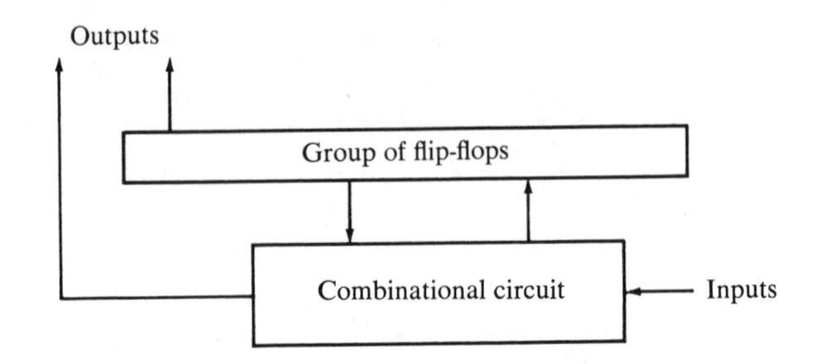

Figure 8-5 Block diagram of an operational register

performed on data stored in registers. The purpose of this section is to introduce a symbolic notation to describe register transfers and elementary operations in a concise and precise manner. From a list of register operations, it is then possible to obtain the Boolean functions for the system.

The register transfer symbols are defined in Table 8-1. Registers are denoted by capital letters and the individual flip-flops of the register by the letter designation with a subscript. Thus, A denotes a register; A_3 denotes flip-flop 3 of register A. The subscript number of each flip-flop must be part of the definition of a register. In this book we shall adopt the convention of numbering the flip-flops of a register in ascending order starting from the right-most position. However, any other convenient numbering of the flip-flops is possible. Figure 8-6 shows a block diagram of an eight-bit register denoted by A. The individual flip-flops (together with their associated combinational gates) are numbered from right to left with subscripts 1 through 8.

Table 8-1 Register transfer symbolic notation

Symbol	*Description*
capital letter	denotes a register
subscript	denotes a flip-flop in a register
parentheses ( )	denotes contents of a register
double line arrow ⇒	denotes transfer of information
brackets [ ]	denotes a portion of a register (see Sec. 10-2)
brackets < >	denotes a memory register specified by an address (see Sec. 8-3)
plus +	denotes arithmetic addition
minus −	denotes subtraction
disjunction ∨	denotes logical OR operation (see Sec. 9-6)
conjunction ∧	denotes logical AND operation (see Sec. 9-6)
exclusive-or ⊕	denotes exclusive-or operation (see Sec. 9-6)
bar ¯	denotes complementation
equality =	denotes equality
colon :	denotes termination of a Boolean control function
comma ,	separates two elementary operations which are performed simultaneously during one clock pulse

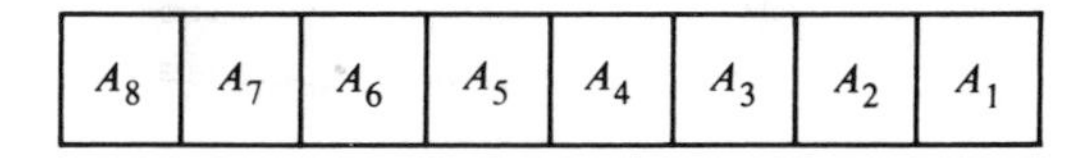

Figure 8-6 Block diagram of an 8-bit register denoted by the letter symbol *A*

Parentheses denote the contents of a register or flip-flop. Thus (A) signifies the contents of register A and (A_i) means the contents of the *ith* flip-flop of register A. The double line arrow denotes the transfer of information between registers. Thus,

$$(A) \Rightarrow B$$

represents a symbolic notation for the parallel transfer of the contents of register A into register B. We shall assume that the contents of register A are not changed during the transfer unless otherwise noted.

The serial transfer of information from register A to register B is done with shift-registers as shown in the block diagram of Fig. 8-7. The right-most flip-flop of register A is labeled A_1 and the left-most flip-flops of registers A and B are labeled A_n and B_n, respectively. When the shift-right command signal S_R is logic-1 and a clock pulse occurs, the contents of registers A and B are shifted once to the right and the value of A_1 is transferred to flip-flops B_n and A_n. This causes the transfer of one bit of information from register A to B and at the same time one bit is recirculated back to register A. This transfer can be expressed by means of symbolic notation as follows:

$$S_R\!: (A_1) \Rightarrow B_n, (A_1) \Rightarrow A_n, (A_{i+1}) \Rightarrow A_i, (B_{i+1}) \Rightarrow B_i$$
$$i = 1, 2, 3, \ldots, n-1$$

The control function S_R is terminated by a colon and designates a Boolean condition; i.e., the register operations listed after the colon are performed only if $S_R = 1$. The elementary operations are separated by a comma and are performed simultaneously during one clock pulse. The subscript i denotes the individual flip-flops of the register. Note that an elementary operation, by definition, is completed during one clock pulse and therefore,

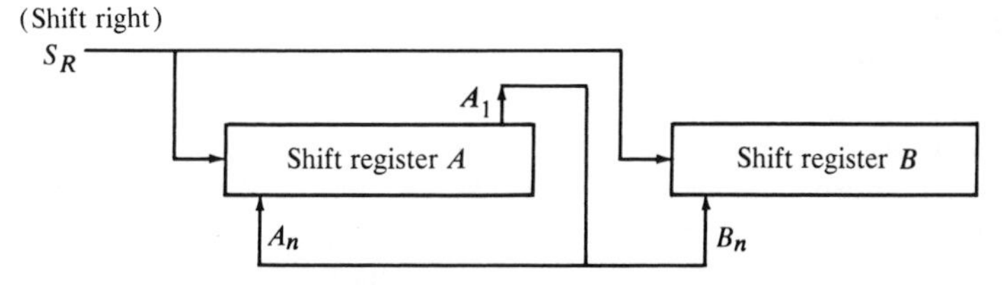

Figure 8-7 Serial transfer by shift registers

the above symbolic statement designates a transfer of only one bit. For a complete transfer of all n bits, the command signal S_R must remain a 1 for a period of n clock pulses. Other examples of elementary operations are listed in Table 8-2 for reference.

A sequence of elementary operations are usually conditioned by control functions. The operations are assumed to follow the listed sequence when the control functions are not included. For example,

$$t_1: (\bar{A}) \Rightarrow A \text{ , } 0 \Rightarrow B$$

$$F_2 t_2: (A) + (B) \Rightarrow A$$

means that when Boolean variable t_1 is logic-1, register A is complemented and register B is cleared. When Boolean function $F_2 t_2 = 1$, the contents of registers A and B are added and the sum is transferred to register A.

On the other hand, the following sequence:

$$(B) \Rightarrow C \text{ , } (B) \Rightarrow T$$

$$(C) + 1 \Rightarrow C \text{ , } (T) - 1 \Rightarrow T$$

does not include control functions and therefore denotes a sequence of four elementary operations performed during two clock pulses. During the first clock pulse, the two elementary operations listed in the first line (separated by a comma) are executed: the contents of register B are transferred to register C and to register T. During the next clock pulse, the two elementary operations listed in the second line are executed: the content of register C is increased by unity and that of register T is decreased by unity.

In Ch. 9 we demonstrate the procedure for obtaining the Boolean functions from the list of register elementary operations. In Chs. 10 and 11, we use register operation sequences to specify the internal operations of digital computers.

8-3 THE MEMORY UNIT

The registers in a digital computer may be classified as either operational or storage. As mentioned previously, an *operational* register is capable of storing binary information in its flip-flops, and in addition, has combinational gates capable of data processing tasks. A *storage* register is used solely for temporary storage of binary information. This information cannot be altered when transferred in and out of the register. A *memory unit* is a collection of storage registers, together with the associated circuits needed to transfer information in and out of the registers. The storage registers in a memory unit are called *memory registers.*

The bulk of the registers in a digital computer are memory registers, to

Table 8-2 Examples of symbolic notation and elementary operations

Symbolic designation	*Description*
A	all bits of register A
A_i	the ith flip-flop of register A
(A_{1-8})	contents of flip-flops 1 through 8 of register A
$(A) \Rightarrow B$	contents of register A are transferred to register B
$0 \Rightarrow C$	clear register C (transfer 0's to all flip-flops)
$(A_{i+1}) \Rightarrow A_i$	contents of flip-flop $i + 1$ of register A are transferred to flip-flop i of register A
$I[M]$	subregister M which is part of register I (see Sec. 10-2)
$<D>$	memory register specified by the address in the D register (see Sec. 8-3)
$(A) + (B) \Rightarrow A$	contents of register A are added to contents of register B and the sum transferred to register A
$(\bar{A}) \Rightarrow A$	complement all flip-flops in register A
$(A) = (B)$	contents of register A are equal to contents of register B
$(A) \Rightarrow C, (C) \Rightarrow A$	swap contents of registers A and C during one clock pulse
P: $(C) + 1 \Rightarrow C$	if Boolean variable P is logic-1, then increment (increase by one) register C

which information is transferred for storage and from which information is available when needed for processing. A comparatively few operational registers are found in the processor unit. When data processing takes place, the information from selected registers in the memory unit is first transferred to the operational registers in the processor unit. Intermediate and final results obtained in the operational registers are transferred back to selected memory registers. Similarly, binary information received from input devices is first stored in memory registers, information transferred to output devices is taken from registers in the memory unit.

The component that forms the binary cells of registers in a memory unit must have certain basic properties, the most important of which are: (a) it must have a reliable two-state property for binary representation, (b) it must be small in size, (c) the cost per bit of storage should be as low as possible, and (d) the time of access to a memory register should be reasonably fast. Examples of memory unit components are: magnetic cores, semiconductor ICs, and magnetic surfaces in tapes, drums, or disks.

A memory unit stores binary information in groups called *words*; each word being stored in a memory register. A word in memory is an entity of n bits that moves in and out of storage as a unit. A memory word may represent an operand, an instruction, a group of alphanumeric characters, or any binary coded information. The communication between a memory unit and its environment is achieved through two control signals and two external registers. The control signals specify the direction of transfer required; that is, whether a word is to be stored in a memory register or whether a word previously stored is to be transferred out of a memory register. One external register specifies the particular memory register chosen out of the thousands available; the other specifies the particular bit configuration of the word in question. The control signals and the registers are shown in the block diagram of Fig. 8-8.

The memory *address register* specifies the memory word selected. Each word in memory is assigned a number identification starting from 0 up to the maximum number of words available. To communicate with a specific memory word, its location number, or *address*, is transferred to the address register. The internal circuits of the memory unit accept this address from the register and open the paths needed to select the word called. An address register with n bits can specify up to 2^n memory words. Computer memory units can range from 1024 words, requiring an address register of 10 bits, to $1{,}048{,}576 = 2^{20}$ words, requiring a 20-bit address register.

The two control signals applied to the memory unit are called *read* and *write*. A write signal specifies a transfer-in function; a read signal specifies a transfer-out function. Each is referenced from the memory unit. Upon accepting one of the control signals, the internal control circuits inside the memory unit provide the desired function. Certain types of storage units, because of their component characteristics, destroy the information stored in a cell when the bit in that cell is read out. Such a unit is said to be a

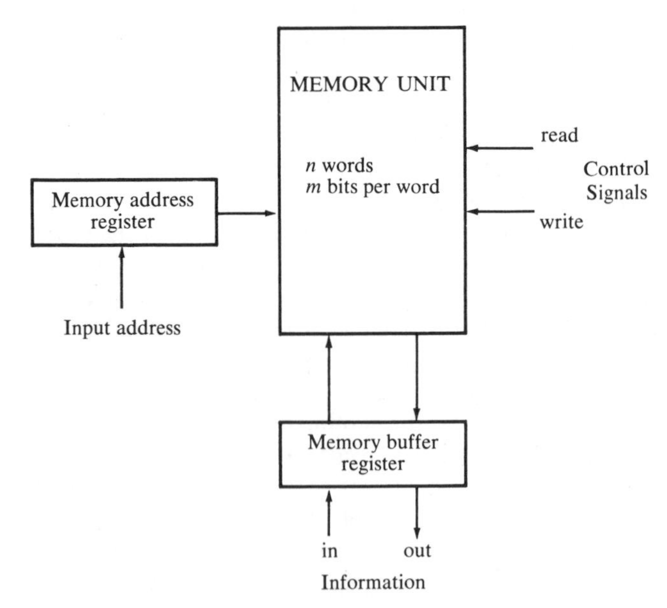

Figure 8-8 Block diagram of a memory unit showing communication with environment

destructive read-out memory as opposed to a nondestructive memory where the information remains in the cell after it is read out. In either case, the old information is always destroyed when new information is written. The sequence of internal control in a destructive read-out memory must provide control signals that will cause the restoration of the word back into its binary cells if the application calls for a non-destructive function.

The information transfer to and from registers in memory and the external environment is communicated through one common register called the memory *buffer register* (other names are: *information register* and *storage register*). When the memory unit receives a *write* control signal, the internal control interprets the contents of the buffer register to be the bit configuration of the word to be stored in a memory register. With a *read* control signal, the internal control sends the word from a memory register into the buffer register. In each case, the contents of the address register specify the particular memory register referenced for writing or reading.

Let us summarize the information transfer characteristics of a memory unit by an example. Consider a memory unit of 1024 words with eight bits per word. To specify 1024 words we need an address of 10 bits, since $2^{10} = 1024$. Therefore, the address register must contain 10 flip-flops. The buffer register must have eight flip-flops to store the contents of words transferred in and out of memory. Let us use the following symbolic notation for the various registers:

D address register
B buffer register
(D) contents of D register
$<D>$ memory register addressed by D
$(<D>)$ the contents of $<D>$

Figure 8-9 shows the initial contents of three registers: the D register, the B register, and the memory register addressed by D:

(D) = 0000101010 = decimal 42 (contents of D register)
$<D>$ = memory register number 42
$(<D>)$ = 01101110 (contents of memory register number 42)
(B) = 10010010 (contents of buffer register)

The following elementary operations are needed to read the contents of memory register number 42:

$$0000101010 \Rightarrow D$$

$$\text{read: } (< D >) \Rightarrow B$$

The binary address is first transferred to the D register. A *read* control signal applied to the memory unit causes a transfer from the specified

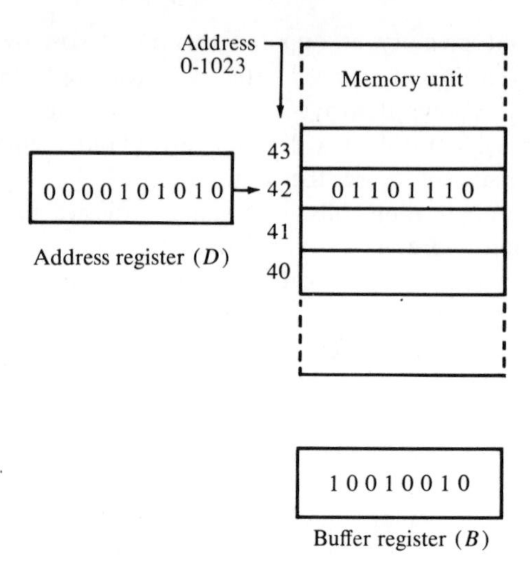

Figure 8-9 Initial values of registers

memory register to the B register. This transfer is depicted in Fig. 8-10(a).

A *write* control signal applied to the memory unit causes a transfer of information as shown in Fig. 8-10(b). The following elementary operations describe the transfer of the eight-bit word 10010010 into memory register 42:

$$0000101010 \Rightarrow D\,, \qquad 10010010 \Rightarrow B$$

$$\text{write: } (B) \Rightarrow < D >$$

The address is transferred to the D register and the contents of the word are transferred into the B register. A *write* control signal transfers the contents of B into the memory register specified by the D register.

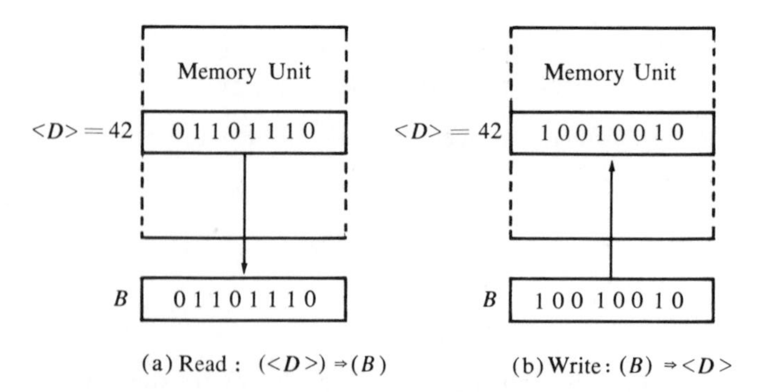

(a) Read : $(<D>) \Rightarrow (B)$ (b) Write: $(B) \Rightarrow <D>$

Figure 8-10 Information transfer during *read* and *write*

In the above example we have assumed a memory unit with nondestructive read-out property. Such memories can be constructed with semiconductor ICs. They retain the information in the memory register when the register is sampled during the reading process so that no loss of information occurs. Another component commonly used in memory units is the magnetic core. A magnetic core characteristically has destructive read-out; i.e., it loses the stored binary information during the reading process. Examples of semiconductor and magnetic core memories are presented in Sec. 8-4.

Because of its destructive read-out property, a magnetic core memory must provide additional control functions to restore the word back into the memory register. A *read* control signal applied to a magnetic core memory unit causes information transfer as depicted in Fig. 8-11. This can be expressed symbolically as follows:

read:

t_1: $(<D>) \Rightarrow B,\ 0 \Rightarrow <D>$ destructive read-out

t_2: $(B) \Rightarrow <D>$ restore contents to memory register

During the first half-cycle designated by t_1, the contents of the memory register are transferred to the B register. Since this is a destructive read-out memory, the contents of the memory register are destroyed. The elementary operation $0 \Rightarrow <D>$ designates the fact that the memory register is automatically cleared; i.e., receives all 0's during the process of reading. Without this additional designation, we would have to assume that the contents of $<D>$ have not changed. To restore the previously stored word into the memory register, we need a second half-cycle t_2. Remember that the B register holds the word just read from memory and the D register holds the address of the memory register, so that the transfer during t_2 automatically restores the lost information.

A *write* control signal applied to a magnetic core memory causes a transfer of information as shown in Fig. 8-12. In order to transfer new

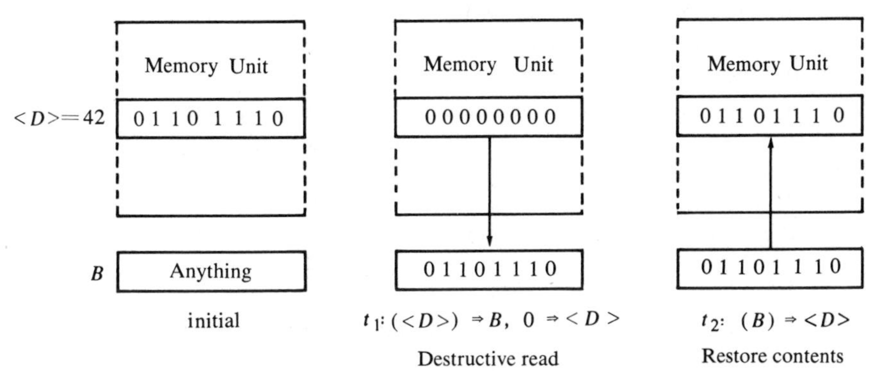

Figure 8-11 Information transfer in a magnetic core memory during *read* command

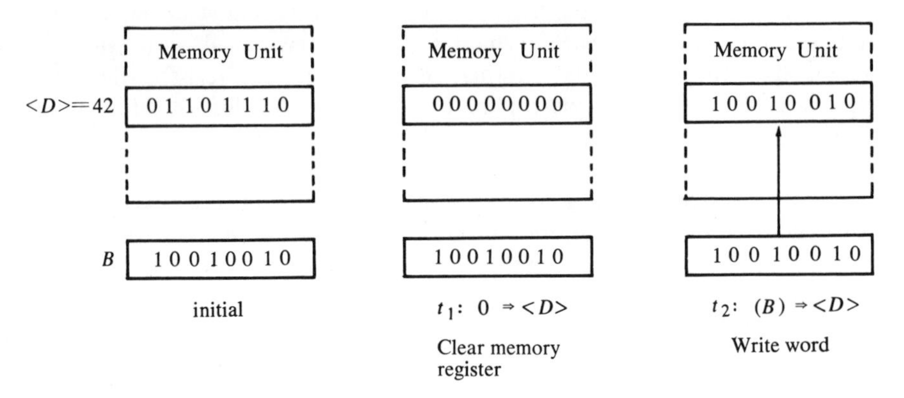

Figure 8-12 Information transfer in a magnetic core memory during *write* command

information into magnetic cores, the old information must be erased first by clearing all cells to zero (see Sec. 8-4). Therefore, the memory register is cleared during the first half cycle and only during the second half-cycle are the contents of the B register transferred to the designated memory register. In symbolic notation we have

write:

t_1: $0 \Rightarrow <D>$ clear memory register
t_2: $(B) \Rightarrow <D>$ transfer word to memory register

The sum of the two half-cycles $t_1 + t_2$ in either the read or write condition is called the *memory cycle* time.

The mode of access of a memory system is determined by the type of components used. In a *random-access* memory, the registers may be thought of as being separated in space, with each register occupying one particular spatial location as in a magnetic core memory. In a *sequential-access* memory, the information stored in some medium is not immediately accessible but is available only at certain intervals of time. A magnetic tape unit is of this type. Each memory location passes the read and write heads in turn, but information is read out only when the requested word has been reached. The *access time* of a memory is the time required to select a word and either read or write it. In a random-access memory, the access time is always the same regardless of the word's particular location in space. In a sequential memory, the access time depends on the position of the word at the time of request. If the word is just emerging out of storage at the time it is requested, the access time is just the time necessary to read or write it. But, if the word happened to be in the last position, the access time also includes the time required for all the other words to move past the

Table 8-3 Access Time and Access Mode of Memory Types.

Storage type	*Typical access time*	*Access mode*
magnetic cores	1 μ sec	random
plated wire	300 nsec	random
semiconductor IC	400 nsec	random
magnetic drum	10-40 msec	sequential
magnetic tape	0.05-500 sec	sequential
magnetic disks	10-80 msec	sequential

terminals. Thus, the access time in a sequential memory is variable. Table 8-3 lists typical access time for various memory types.

Memory units whose components lose stored information with time or when the power is turned off are said to be *volatile*. A semiconductor memory unit is of this category since its binary cells need external power to maintain the needed signals. By contrast, a nonvolatile memory unit, such as magnetic core or magnetic disk, retains its stored information after removal of power. This is because the stored information in magnetic components is manifested by the direction of magnetization, which is retained when power is turned off. A nonvolatile property is desirable in digital computers because many useful programs are left permanently in the memory unit. When power is turned off and then on again, the previously stored programs and other information are not lost but continue to reside in memory.

8-4 EXAMPLES OF RANDOM-ACCESS MEMORIES

The internal construction of two different types of random-access memories are presented diagramatically in this section. The first is constructed with flip-flops and gates and the second with magnetic cores. To be able to include the entire memory unit in one diagram, it is necessary that a limited storage capacity be used. For this reason, the memory units presented here have a small capacity of 12 bits arranged in four words of 3 bits each. Commercial random-access memories may have a capacity of thousands of words and each word may range somewhere between 8 and 64 bits. The logical construction of large capacity memory units would be a direct extension of the configuration shown here.

The logic diagram of a memory unit that uses flip-flops and gates is shown in Fig. 8-13. The entire unit may be constructed physically with ICs deposited in one semiconductor chip. The binary cell that stores one bit of information consists of an *RS* flip-flop, three AND gates, and an inverter. The input gates allow information to be transferred into the flip-flop when

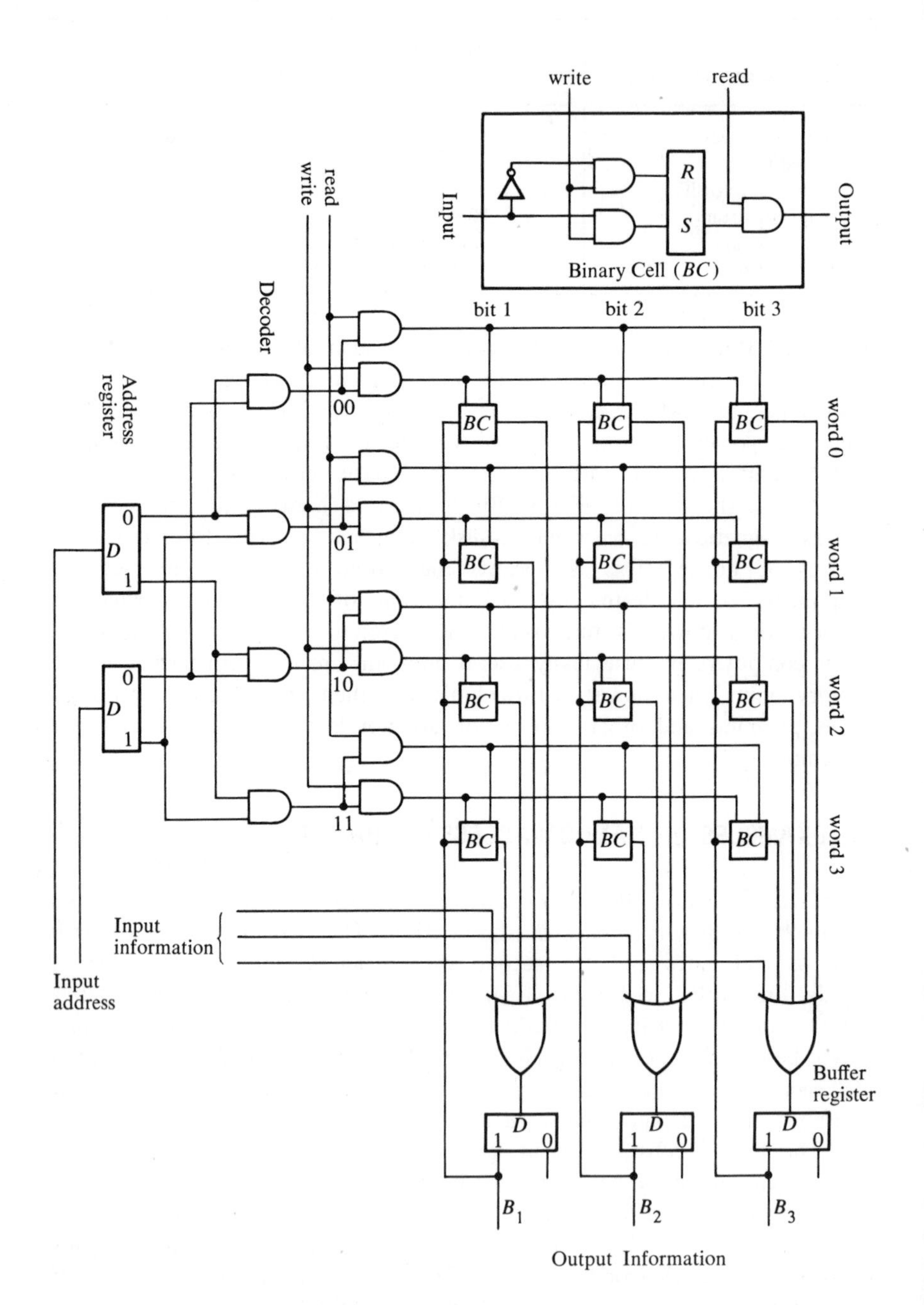

Figure 8-13 Semiconductor I. C. memory unit

the *write* signal is logic-1. An output gate samples the output of the flip-flop when the *read* signal is logic-1. Each small box labeled BC in the diagram includes within it the circuit of a binary cell. The four lines going into each BC box designate the three inputs and one output as specified in the detailed diagram of the binary cell.

To distinguish between four words we need a two-bit address. Therefore, the address register must have two flip-flops. To store a word of three bits, we need a buffer register with three flip-flops. The input information to be stored in memory is first transferred into the buffer register. The output information read from memory is available in the buffer register.

The address of the word held in the address register goes through a decoder circuit and each of the four outputs of the decoder is applied to the inputs of two gates. One of these gates receives the read signal and the other, the write signal. When the read signal is logic-1, the gates that go to the read input of the binary cells become logic-1 and the contents of the three cells of the word specified by the address register are transferred to the buffer register. When the write signal is logic-1, the gates that go to the write input of the binary cells become logic-1 and the contents of the buffer register are transferred to the three cells of the word specified by the address register. The two operations perform the required *read* and *write* transfers as specified in Sec. 8-3.

A magnetic core memory uses magnetic cores to store binary information. A magnetic core is a doughnut-shaped toroid made of magnetic material. In contrast to a semiconductor flip-flop that needs only one physical quantity such as voltage for its operation, a magnetic core employs three physical quantities: current, magnetic flux, and voltage. The signal that excites the core is a *current* pulse in a wire passing through the core. The binary information stored is represented by the direction of *magnetic flux* within the core. The output binary information is extracted from a wire linking the core in the form of a *voltage* pulse.

The physical property that makes a magnetic core suitable for binary storage is its hysteresis loop, as shown in Fig. 8-14(c). This is a plot of current vs. magnetic flux and has the shape of a square loop. With zero current, a flux which is either positive (counter-clockwise direction) or negative (clockwise direction) remains in the magnetized core. One direction, say counter-clockwise magnetization, is used to represent a 1 and the other to represent a 0.

A pulse of current applied to the winding through the core can shift the direction of magnetization. As shown in Fig. 8-14(a), current in the downward direction produces flux in the clockwise direction causing the core to go to the 0 state. Fig. 8-15(b) shows the current and flux directions for storing a 1. The path that the flux takes when the current pulse is applied is indicated by arrows in the hysteresis loop.

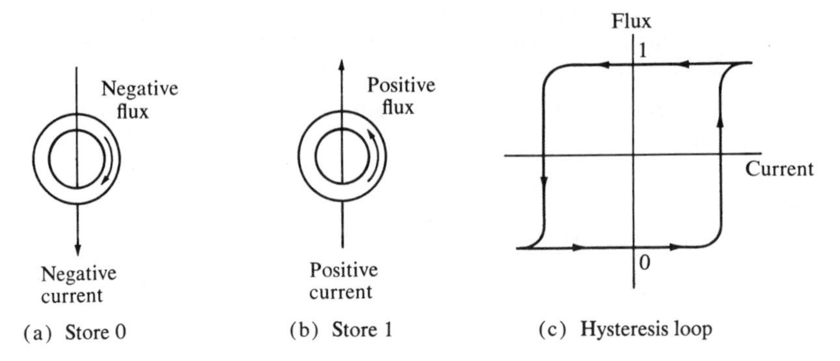

Figure 8-14 Storing a bit into a magnetic core

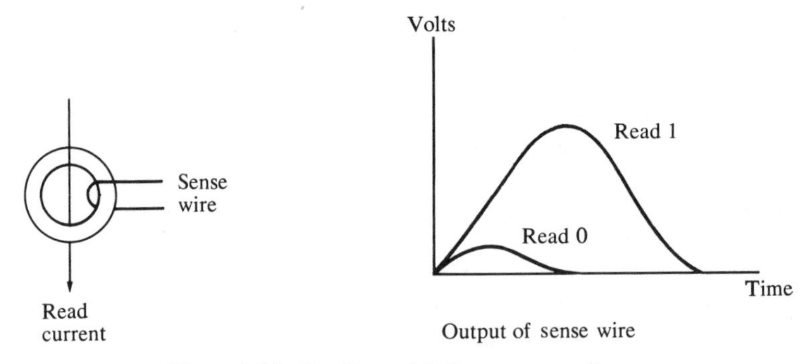

Figure 8-15 Reading a bit from a magnetic core

Reading out the binary information stored in the core is complicated by the fact that flux cannot be detected when it is not changing. However, if flux is changing with respect to time, it induces a voltage in a wire that links the core. Thus, read-out could be accomplished by applying a current in the negative direction as shown in Fig. 8-15. If the core is in the 1 state, the current reverses the direction of magnetization and the resulting change of flux will produce a voltage pulse in the sense wire. If the core is already in the 0 state, the negative current will leave the core magnetized in the same direction, causing a very slight disturbance of magnetic flux which results in a very small output voltage in the sense wire. Note that this is a destructive read-out since the read current always returns the core to the 0 state. The previously stored value is lost.

Figure 8-16 shows the organization of a magnetic core memory containing four words with three bits each. Comparing it with the semiconductor memory unit of Fig. 8-13 we note that the buffer register, address register, and decoder are exactly the same. The binary cell now is a magnetic core and the wires linking it. The excitation of the core is accomplished by means of a current pulse generated in a driver (abbreviated DR). The output information goes through a sense amplifier (abbreviated

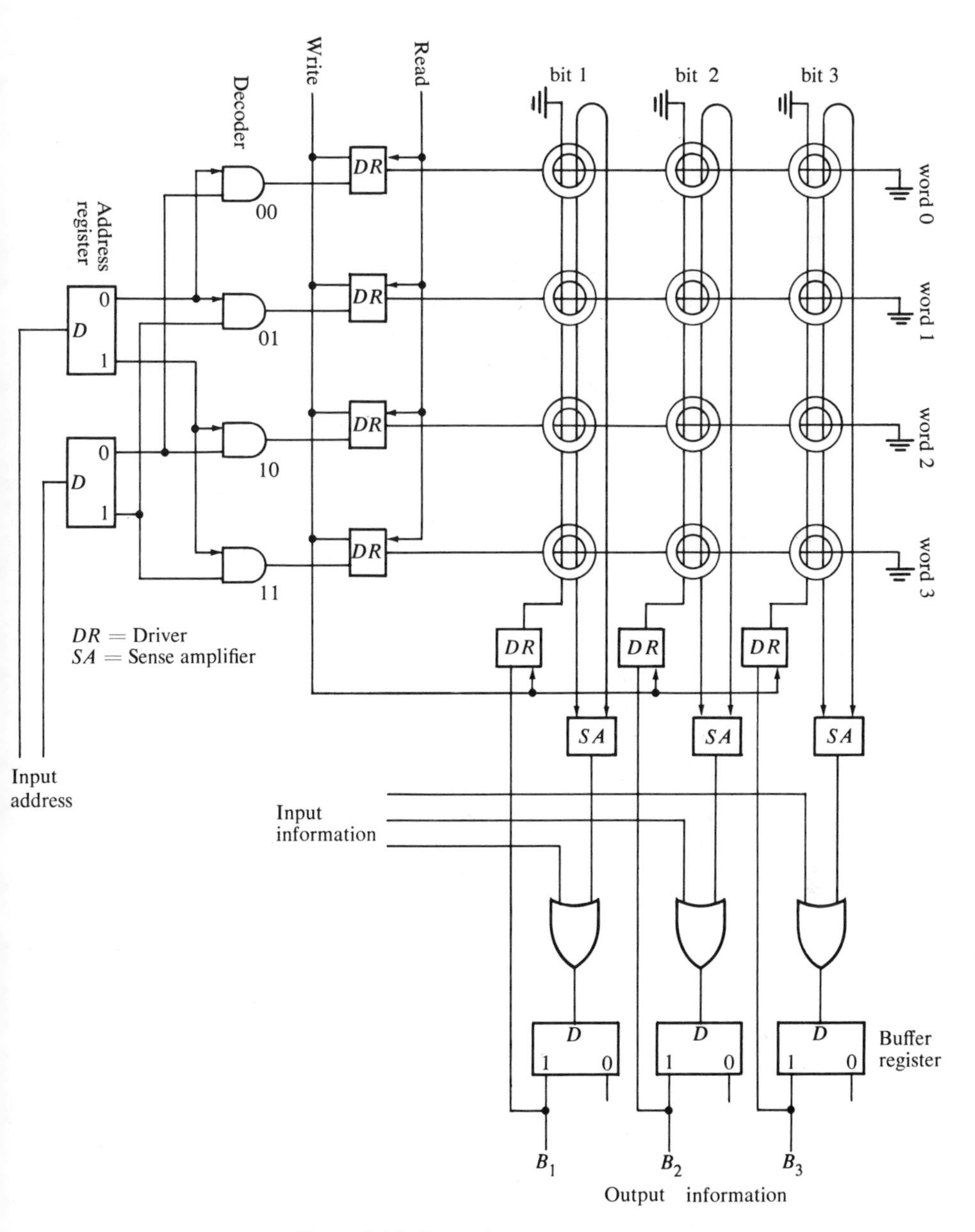

Figure 8-16 Magnetic core memory unit

SA) whose outputs set corresponding flip-flops in the buffer register. Three wires link each core. The word-wire is excited by a word-driver and goes through the three cores of a word. A bit-wire is excited by a bit-driver and goes through four cores in the same bit position. The sense-wire links the same cores as the bit-wire and is applied to a sense amplifier that shapes the voltage pulse when a 1 is read and rejects the small disturbance when a 0 is read.

During a *read* operation, a word-driver current pulse is applied to the cores of the word selected by the decoder. The read current is in the negative direction (Fig. 8-15) and causes all cores of the selected word to go to the 0 state, irrespective of their previous state. Cores which previously contained a 1 switch their flux and induce a voltage into their sense-wire. The flux of cores which already contained a 0 is not changed. The voltage pulse on a sense-wire of cores with a previous 1 is amplified in the SA; sets the corresponding flip-flop in the buffer register.

During a *write* operation, the buffer register holds the information to be stored in the word specified by the address register. We assume that all cores in the selected word are initially cleared; i.e., all are in the 0 state so that cores requiring a 1 need to undergo a change of state. A current pulse is generated simultaneously in the word-driver selected by the decoder and in the bit-driver, whose corresponding buffer register flip-flop contains a 1. Both currents are in the positive direction, but their magnitude is only half that needed to switch the flux to the 1 state. This half current by itself is too small to change the direction of magnetization. But the sum of two half currents is enough to switch the direction of magnetization to the 1 state. A core switches to the 1 state only if there is a coincidence of two half currents from a word-driver and a bit-driver. The direction of magnetization of a core does not change if it receives only half current from one of the drivers. The result is that the magnetization of cores is switched to the 1 state only if the word and bit wires intersect; that is, only in the selected word and only in the bit position in which the buffer register is a 1.

The read and write operations described above are incomplete. This is because the information stored in the selected word is destroyed by the reading process and the write operation works properly only if the cores are initially cleared. As mentioned in Sec. 8-3, a read operation must be followed by another cycle that restores the values previously stored in the cores. A write operation is preceeded by a cycle that clears the cores of the selected word.

The restore operation during a read cycle is equivalent to a write operation which, in effect, rewrites the previously read information from the buffer register back into the word selected. The clear operation during a write cycle is equivalent to a read operation which destroys the stored information but prevents the read information from reaching the buffer register by inhibiting the SA. Restore and clear cycles are normally initiated by the memory internal control, so that the memory unit appears to the outside as having a nondestructive read-out property.

Many types of memory units other than the two presented in this section have been used in digital computers. Information about other types of memories including an extensive bibliography can be found in references 4-6.

8-5 DATA REPRESENTATION IN REGISTERS

The bit configuration found in registers represents either data or control information. Data are operands and other discrete elements of information operated on to achieve required results. Control information is a bit or group of bits that specify the operations to be done. A unit of control information stored in digital computer registers is called an instruction and is a binary code that serves to specify the operations to be performed on the stored data. Instruction codes and their representation in registers are presented in Sec. 10-1. Some commonly used types of data and their representation in registers are presented in this section.

Binary Numbers

A register with n flip-flops can store a binary number of n bits; each flip-flop represents one binary digit. This represents the magnitude of the number but does not give information about its sign or the position of the binary point. The sign is needed for arithmetic operations, as it shows whether the number is positive or negative. The position of the binary point is needed to represent integers, fractions, or mixed interger-fraction numbers.

The sign of a number is a discrete quantity of information having two values: plus and minus. These two values can be represented by a code of one bit. The convention is to represent a plus with a 0 and a minus with a 1. To represent a sign binary number in a register we need $n + 1$ flip-flops; n flip-flops for the magnitude and one for storing the sign of the number.

The representation of the binary point is complicated by the fact that it is characterized by a *position* between two flip-flops in the register. There are two possible ways of specifying the position of the binary point in a register: by giving it a *fixed-point* position or by employing a *floating-point* representation. The fixed-point method assumes that the binary point is always fixed in one position. The two positions most widely used are: (a) a binary point in the extreme left of the register to make the stored number a fraction or (b) a binary point in the extreme right of the register to make the stored number an integer. In either case, the binary point is not physically visible but is assumed from the fact that the number stored in the register is treated as a fraction or as an integer. The floating-point representation uses a second register to store a number that designates the position of the binary point in the first register. Floating-point representation is explained in more detail below.

When a fixed-point binary number is positive, the sign is represented by a 0 and the magnitude by a positive binary number. When the number is negative, the sign is represented by a 1 and the magnitude may be represented in any one of three different ways. These are:

1. sign-magnitude
2. sign-1's complement
3. sign-2's complement

In the sign-magnitude representation, the magnitude is represented by a positive binary number. In the other two representations, the number is either in 1's or 2's complement. If the number is positive, the three representations are the same.

As an example, the binary number 9 is written below in the three representations. It is assumed that a seven-bit register is available to store the sign and the magnitude of the number.

	+9	-9
sign-magnitude	0 001001	1 001001
sign-1's complement	0 001001	1 110110
sign-2's complement	0 001001	1 110111

A positive number in any representation has a 0 in the left-most bit for a plus followed by a positive binary number. A negative number always has a 1 in the left-most bit for a minus, but the magnitude bits are represented differently. In the sign-magnitude representation, these bits are the positive number; in the 1's-complement representation these bits are the complement of the binary number; and in the 2's complement representation, the number is in its 2's complement form.

The addition and subtraction of two numbers in sign-magnitude representation is identical to paper and pencil arithmetic, but the machine implementation of this calculation is somewhat involved and inefficient. On the other hand, the rule for addition and subtraction of two numbers in complement representation is much simpler to implement. The algorithm and implementation of sign-magnitude addition and subtraction can be found in Ch. 12. Binary arithmetic with sign-complement representation is treated in Sec. 9-9.

Decimal Numbers

The representation of decimal numbers in registers is a function of the binary code used to represent a decimal digit. A four-bit decimal code, for example, requires four flip-flops for each decimal digit. The representation of + 4385 in BCD requires at least 17 flip-flops; 1 flip-flop for the sign and 4 for each digit. This number is represented in a register with 25 flip-flops as follows:

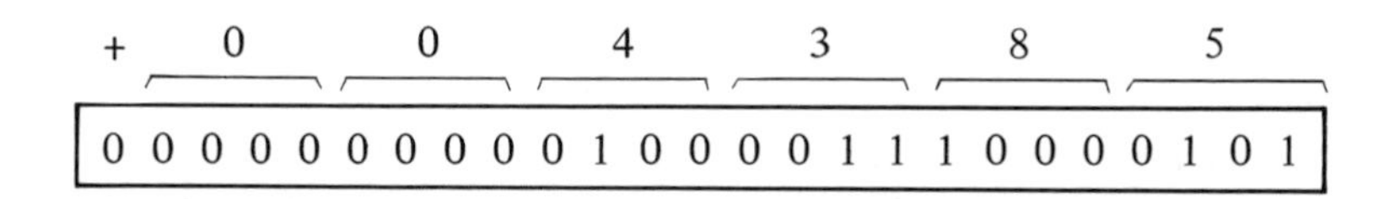

By representing numbers in decimal, we are wasting a considerable amount of storage space since the number of flip-flops needed to store a decimal number in a binary code is greater than the number of flip-flops needed for its equivalent binary representation. Also, the circuits required to perform decimal arithmetic are much more complex. However, there are some advantages in the use of decimal representation, mostly because computer input and output data are generated by people that always use the decimal system. A computer that uses binary representation for arithmetic operations requires data conversion from decimal to binary prior to performing calculations. Binary results must be converted back to decimal for output. This procedure is time consuming; it is worth using provided the amount of arithmetic operations is large, as is the case with scientific applications. Some applications such as business data processing require small amounts of arithmetic calculations. For this reason, some computers perform arithmetic calculations directly on decimal data (in binary code) and thus eliminate the need for conversion to binary and back to decimal. Large-scale computer systems usually have hardware for performing arithmetic calculations both in binary and in decimal representation. The user can specify by programmed instructions whether he wants the computer to perform calculations on binary or decimal data. A decimal adder is introduced in Sec. 12-2.

There are three ways to represent negative fixed-point decimal numbers. They are similar to the three representations of a negative binary number except for the radix change:

1. sign-magnitude

2. sign-9's complement

3. sign-10's complement

For all three representations, a positive decimal number is represented by a 0 (for plus) followed by the magnitude of the number. It is in regard to negative numbers that the representations differ. The sign of a negative number is represented by a 1 and the magnitude of the number is positive in sign-magnitude representation. In the other two representations the magnitude is represented by the 9's or 10's complement.

The sign of a decimal number is sometimes taken as a four-bit quantity to conform with the four-bit representation of digits. For example, one very popular computer uses the code 1100 for plus and 1101 for minus.

Floating-Point Representation

Floating-point representation of numbers needs two registers. The first represents a signed fixed-point number and the second, the position of the radix point. For example, the representation of the decimal number +6132.789 is as follows:

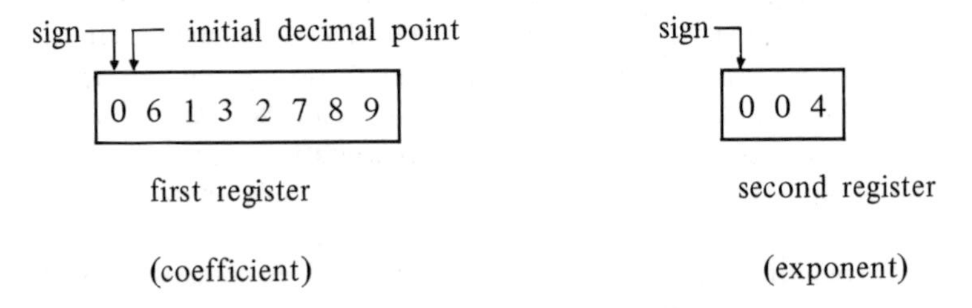

first register

(coefficient)

second register

(exponent)

The first register has a 0 in the most significant flip-flop position to denote a plus. The magnitude of the number is stored in a binary code in 28 flip-flops, with each decimal digit occupying 4 flip-flops. The number in the first register is considered a fraction, so the decimal point in the first register is fixed at the left of the most significant digit. The second register contains the decimal number +4 (in binary code) to indicate that the *actual* position of the decimal point is four decimal positions to the right. This representation is equivalent to the number expressed by a fraction times 10 to an exponent; that is + 6132.789 is represented as $+.6132789 \times 10^{+4}$. Because of this analogy, the contents of the first register are called the *coefficient* (and sometimes *mantissa* or *fractional part*) and the contents of the second register, the *exponent* (or *characteristic*).

The position of the actual decimal point may be outside the range of digits of the coefficient register. For example, assuming sign-magnitude representation, the following contents

coefficient

exponent

represent the number $+.2601000 \times 10^{-4} = +.000026010000$, which produces four more 0's on the left. On the other hand the following contents

represent the number $-.2601000 \times 10^{12} = -260100000000$, which produces five more 0's on the right.

In the above examples, we have assumed that the coefficient is a fixed-point fraction. Some computers assume it to be an integer, so the initial decimal point in the coefficient register is to the right of the least significant digit.

Another arrangement used for the exponent is to remove its sign bit altogether and consider the exponent as being "biased." For example, numbers between 10^{+49} and 10^{-50} can be represented with an exponent of two digits (without sign bit) and a bias of 50. The exponent register always contains the number $E + 50$, where E is the actual exponent. The subtraction of 50 from the contents of the register gives the desired exponent. This way, positive exponents are represented in the register in the range of numbers from 50 to 99. The subtraction of 50 gives the positive values from 00 to 49. Negative exponents are represented in the register in the range of 00 to 49. The subtraction of 50 gives the negative values in the range of -50 to -1.

A floating-point binary number is similarly represented with two registers, one to store the coefficient and the other, the exponent. For example, the number +1001.110 can be represented as follows:

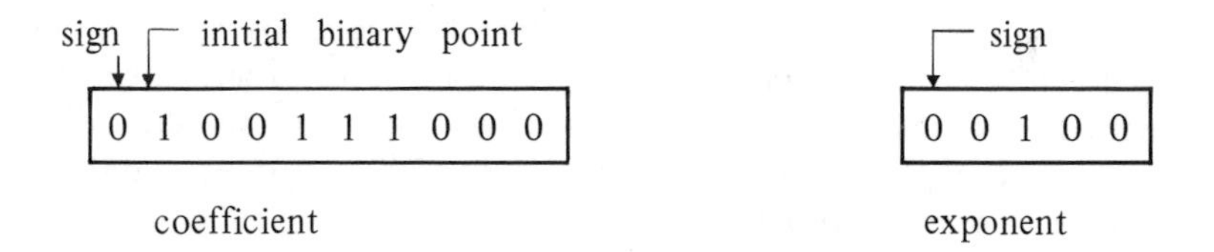

The coefficient register has 10 flip-flops; 1 for sign and 9 for magnitude. Assuming that the coefficient is a fixed-point fraction, the actual binary point is four positions to the right, so the exponent has the binary value +4. The number is represented in binary as $.10011000 \times 10^{100}$ (remember that 10^{100} in binary is equivalent to decimal 2^4).

Floating-point is always interpreted to represent a number in the following form:

$$c \cdot r^e$$

where c represents the contents of the coefficient register and e the contents of the exponent register. The radix (base) r and the radix-point position in the coefficient are always assumed. Consider, for example, a computer that assumes integer representation for the coefficient and base 8 for the exponent. The octal number $+17.32 = +1732 \times 8^{-2}$ will look like this:

When the octal representation is converted to binary, the binary value of the registers becomes:

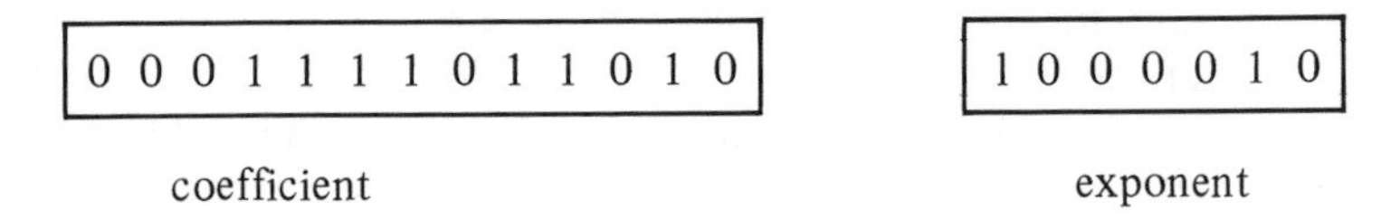

A floating-point number is said to be *normalized* if the most significant position of the coefficient contains a nonzero digit. In this way, the coefficient has no leading zeros and contains the maximum possible number of significant digits. Consider, for example, a coefficient register that can accomodate five decimal digits and a sign. The number $+.00357 \times 10^3 = 3.57$ is not normalized because it has two leading zeros and the unnormalized coefficient is accurate to three significant digits. The number can be normalized by shifting the coefficient two positions to the left and decreasing the exponent by two to obtain: $+.35700 \times 10^1 = 3.5700$, which is accurate to five significant digits.

Arithmetic operations with floating-point number representation are more complicated than arithmetic operations with fixed-point numbers and their execution takes longer and requires more complex hardware. However, floating-point representation is more convenient because of the scaling problems involved with fixed-point operations. Many computers have a built-in capability to perform floating-point arithmetic operations. Those that do not have this hardware are usually programmed to operate in this mode.

Adding or subtracting two numbers in floating-point representation requires first an alignment of the radix point since the exponent part must be made equal before adding or subtracting the coefficients. This alignment is done by shifting one coefficient while its exponent is adjusted until it is equal to the other exponent. Floating-point multiplication or division requires no alignment of the radix-point. The product can be formed by multiplying the two coefficients and adding the two exponents. Division is accomplished from the division with the coefficients and the subtraction of the divisor exponent from the exponent of the dividend.

Double Length Words

The registers in a digital computer are of finite length. This means that the maximum number that can be stored in a register is finite. If the precision needed for calculations exceeds the maximum capacity of one register, the user can double the bit capacity by employing two registers to store an operand. One register would contain the most significant digits and the other, the least significant digits.

For example, consider a computer whose memory and processor registers contain 16 bits. One bit is needed to store the sign of a fixed-point number, so the range of integers that can be accomodated in a register is between $\pm 2^{15} = \pm 32{,}768$. This number may be too small for a particular application. By using two registers it is possible to increase the range of integers to $\pm 2^{31}$.

The same reasoning applies to floating-point numbers when greater length coefficients are needed for greater resolution. Using a computer with 16-bit words as before, a floating-point number can be stored in two words. The exponent may occupy the first seven bits of the first word giving an exponent value of ±63. The coefficient would then occupy 9 bits of the first word and 16 bits of the second word for a total of 25 bits. The number of bits in the coefficient can be extended to 41 by using a third word.

Double length words and sometimes floating-point operands are stored in two or more consecutive memory registers. They need two or more cycles to be read or written in memory. Some computers have available double length registers and/or special floating-point registers in their processor unit. These registers can usually execute the required arithmetic operations directly with double precision or floating-point numbers. Computers with a single length register can be programmed to operate with double length words. In any case, operands that exceed the number of bits of a memory word must be stored in two or more consecutive memory registers.

Some computers use variable length words and allow the user to specify the number of bits that he wants to use for the operands. This implies a memory structure composed of variable length cells instead of fixed length words. This can be done by making the smallest data component addressable and using a memory address that specifies the first and last cell of the operand.

Character Strings

The types of data considered thus far represent numbers that a computer uses as operands for arithmetic operations. However, a computer is not only a machine that stores numbers and does high-speed arithmetic. Very often, a computer manipulates symbols rather than numbers. Most programs written by computer users are in the form of characters; i.e., a set of symbols comprised of letters, digits, and various special characters. A computer is capable of accepting characters (in a binary code), storing them in memory, performing operations on and transferring characters to an output device. A computer can function as a character-string manipulating machine. By a *character string* is meant a finite sequence of characters written one after another.

Characters are represented in computer registers by a binary code. In Table 1-5 we listed three different character codes in common use. Each member of the code represents one character and consists of either six, seven, or eight bits, depending on the code. The number of characters that can be stored in one register depends on the length of the register and the number of bits used in the code. For example, a computer with a word length of 36 bits that uses a character code of 6 bits can store six characters per word. Character strings are stored in memory in consecutive locations. The first character in the string can be specified from the address of the first word. The last character of the string may be found from the address of the last word, or by specifying a character count, or by a special mark designating end of character string. The manipulation of characters is done in the registers of the processor unit, with each character representing a unit of information.

Logical Words

Certain applications call for manipulating the bits of a register with logical operators. Logical operations such as complement, AND, OR, exclusive-or, etc., can be performed with data stored in registers. However, logical operations must consider each individual bit of the register separately, since they operate on two-valued variables. In other words, the logical operations must consider each bit in the register as a Boolean variable having the value of 1 or 0.

For example, to complement the contents of a register we need to complement each bit of the word stored in it. As a second example, the OR operation between the contents of two registers A and B is shown below:

(A)	1100	contents of A
(B)	1010	contents of B
$(A) \vee (B)$	1110	result of OR operation

The result is obtained by performing the OR operation with every pair of corresponding bits.

The datum available in a register during logical operations is called a *logical word*. A logical word is interpreted to be a *bit string* as opposed to character strings or numbers. Each bit in a logical word functions in exactly the same way as every other bit; in other words, the unit of information of a logical word is a bit.

PROBLEMS

8-1. Show the configuration of a bus transfer system that contains four registers of three bits each. Use multiplexers and demultiplexer circuits.

8-2. A digital system has 16 registers, each with 32 bits. It is necessary to provide transfer paths for data from each register to each other register. (a) How many lines are needed for direct transfers? (b) How many lines are needed for bus transfer? (c) How many lines are needed if the 16 registers form a memory unit? Compare the three configurations with respect to the time it takes to transfer data between two given registers.

8-3. A three-bit register A has one input x and one output y. The register operations can be described symbolically as follows:

$$P: x \Rightarrow A_1, (A_3) = y, (A_i) \Rightarrow A_{i+1} \qquad i = 1, 2$$

Draw the state table for the sequential circuit and its corresponding state diagram. What is the function of the circuit?

8-4. A digital system has three flip-flops; i.e., three one-bit registers labeled A, B, C. The state of the flip-flops changes according to the following elementary operations:

$$\begin{aligned} x&: \quad 0 \Rightarrow A, \quad 1 \Rightarrow B, \quad 1 \Rightarrow C \\ y&: \quad (\bar{B}) \Rightarrow A, \quad (\bar{C}) \Rightarrow B, \quad (B) \Rightarrow C \\ zA&: \quad (\bar{A}) \Rightarrow A, \quad (\bar{B}) \Rightarrow B, \quad (A) \Rightarrow C \end{aligned}$$

Give the sequence of states of the flip-flops for the following sequence of control signals.

(a) x, y, z, y, z, z, y, x.

(b) x, y, y, z, y, y, z.

8-5. List the elementary operations needed to transfer bits 1 - 8 of register A to bits 9-16 of register B and bits 1-8 of register B to bits 9-16 of register A, all during one clock pulse.

8-6. Express the transfers depicted in Fig. 1-3, Sec. 1-7 with elementary operations. Include address and buffer registers for the memory unit.

8-7. Express the function of the shift register of Fig. 6-23, Sec. 6-6, with symbolic notation.

8-8. It is required to construct a memory with 256 words, 16 bits per word, organized as in Fig. 8-16. Cores are available in a "matrix" of 16 rows and 16 columns.

(a) How many matrices are needed?

(b) How many flip-flops are in the address and buffer registers?

(c) How many cores receive current during a *read* cycle.

(d) How many cores receive at least half current during a *write* cycle?

8-9. A memory unit has a capacity of 40,000 words of 20 bits each. How many decimal digits can be stored in a memory word? How many flip-flops are needed for the address register if the words are numbered in decimal?

8-10. List the elementary operations for the following transfers using (1) a nondestructive memory and (2) a destructive read-out memory.

(a) To store binary 125 in location 63.

(b) To read memory register number 118.

(c) To transfer contents of memory register 75 to memory register number 90.

8-11. The magnetic core memory of Fig. 8-16 is called a *linear selection* memory and has a two-dimensional (2D) organization. There is another magnetic core memory organization called *coincident-current selection*, which has a three-dimensional (3D) organization. A third memory organization, a combination of the two types, is called a 2½D selection. Consult outside sources and describe the other two memory organizations.

8-12. What modifications or additions are required in a memory unit to add the following capabilities:

(a) To be able to transfer incoming and outgoing words serially.

(b) To be able to select consecutive locations in memory.

(c) To be able to address memory locations with decimal numbers in BCD.

(d) To be able to select one particular bit of the selected word.

8-13. Represent binary +27 and −27 in three representations using a register of 10 bits.

8-14. Represent the numbers +315 and −315 in binary using the following representations:

(a) sign-magnitude,

(b) 1's complement,

(c) 2's complement, and

(d) sign-magnitude floating-point normalized.

8-15. Represent the numbers +315 and −315 in decimal using BCD in the following representations:

(a) sign-magnitude,

(b) 9's complement,

(c) 10's complement,

(d) sign-magnitude floating-point normalized.

8-16. Given two binary numbers X and Y in floating-point sign-magnitude representation. The numbers are normalized and placed in registers and an addition is to be performed; i.e., $X + Y$.

(a) Which of the two coefficients should be shifted, in what direction, and by how much?

(b) How can we determine the value of the exponent of the sum?

(c) After the addition of the two coefficients, the sum may be unnormalized. Specify what should be done to the coefficient and the exponent in order to normalize the sum.

8-17. What is the largest and smallest positive quantity which can be represented when the coefficient is normalized:

(a) by a 36-bit floating-point binary number having 8 bits plus sign for the exponent and fraction representation for the coefficient?

(b) by a 48-bit floating-point binary number having 11 bits plus sign for the exponent and integer representation for the coefficient?

REFERENCES

1. Reed, I. S. "Symbolic Design of Digital Computers," *MIT Lincoln Lab. Tech.*, Mem. 23 (January 1953).

2. Chu, Y. *Digital Computer Design Fundamentals.* New York: McGraw-Hill Book Co., 1962.

3. Bartee, T. C., I. L. Lebow, and I. S. Reed. *Theory and Design of Digital Machines.* New York: McGraw-Hill Book Co., 1962.

4. Scott, N. R. *Electronic Computer Technology.* New York: McGraw-Hill Book Co., 1970 CH. 10.

5. Mayerhoff, A. J. *Digital Applications of Magnetic Devices.* New York: John Wiley & Sons, Inc., 1960.

6. Special Issue on High-Speed Memories, *IEEE Trans. on Electronic Computers*, Vol. EC-15 (August 1966).

9 BINARY ADDITION AND THE ACCUMULATOR REGISTER

9-1 INTRODUCTION

General purpose digital computers as well as many special purpose systems quite often incorporate a multipurpose register into the arithmetic section or the central processing unit. Some digital computers have more than one such register. These registers are called by various names, but the name *accumulator* is the one most widely used. The name is derived from the arithmetic addition process encountered in digital computers. The process of arithmetic addition of two or more numbers is carried out by initially storing these numbers in the memory unit and clearing the accumulator register. The numbers are then read from memory one at a time and added to the register in consecutive order. The first number is added to zero and the sum transferred to the register. The second number is added to the contents of the register and the newly formed sum replaces its previous value. This process is continued until all numbers are added and the sum formed. Thus, the register "accumulates" the sum in a step-by-step manner by performing sequential additions between a new number read from memory and the previously accumulated sum.

The accumulator is the most used register in the arithmetic unit. In addition to the function described above, this register performs various other operations either specified by programmed instructions or required to implement other machine instructions. The accumulator stores the operands used in arithmetic and logical operations and retains the results of computations formed. When two numbers are multiplied, the multiplier and multiplicand are read from memory into two registers and multiplication is performed by forming a series of partial sums in the accumulator. Similarly, subtraction, division, and logical operations are performed by the accumulator logic circuits. Other operations such as complementing, counting, and shifting are also encountered.

The purpose of this chapter is to present a simple multipurpose accumulator register, together with its associated combinational circuits. The register proper will be designated by the letter A and will consist of n flip-flops $A_1, A_2, \ldots, A_n$, numbered in ascending order starting from the right-most element. We shall assume "the accumulator" to mean the A register plus its associated logic circuits and the A register will be called "the accumulator register." We concentrate the discussion first on the subject of addition of two binary numbers, since this is the most involved part of the logic design of an accumulator. In Sec. 9-8 a multipurpose register having other elementary operations besides addition is specified.

One of the primary functions of an accumulator is the addition of two numbers. The following sections present a few alternative ways to accomplish this task. One of these alternatives will be chosen for inclusion in the final design of the register. In order to realize the machine performance of binary addition, we must visualize the existence of two registers, one storing the augend and the other the addend. The outputs of these registers are applied to logic circuits that produce the sum; the sum is then transferred into a register for storage. The following numerical example will clarify the process:

Previous carry	00110	C_i
Augend	01011	A_i
Addend	10011	B_i
Sum	11110	S_i
Carry	00011	C_{i+1}

The augend is in register A (accumulator register) and the addend in register B. The bits are added starting from the least significant position to form a sum bit and carry bit. The previous carry bit C_i for the least significant position must be 0 and the carry bit C_{i+1} for this example is a 1. The value of C_{i+1} in a given significant position is copied into C_i one higher significant position to the left. The sum bits are thus generated starting from the right-most position and are available as soon as the corresponding previous carry bit C_i is known. The sum can be transferred to a third register; however, when the process employs an accumulator, the sum is transferred to the accumulator register, destroying the previously stored augend.

9-2 PARALLEL ADDITION

Binary addition is performed in parallel when all bits of the augend and addend registers are available as outputs. The combinational circuit among the registers that forms the sum depends on the type of flip-flops employed in the accumulator register. Three different combinational circuits are derived in this section. The first employs D flip-flops and the other two, T flip-flops. Circuits using RS or JK flip-flops are equivalent to those employing D or T flip-flops, respectively.

Addition with Full-Adders

The simplest and most forward way to add two binary numbers in parallel is to use one full-adder (abbreviated FA) circuit for each pair of significant bits.* This is shown schematically in Fig. 9-1 for a four-bit adder. The implementation of adders with more than four bits is accomplished by extending the registers with more flip-flops and corresponding

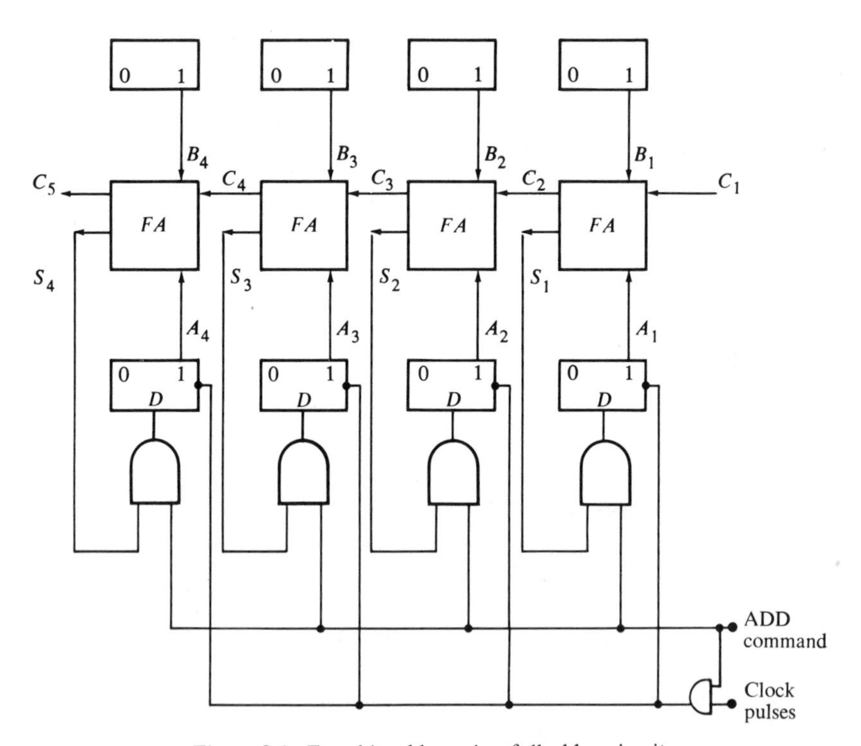

Figure 9-1 Four-bit adder using full-adder circuits

*The logic diagram of a full-adder circuit was derived in Sec. 4-3.

FA circuits. The contents of registers A and B are added in the four FA circuits; their sum is formed in S_1 to S_4. Note that the input carry to the first stage must be connected to a signal corresponding to logic-0 and that the carry-out from the last stage can be used for an input to a fifth stage or as an overflow bit in a four-bit adder. The sum outputs S_1 to S_4 are transferred to the A register through the D inputs of the flip-flops when an ADD command signal is enabled during one clock pulse. The sum transferred will obviously destroy the previously stored augend so that the A register acts as an accumulator register.

Some minor variations are possible. The first stage could use a half-adder circuit instead of a full-adder, which will result in the savings of a few gates. *RS* or *JK* flip-flops could be used, but the complement values of S_1-S_4 must be generated for the clear inputs. Only the normal outputs of the flip-flops are shown going into the FA circuits. If the complement outputs are also used, the inverter circuits which produce the complements inside the FA circuits can be eliminated. The diagram chosen for Fig. 9-1 assumes the availability of a single IC chip, which includes all four FA circuits with the carry leads internally connected. Such an IC is commonly named "four-bit adder." It has nine input terminals, five output terminals, at least one terminal for the supply voltage, and one for ground.

The binary adder in Fig. 9-1 is constructed with four identical stages, each consisting of a pair of flip-flops and their corresponding full-adder circuit. It is interesting to note that each stage is similar to all others, including similar interconnections among neighborhood stages. It is therefore convenient to partition such circuits into n similar circuits, each having the same general configuration. The ability to partition a large system into a number of smaller and possibly similar subsystems has many advantages. It is convenient for the design process because the design of only one typical stage is sufficient for obtaining all the information needed to design the entire register logic. It also brings about standardization of common circuit configurations which can be eventually incorporated into an IC chip. In fact, binary adders as well as accumulator registers are comparatively easier to design and manufacture because they can be partitioned into smaller subregister units. Practical adders and accumulators range somewhere between 12 and 64 bits long. The task of designing a sequential circuit with 64 flip-flops is formidable, but the partition into 64 identical stages reduces the design process down to the point where a sequential circuit with only one flip-flop is considered. We shall utilize this partitioning property in the remainder of this chapter and concentrate our efforts on the design of only one typical stage. The entire register and its associated logic will then consist of a number of such typical stages, where that number is equal to the number of bits of the register.

Addition with Complementing Flip-Flops

Flip-flops having the property of complementation are very useful for reducing the number of gates required to form the sum. Such flip-flops are the T and JK types. This advantage can be deduced from inspection of the Boolean functions that describe the FA circuit. For a typical stage i, the sum and carry are expressed by the following Boolean functions*:

$$S_i = A_i'B_i'C_i + A_i'B_iC_i' + A_iB_i'C_i' + ABC \tag{9-1}$$

$$= A_i \oplus B_i \oplus C_i \tag{9-2}$$

$$C_{i+1} = A_iB_i + A_iC_i + B_iC_i \tag{9-3}$$

The sum bit is obtained from the exclusive-or operation of the three input variables. Also note that the augend is in the A register and that the sum is also transferred into the A register. Now, an inherent property of a complementing flip-flop is that its next state produces the exclusive-or of its input variable and the previous state value. Consider the following state table of a single flip-flop A and a variable X applied to its T input.

Present state	*T input*	*Next state*
$A(t)$	X	$A(t+1) = A(t) \oplus X$
0	0	0
0	1	1
1	0	1
1	1	0

It is clear from this table that the next state of A is the exclusive-or of the present state of A and the input X.

From this observation, it is possible to deduce that if $B_i \oplus C_i$ is applied to the T input of an A_i flip-flop, the next state of A_i will hold the value obtained from the exclusive-or of B_i, C_i, and the previous state of A_i. This is exactly the value of the sum bit as given in Eq. 9-2.

It is possible to arrive at the same conclusion by a straightforward design process. A typical adder stage consists of one flip-flop A_i that changes its state during the addition process. The single stage will have an input addend bit B_i and a previous carry C_i. (Note that B_i is a flip-flop that does not change state during the addition process and therefore can be considered as an input terminal). A state table for this simple sequential circuit is tabulated in Fig. 9-2(a). A column is included in the table for the flip-flop

*Equations 9-1 and 9-3 were derived in Sec. 4-3. Equation 9-2 follows directly from Equation 9-1.

Present state	Inputs		Next state	Flip-flop input
A_i	B_i	C_i	$A_i = S$	TA_i
0	0	0	0	0
0	0	1	1	1
0	1	0	1	1
0	1	1	0	0
1	0	0	1	0
1	0	1	0	1
1	1	0	0	1
1	1	1	1	0

(a)

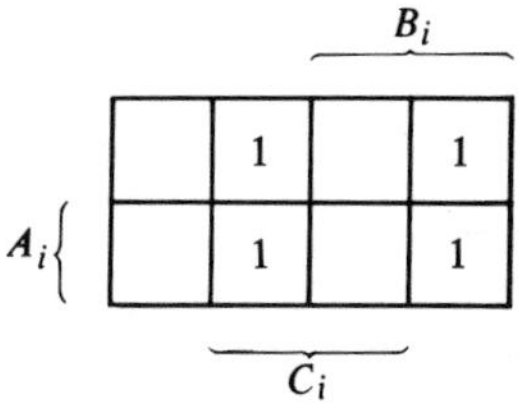

	1		1
	1		1

A_i; B_i; C_i

$TA_i = B_i C'_i + B' C_i = B_i \oplus C_i$

(b)

Figure 9-2 Excitation table and map for a parallel adder stage

T input and its necessary excitation. The combinational circuit for this two-state sequential circuit can be derived immediately from the map shown in Fig. 9-2(b). The required flip-flop input function is

$$TA_i = B_i C'_i + B'_i C_i = B_i \oplus C_i$$

which is the same condition derived previously. The carry output C_{i+1} has not been shown in the table. It is a function of the present state and the inputs and can easily be shown to be the same as Eq. 9-3. The logic diagram of a typical stage is drawn in Fig. 9-3. The part of the circuit that produces the sum has only two AND gates and one OR gate, compared to the four AND gates and one OR gate required for the implementation of Eq. 9-1 in sum of products form. A *JK* flip-flop can be used instead of a *T*, with the two inputs *J* and *K* connected to receive the same signal. The circuit of Fig. 9-3 with a *JK* flip-flop is the one chosen for the final version of the accumulator to be completed in Sec. 9-8. For an *n*-bit register, *n* identical circuits are required with the output carry C_{i+1} of one stage connected to the input carry C_i of the next stage on its left, except for the least significant position, which requires an input C_i of 0, or a circuit that corresponds to a half-adder.

Two-Step Adder

A further reduction in the number of gates per stage is realized if the time of execution is extended to a period equal to two clock pulses by requiring two consecutive command signals to complete the addition. Consider two signals *AD*1 and *AD*2 as shown in Fig. 9-4. Compared with the single command signal used for the circuit in Fig. 9-3, the two-step addition scheme to be developed here requires two clock pulses for completion. On the other hand, the two-step adder requires fewer combinational gates for its implementation.

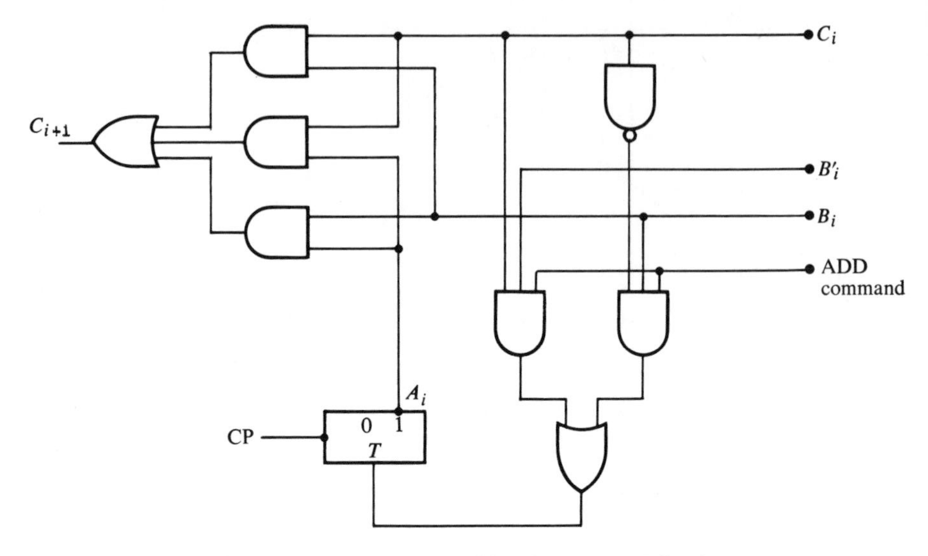

Figure 9-3 Typical stage of adder with a T flip-flop

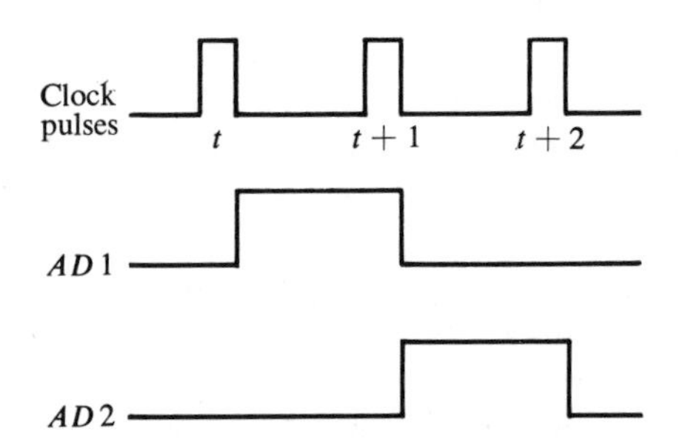

Figure 9-4 Two consecutive Add signals for a two-step adder

The combinational logic necessary for a two-step adder will be derived intuitively. The next state of a T flip-flop is the exclusive-or of its present state and its input. The sum bit is obtained from the exclusive-or of B_i, C_i, and the present state of A_i. Now, if B_i is applied during the time $t + 1$, when the first command signal $AD1$ is enabled, one obtains for the next state $A_i(t + 1)$, the value of $A_i \oplus B_i$. Then during the time $t + 2$, when command signal $AD2$ is enabled, C_i is applied to the input of the flip-flop producing a final next state $A_i(t + 2)$ equal to $A_i \oplus B_i \oplus C_i$, which is equal to the sum bit.

The carry bit must be generated in all the stages of the adder prior to the application of the $t + 2$ pulse. However, after signal $AD1$ has been applied, the state of the flip-flop has the value of $A_i \oplus B_i$. Therefore, the carry to the next stage C_{i+1} must be formed from the $(t + 1)$ value of the flip-flop output. To determine the logic required for generating the carry, we need the table of Fig. 9-5(a). In this table, the value of $A_i \oplus B_i$ is first determined. The column under C_{i+1} is determined from the bits of $A_i(t)$, B_i, and C_i. However, the combinational logic required for output C_{i+1} is obtained from the columns under B_i, $A_i(t+1)$, and C_i because only these values are available after command signal $AD1$ and during command signal $AD1$. Remembering that the value of A_i after $t+1$ is the exclusive-or of the addend and augend bits, then from the map of Fig. 9-5b, we obtain the Boolean function for the output carry to be:

$$C_{i+1} = B_i A_i' + A_i C_i$$

where A_i is the state of the flip-flop after command signal $AD1$ has been removed and during the application of command signal $AD2$.

Figure 9-6 shows the logic diagram of a two-step adder. It can be seen that this circuit requires fewer gates per stage as compared with Fig. 9-3.

(a)

$A_i(t)$	B_i	C_i	$A_i(t+1) = A_i \oplus B_i$	C_{i+1}
0	0	0	0	0
0	0	1	0	0
0	1	0	1	0
0	1	1	1	1
1	0	0	1	0
1	0	1	1	1
1	1	0	0	1
1	1	1	0	1

(b)

C_{i+1} (map; C_i over the two right columns, B_i on the bottom row, $A_i \oplus B_i = A_i(t+1)$ under the two middle columns)

			1	
B_i	1		1	1

Figure 9-5 Derivation of output carry in one stage of a two-step adder

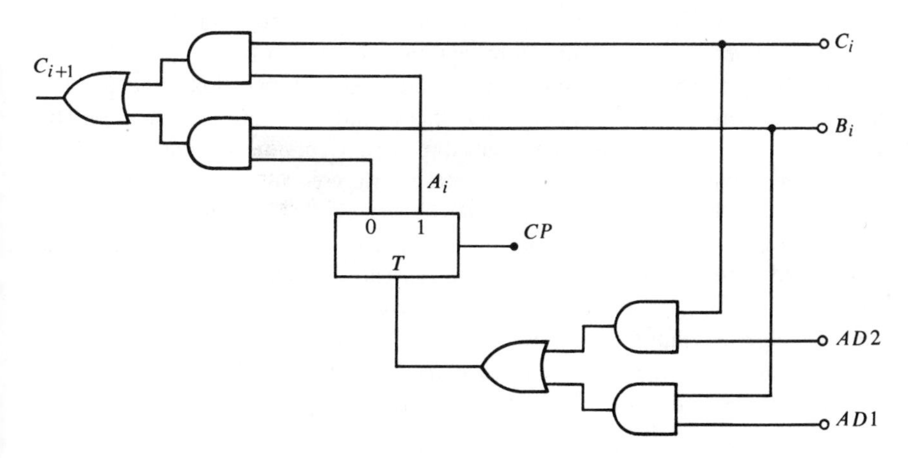

Figure 9-6 One stage of a two-step adder

The price one pays for this saving is an increase in the execution time. The process of addition in the two-step adder of Fig. 9-6 is as follows: First the command signal $AD1$ is applied. After $t + 1$, the carry through all the stages is allowed to propagate. Then command signal $AD2$ is applied, forming the sum in the A register after clock pulse $t + 2$.

9-3 CARRY PROPAGATION

The addition of two numbers in parallel implies that all the bits of the augend and addend are available for computation at the same instant of time. This is the difference between parallel and serial operations; in the latter the bits of the augend and addend are available for computation only one at a time. Before leaving the subject of parallel adders and going into serial addition, the timing problem involved with the carry propagation will be explained.

Electronic circuits take a certain amount of time from the instant the input signal is applied to the instant when the output settles into its appropriate logic value. This interval is defined as the propagation time of the circuit. At any instant, a digital circuit may be in one of three conditions: in one constant value which represents logic-1, in another constant value which represents logic-0, or in transition between the two values. For example, an OR gate with two inputs having the values of 0 and 1 will have in its output the value of 1. Now, if the 0 input is changed to a 1, the output remains a 1 and the propagation time is zero. If, on the other hand, the 1 input changes to a 0, the output will go through a transition from 1 to 0. The maximum time of this transition period is the maximum propagation time of the OR gate. When logic gates are interconnected and values of inputs cannot be predicted, the propagation time between the inputs and the outputs cannot be predicted exactly. However, we can predict the maximum propagation time by evaluating the longest possible transition through the various gates. The propagation time is variable and depends on the circuit components and the input combination. In clocked sequential systems, one must always wait for the maximum possible propagation time before the output signal is triggered into a flip-flop.

Let us consider the operation of a parallel adder as shown in Fig. 9-1 or 9-3. Initially, one number is in the A register and the other number is transferred into the B register. As soon as the outputs of the flip-flops in register B settle down to their appropriate values, all the outputs A_i and B_i are available as inputs to the combinational circuit. The outputs of the combinational circuit at any time have the value either of logic-0 or logic-1, or are in transition between these two states. These outputs settle down to a constant value only after the signal propagates through the appropriate gates. In other words, the outputs of the combinational circuit represent the sum bits only after the signals have been propagated. Consider the sum bit from a single stage, say stage 15. A_{15} and B_{15} are available in their final form, but C_{15} will not settle to its final form until the correct C_{14} is available in its final form. Similarly, C_{14} has to wait for C_{13} and so on down to C_2. Thus only after the carry is propagated can we apply the ADD command pulse so that at the occurrence of the next clock pulse the correct sum bits are transferred into the A register.

Let us look at a specific example. Assume that two 36-bit numbers have to be added. Therefore, registers A and B must have 36 flip-flops each. Looking at one typical stage from Fig. 9-3, we see that the carry to the next stage C_{i+1} must propagate through two levels. The first level consists of three AND gates; the second, of one OR gate. Let us assume that the maximum propagation time through each gate is 20 nsec (1 nano-second = 10^{-9} seconds). From the time that the previous carry C_i settled down until the next carry C_{i+1} settles down, the signals must propagate through one level of AND gates and one OR gate with a maximum propagation time of 40 nsec. The carry must propagate through all 36 stages, giving a maximum delay of 1440 nsec.* Thus, the ADD command signal must wait at least 1.44 μsec from the time the signals of register B settle down before it can be applied. Otherwise, the outputs of the combinational circuits may not represent the correct sum bits. Assume that the master-clock has a frequency of 2 MHz, so that a clock pulse occurs every 500 nsec. If clock pulse t_0 (at time 0) was used to transfer the addend to register B, and assuming that the flip-flops have a maximum propagation time of 50 nsec, the total delay encountered is 1490 nsec. This is equivalent to a delay of three clock pulses and the ADD signal should not occur prior to clock pulse t_3.

The carry propagation time is a limiting factor on the speed by which two numbers can be added in a parallel digital system. Since all other arithmetic operations are implemented through successive additions, the time consumed during the addition process is very critical. One of the major problems encountered in the design of an accumulator is the reduction of the time for addition. One solution is to design digital circuits with reduced propagation time. For instance, the reduction in the maximum propagation time through a gate from 20 to 10 nsec reduces the carry propagation time from 1440 to 720 nsec. But physical circuits have a limit to their capability. Logic designers, however, have come out with ingenious schemes to

*The propagation in the last stage is through the gates that form the sum; we neglect the propagation time through the inverter of Fig. 9-3.

reduce the time for addition, not necessarily by means of faster circuits but by the incorporation of additional logic circuits.**

One way of reducing the carry propagation time is by a method known as *carry look-ahead*. In this method as in all other methods, increase in speed is traded with increase in equipment and complexity. We have discussed previously the advantages of partitioning a digital system into smaller and possibly similar subunits. So far in our discussion the accumulator register and its associated logic were partitioned into n typical stages, each having one flip-flop and associated combinational logic. Now consider a partition of an n-bit register into j groups of k bits each so that $j \cdot k = n$. As a specific example, let $n = 36$, $j = 12$, $k = 3$. Each group will consist of three flip-flops and their associated combinational logic for a total of 12 groups. The carry for the least significant group is given by the following Boolean functions:

$$C_1 = 0 \tag{9-4}$$
$$C_2 = A_1 B_1 \tag{9-5}$$
$$C_3 = A_2 B_2 + (A_2 + B_2)\, C_2 \tag{9-6}$$
$$C_4 = A_3 B_3 + (A_3 + B_3)\, C_3 \tag{9-7}$$

The input carry to the second group is C_4. Now, instead of waiting for C_4 to propagate through the first group, which will require five levels of gates, it can be generated from a combinational circuit with inputs A_1, A_2, A_3, and B_1, B_2, B_3. The Boolean function for this circuit is easily obtained by substituting C_2 from Eq. 9-5 into Eq. 9-6, and then substituting C_3 from Eq. 9-6 into Eq. 9-7. The final result, after rearranging, is a Boolean expression which is a function of the output variables of the flip-flops in group 1.

$$\begin{aligned} C_4 = {} & A_3 B_3 + A_2 A_3 B_2 + A_1 A_2 A_3 B_1 + A_1 A_3 B_1 B_2 + \\ & A_2 B_2 B_3 + A_1 A_2 B_1 B_3 + A_1 B_1 B_2 B_3 \end{aligned}$$

Since the Boolean function is expressed in the sum of products form, it can be implemented with one level having seven AND gates followed by one OR gate. The maximum propagation time for this curcuit is 40 nsec (using the figure from the previous example). The carry input to the third group can similarly be obtained from a two-level realization of a Boolean expression as a function of C_4, A_4, A_5, A_6, and B_4, B_5, B_6. Thus each group produces not only its own carries but also the carry for the next group ahead of time. By this scheme the maximum total carry propagation is 440 nsec for the carry to reach the input of the last (12th) group plus 120 nsec for the carry to propagate out of the last group, for a total of 560 nsec. This is compared with the delay of 1440 nsec obtained for single-stage

**The detailed description of all possible schemes is beyond the scope of this book. The interested reader is referred to the literature (references 1-5) for details.

partitioning. Obviously, if the register is partitioned into groups with more flip-flops, the carry propagation may be reduced even further, but again one must pay the price of more equipment. As mentioned previously this is not the only scheme available. Even this scheme can be improved by partitioning into groups within groups.

9-4 SERIAL ADDITION

A digital system is said to operate in a serial fashion when information is transferred and manipulated one bit at a time. The contents of one register are transferred to another by shifting the bits out of one register and into the other. Similarly, any operation to be done on a single register or between two registers is manipulated one bit at a time and the result transferred serially into a shift register. As a general rule, digital systems that operate in a serial mode require far less equipment than systems operating in a parallel mode, but their speed of operation is slower.

A block diagram of a serial adder is shown in Fig. 9-7. It consists of a shift-right accumulator register and a logic circuit capable of adding two bits and a previous carry. The augend is stored in the accumulator register and the addend transferred from a serial memory unit which, for this application, can be thought of as just another shift register. The sum formed is transferred into the accumulator and replaces the previously stored augend.

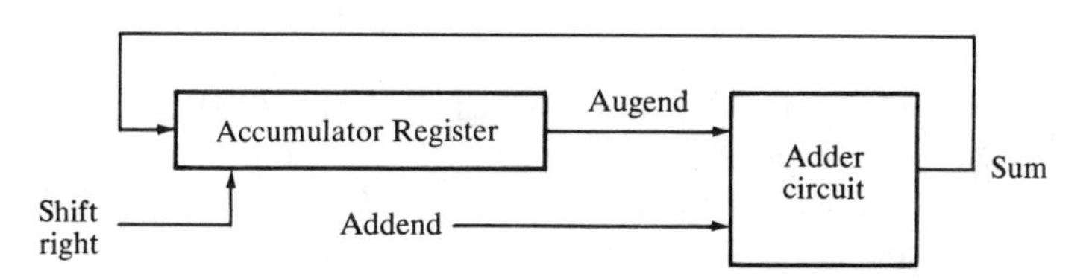

Figure 9-7 Block diagram of a serial adder

At the start of the addition process, the two least significant bits of the augend and addend are presented as inputs to the adder circuits. As the accumulator register is shifted to the right, the sum bit formed in the adder circuit enters the left-most flip-flop of the accumulator, and the next bit of the addend appears from the serial memory unit. The accumulator register proper is a shift-right register and its implementation is straightforward. The block marked "adder-circuit" may be thought of as a FA circuit that produces the sum but must also possess, in addition, one storage element for temporary storage of the carry bit. Since the adder circuit must include a memory element, it must be sequential. This sequential circuit may be derived intuitively when we realize that the carry-out of the FA can be stored in a flip-flop whose output can be used as the previous carry for the

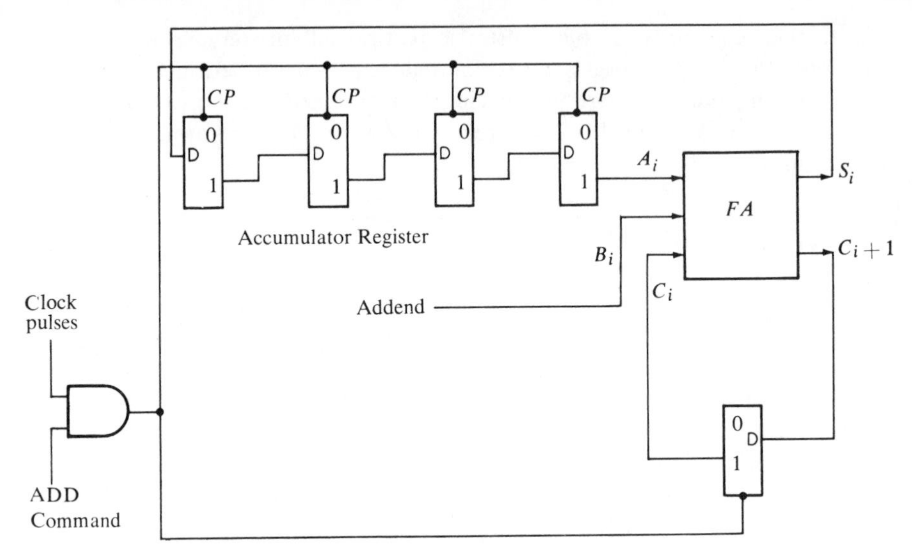

Figure 9-8 Four-bit serial adder

next significant bits. Thus the circuit of Fig. 9-8 is obtained with the "adder circuit" of Fig. 9-7 implemented as a sequential circuit consisting of a single D flip-flop together with a combinational FA circuit.

The operation of the serial adder of Fig. 9-8 will now be described. The accumulator initially holds the augend, and the addend is transferred from a serial memory unit or from another shift-register. At the start of the addition process, the least significant bits are presented to the input of the FA circuit, the carry flip-flop C is cleared, and the ADD command signal enabled. The signal propagates through the combinational circuit and produces a sum bit and a carry bit. The sum bit is returned to the most significant position of the accumulator register and the carry bit is applied to flip-flop C. When a clock pulse arrives, the sum bit is transferred into the left-most flip-flop of the accumulator, the carry into flip-flop C, the augend shifted one position to the right, and the next bit of the addend arrives from memory. The output of flip-flop C now holds the previous carry. When the next clock pulse occurs, the new sum bit to enter the left-most position of the accumulator is the one corresponding to the second significant position. This process continues until the fourth clock pulse transfers the last bit of the sum into the accumulator. At this time, the ADD command signal is disabled. Thus the addition is accomplished by passing each pair of bits together with the previous carry through a single FA circuit and transferring the sum to the accumulator.

The question now arises, is it possible to reduce the number of gates in the FA of Fig. 9-8 by using a different type of flip-flop? The section of the circuit that produces the sum cannot be simplified because the flip-flop that holds the addend bit is different from the one that receives the sum bit. However, the combinational circuit that produces the carry is applied to a flip-flop that holds the previous carry and may be simplified if some other type of flip-flop is used. The *D* flip-flop was originally chosen because it resulted in the most straightforward design. It is possible to show that if a *JK* or *RS* flip-flop is used, a reduction in the number of combinational gates will ensue. The input functions of a *JK* flip-flop can be derived directly from a state diagram. As mentioned previously, the portion of the serial adder that forms the sum bit is a sequential circuit with at least one flip-flop. Designating the carry flip-flop by the variable C, and noting that the inputs to the circuit are the augend bit A_1 and the addend bit B_1, the state table of Fig. 9-9(a) is obtained. The next state of C is the carry for the next significant bits and is derived from the present state of C (input carry) and the two inputs. A column for the output S could be included in this table but was deleted because the sum output will obviously be the same as Eq. 9-1. The excitation table for the two inputs of the flip-flop is then obtained from inspection of the present-state and the next-state. The input functions are simplified in the maps of Fig. 9-9(b), from which we obtain

$$JC = A_1 B_1$$
$$KC = A_1' B_1'$$

Present state	inputs		Next state	Flip-flop inputs	
C	A_1	B_1	C	JC	KC
0	0	0	0	0	X
0	0	1	0	0	X
0	1	0	0	0	X
0	1	1	1	1	X
1	0	0	0	X	1
1	0	1	1	X	0
1	1	0	1	X	0
1	1	1	1	X	0

(a)

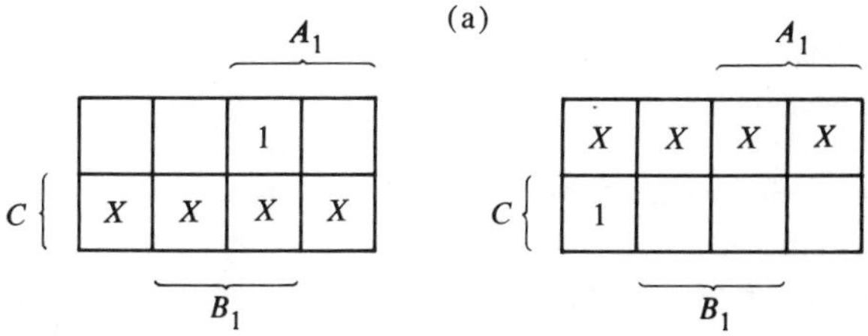

Map for JC (A_1 over the third and fourth columns, B_1 under the second and third columns):

			1	
C	X	X	X	X

$JC = A_1 B_1$

Map for KC (A_1 over the third and fourth columns, B_1 under the second and third columns):

	X	X	X	X
C	1			

$KC = A_1' B_1'$

(b)

Figure 9-9 Excitation table and maps for the *JK* carry flip-flop in a serial adder

Thus, only two AND gates are required to store the next carry in a *JK* (or *RS*) flip-flop as compared to four gates for the *D* flip-flop. This result could be obtained intuitively from inspection of the state table of Fig. 9-9(a) by noting that the next state is 1 when both inputs are 1, that the next state is 0 when both inputs are 0, and that the carry flip-flop does not change states otherwise.

9-5 PARALLEL VERSUS SERIAL OPERATIONS

Before we continue with the design of a multipurpose parallel accumulator register, let us pause to consider the difference between parallel and serial operations. In a parallel adder, all inputs and outputs of the register are available for manipulation at the same time. The register in the serial adder communicates with external circuits only through its left-most and right-most flip-flops and the addition must be processed with each pair of significant bits one at a time. This property is inherent in the two processes, not only in adder circuits but in other operations as well. The registers in serial digital systems are shift registers. The external circuit that performs the operation receives the bits sequentially at its inputs. The time interval between adjacent bits is called the "bit-time," and the time required to present the whole number is called the "word-time." These timing sequences are generated either by the serial memory unit or by the control section of the system. In a parallel operation, the ADD command signal is enabled during one pulse interval to transfer the sum bits into the accumulator upon the occurrence of a single clock pulse. In the serial adder, the ADD command signal must be enabled during one whole word-time, and a pulse applied every bit-time to transfer the generated sum bits one at a time. Moreover, the external circuits that perform operations and change contents of registers are always combinational in parallel systems. In serial computers, these external circuits would be sequential (requiring extra flip-flops) if their outputs depended not only on the present values of the inputs but also on values of previous inputs. This has been shown in the serial adder, where the output sum bit is a function of the previous carry, which in turn is a function of the previous inputs.

A comparison between the parallel adder shown in Fig. 9-1 and the serial adder of Fig. 9-8 will give some indication of the relative cost and speed between the modes of operation. Consider again a 36-bit accumulator and a master-clock pulse generator with a frequency of 2 MHz. Assume that the maximum propagation time through any gate is 20 nsec and that the settling time of a flip-flop is 50 nsec. For the parallel adder, we had to wait three clock pulses, or 1.5 μsec, to complete the addition. The serial adder required the application of 36 clock pulses for a total time of 18 μsec to complete the addition, a speed advantage of the parallel over the serial adder of 12 to 1. On the other hand the number of FA circuits is 36 to 1, thus the added equipment is greater than the speed advantage. Obviously, we can increase the speed of the parallel adder considerably if faster circuits are used and the carry propagation reduced. Circuits that add two 36-bit numbers in 100 nano-seconds can be thus attained. The speed of the serial adder can also be increased by increasing the frequency of the master-clock generator. The limit on this frequency is a function of the maximum propagation time of the flip-flops plus the combinational gates. In this example, the next sum bit will be available after 50 nsec for the flip-flops to settle down plus 40 nsec for the sum and carry to propagate through two levels of gates, for a total of 90 nsec. This gives a maximum frequency for the master-clock of 11 MHz, a bit-time of 90 nsec, and a word-time for 36 bits of 3.24 μsec. From this example one can clearly see that although, as a general rule, serial operations tend to be slower than parallel operations, there may be some operations that are done faster serially if one uses high-speed circuits compared to slower circuits used in the parallel counterpart.

9-6 ELEMENTARY OPERATIONS

The most important function of an accumulator is to add two numbers. Other arithmetic operations can be processed using this basic operation in conjunction with other elementary operations. Subtraction can be done by addition and complementation, multiplication is reduced to repeated additions and shifting, and division can be processed by repeated subtractions and shifting. The 2's complement of a number stored in a register can be obtained by complementing the register and then incrementing it by 1. Logical operations on the individual bits of a register are useful for extracting and packing parts of words. A digital computer incorporates these elementary operations together with other operations in its instruction list. Instructions such as multiply and divide are usually processed internally in computer registers by repeated applications of elementary operations. However, the availability of the elementary operations to the user allows him to specify his own sequence of repeated operations and thus extend the processing capabilities of the machine. To save equipment, some small computers do not include multiplication or division instructions as part of their hardware. In this case, it is up to the user to specify in his program the sequence of elementary operations needed for their execution.

A set of elementary operations for an accumulator register is listed in Table 9-1. The command signals p_1 to p_9 are generated by control logic circuits and should be viewed as inputs to the accumulator for the purpose of initiating the corresponding elementary operation. The nine control signals are assumed to be mutually exclusive; i.e., one and only one signal is enabled at any interval between two clock pulses.

Table 9-1 List of elementary operations for an accumulator

Control signal	*Operation*	*Symbolic designation*	
p_1	arithmetic addition	$(A) + (B) \Rightarrow A$	
p_2	clear	$0 \Rightarrow A$	
p_3	complement	$(\bar{A}) \Rightarrow A$	
p_4	logical AND	$(A) \wedge (B) \Rightarrow A$	
p_5	logical OR	$(A) \vee (B) \Rightarrow A$	
p_6	exclusive-or	$(A) \oplus (B) \Rightarrow A$	
p_7	shift-right	$(A_{i+1}) \Rightarrow A_i$	$i = 1, 2, \ldots, n\text{-}1$
p_8	shift-left	$(A_{i-1}) \Rightarrow A_i$	$i = 2, 3, \ldots, n$
p_9	increment	$(A) + 1 \Rightarrow A$	
z	check if zero	if $(A) = 0$; then $z = 1$	

There are two types of operations listed in the table. The unary operations–clear, complement, shift, and increment–require only one operand, the one stored in the A register. The binary operations-arithmetic addition and the three logical operations–require two operands, one stored in A and the other in B. All the operations will change the state of the A register except the last function listed in the table (an output function). Most of the elementary operations are self-explanatory. Arithmetic addition of two numbers has been discussed extensively in previous sections. The clear operation clears all flip-flops to 0. The complement operation changes the states of all individual flip-flops. Shift registers have been discussed previously. The increment operation increases the contents of the accumulator register by one. The logical operations have their usual meaning but they operate on individual bits of the register. For example, if the contents of a four-bit accumulator is 0011 and the contents of the B register is 0101, the AND operation changes the contents of the accumulator to 0001, the OR operation to 0111, and the exclusive-or to 0110. Note that the most elementary operation of a simple transfer from register B into the accumulator is not included in the table. This simple transfer can be implemented by clearing the accumulator and then ORing to it the contents of B.

A symbolic notation is given to each operation in the table. A double arrow designates a transfer into a register. The letter A is used for the accumulator register and the letter B for the register that stores the operand received from an external unit. A letter without a subscript refers to all n flip-flops of the register, while a subscripted letter refers to a single flip-flop. When enclosed in parentheses, the letter refers to the contents of the register or subregister. In order to distinguish between arithmetic plus and logical OR, the logical operation is given the $\vee$ symbol. The $\wedge$ symbolizes logical AND operation. The overbar symbolizes a bit-by-bit complementation operation. The symbolic notation of Table 9-1 conforms with the notation introduced in Sec. 8-2 and defined in Table 8-1.

A block diagram of the digital system that implements the elementary operations listed in the table is shown in Fig. 9-10. The accumulator includes a parallel A register together with its associated combinational circuit. The register proper consists of n flip-flops $A_1, A_2, \ldots, A_n$, numbered in ascending order starting from the right-most position. The inputs to the accumulator are: (1) a continuous clock pulse train for synchronization, (2) the nine command signals p_1 to p_9 with only one p_j being enabled at any clock pulse period, (3) n inputs $B_1, B_2, \ldots, B_n$ from register B, (4) n outputs $A_1, A_2, \ldots, A_n$ from register A, and (5) a single output z, whose output is logic-1 when the contents of the accumulator register is 0. The parallel accumulator will be partitioned into n similar stages with each stage consisting of one flip-flop A_i and its associated combinational circuit. In subsequent discussions only one typical stage will be considered with the understanding that an n-bit accumulator consists of n such typical stages with similar interconnections among neighbors. The extreme right-most and left-most stages have no neighbors and require special attention.

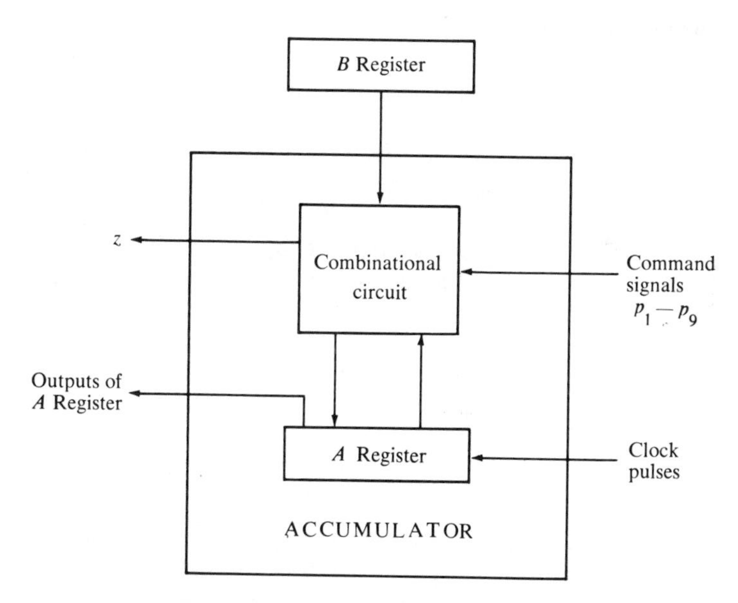

Figure 9-10 Block diagram of accumulator

9-7 ACCUMULATOR DESIGN

The implementation of each elementary operation listed in Table 9-1 is carried out in this section. The individual combinational circuits so obtained are incorporated into a unified system in the next section. The specific combinational circuits obviously depend on the type of flip-flop chosen for the A register. We shall employ JK flip-flops in the design that follows and leave the design with other types of flip-flops for problems at the end of the chapter.

For each elementary operation, there corresponds one and only one command signal p_j that activates it. This signal is generated by an external control unit and is considered as an input entering each stage of the accumulator. The presence of this input (i.e., when it is logic-1) dictates the required change of state of the A register upon the occurrence of the next clock pulse. Its absence (when it is logic-0) dictates no change of state. Therefore, the state diagrams to be used subsequently do not list the p_j input with the understanding that the next state is conditional upon its presence. Moreover, a binary operation takes the present values of A_i and B_i and stores the result of the operation in A_i. Only the A_i flip-flop changes its state in response to a command signal; the corresponding bit of B_i remains unaltered. Therefore, B_i is considered as an input to a single stage and is listed as such in the state diagrams.

p_1: Arithmetic Addition.

This operation has been discussed extensively in previous sections. We shall select the binary adder circuit of Fig. 9-3 with JK flip-flops instead of T. Since the JK flip-flop behaves as a complementing flip-flop when both inputs are excited simultaneously, it follows that both J and K must receive the same input received by the T flip-flop of Fig. 9-3. The input ADD command signal can now be replaced by the symbol p_1. The input functions to the flip-flop and the output carry to the next stage are:

$$JA_i = (B_iC_i' + B_i'C_i)\, p_1 \tag{9-8a}$$

$$KA_i = (B_iC_i' + B_i'C_i)\, p_1 \tag{9-8b}$$

$$C_{i+1} = A_iB_i + A_iC_i + B_iC_i \tag{9-9}$$

p_2: Clear.

The clear command signal clears all flip-flops of the A register; in other words, it changes the contents of the register to 0. To cause this transition in a JK flip-flop we need only apply command signal p_2 to the K input during one clock pulse period. The logic diagram for this operation shown in Fig. 9-11 is obvious and required no design effort. The input functions are:

$$JA_i = 0 \tag{9-10a}$$

$$KA_i = p_2 \tag{9-10b}$$

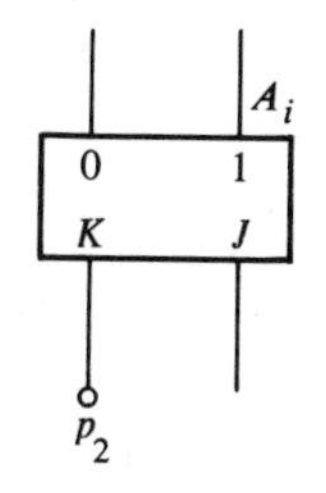

Figure 9-11 Clear

p_3: Complement

The complement command signal changes the state in each flip-flop of the A register. To cause this transition in a JK flip-flop we need to apply command signal p_3 to both inputs. The logic diagram is shown in Fig. 9-12. The input functions are:

$$JA_i = p_3 \tag{9-11a}$$

$$KA_i = p_3 \tag{9-11b}$$

p_4: Logical AND

This operation forms the bit-by-bit logical AND between A_i and B_i and transfers the result to A_i. The state table and input excitation for this operation are given in Fig. 9-13(a). The input functions are derived in the maps of Fig. 9-13(b):

$$JA_i = 0 \tag{9-12a}$$

$$KA_i = B_i'p_4 \tag{9-12b}$$

and the logic diagram is shown in Fig. 9-13(c). The procedure for obtaining the state diagram, input excitation, the simplified input Boolean functions, and the logic diagram is straightforward and follows the procedures outlined in Ch. 7.

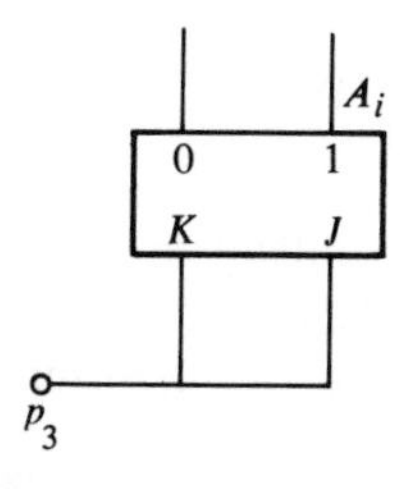

Figure 9-12 Complement

Present state	Input	Next state	Flip-flop inputs	
A_i	B_i	A_i	JA_i	KA_i
0	0	0	0	X
0	1	0	0	X
1	0	0	X	1
1	1	1	X	0

(a)

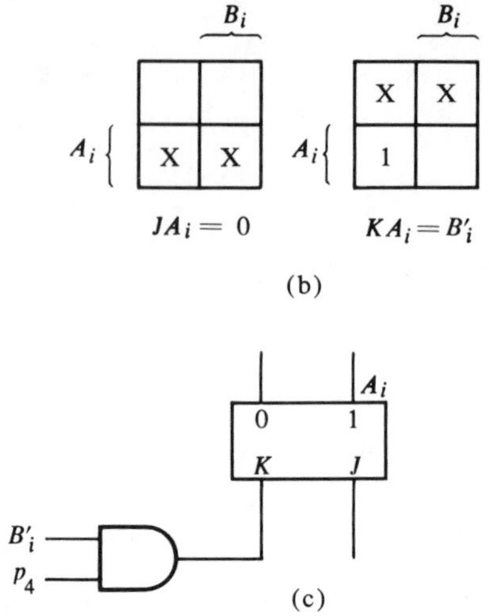

(b)

(c)

Figure 9-13 Logical AND

p_5: Logical OR

This operation forms the bit-by-bit logical OR between A_i and B_i and transfers the result to A_i. Figure 9-14 shows the state table, input excitation, the map minimization, and the logic diagram for the OR elementary operation. The input Boolean functions are:

$$JA_i = B_i p_5 \quad (9\text{-}13a)$$

$$KA_i = 0 \quad (9\text{-}13b)$$

p_6: Exclusive-or

This operation forms the bit-by-bit logical exclusive-or between A_i and B_i and transfers the result to A_i. The pertinent information for this operation is shown in Fig. 9-15. The Boolean input functions are:

$$JA_i = B_i p_6 \quad (9\text{-}14a)$$

$$KA_i = B_i p_6 \quad (9\text{-}14b)$$

p_7: Shift-right

This operation shifts the contents of the A register one place to the right. Clearly, this operation transfers the content of A_{i+1} into A_i for $i = 1, 2, \ldots, n - 1$, with the transfer into A_n unspecified.* The

*The data transferred to A_n when shifting right or to A_1 when shifting left determines the type of shift used. This is discussed in Sec. 9-9.

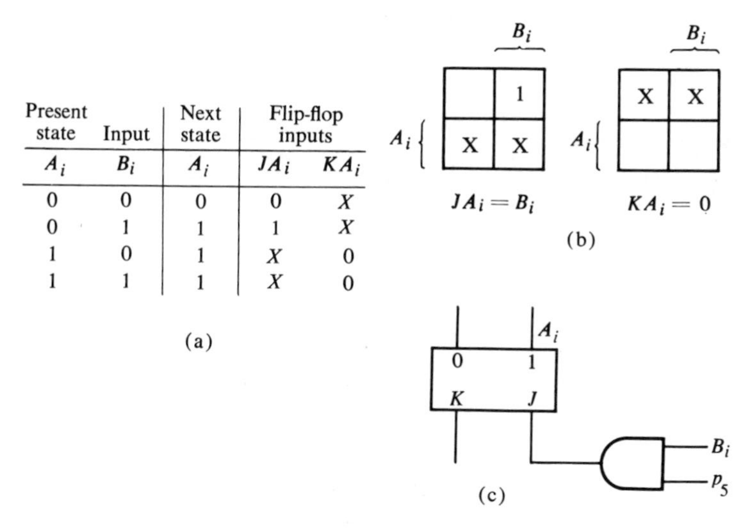

Present state	Input	Next state	Flip-flop inputs	
A_i	B_i	A_i	JA_i	KA_i
0	0	0	0	X
0	1	1	1	X
1	0	1	X	0
1	1	1	X	0

Figure 9-14 Logical OR

Present state	Input	Next state	Flip-flop inputs	
A_i	B_i	A_i	JA_i	KA_i
0	0	0	0	X
0	1	1	1	X
1	0	1	X	0
1	1	0	X	1

(a)

$JA_i = B_i$ $KA_i = B_i$

(b)

(c)

Figure 9-15 Exclusive-or

excitation table for this operation is shown in Fig. 9-16(a). Note that A_{i+1} is an input to stage i and that the next state is equivalent to the input. The map minimization is obtained in Fig. 9-16(b) and the logic diagram in Fig. 9-16(c), where the shift-right is executed only when command signal p_7 is enabled. The input functions are:

$$JA_i = A_{i+1}p_7 \quad (9\text{-}15a)$$

$$KA_i = A'_{i+1}p_7 \quad (9\text{-}15b)$$

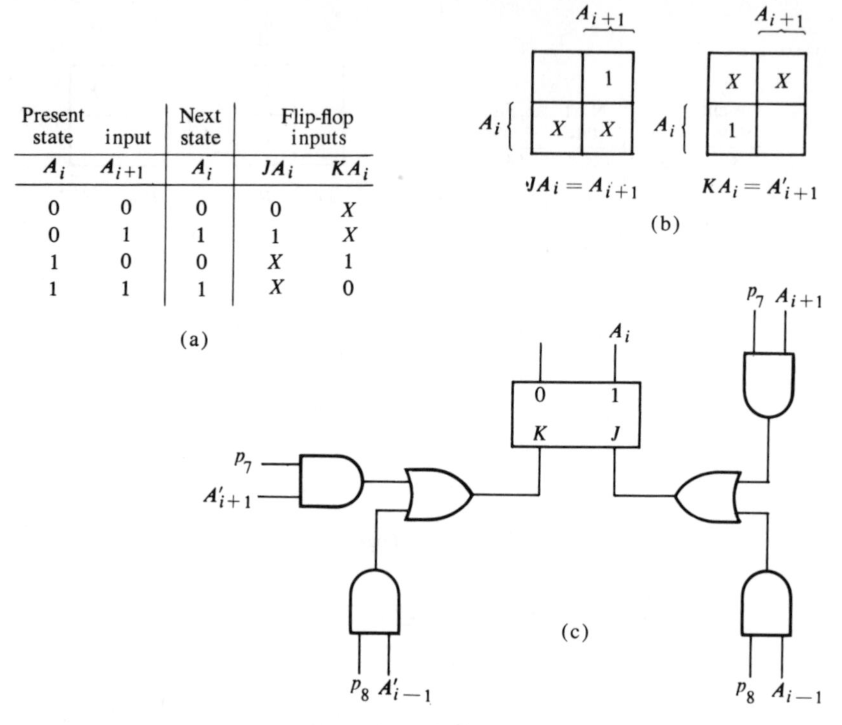

Present state	input	Next state	Flip-flop inputs	
A_i	A_{i+1}	A_i	JA_i	KA_i
0	0	0	0	X
0	1	1	1	X
1	0	0	X	1
1	1	1	X	0

Figure 9-16 Shift-right

p_8: Shift-left

This operation shifts the contents of the A register one place to the left. The design procedure here is the same as the shift-right operation with A_{i-1} replacing A_{i+1} and will not be repeated. The logic diagram is shown in Fig. 9-16(c), with command signal p_8 initiating the shift-left operation. The input functions are similar to Eq. 9-15:

$$JA_i = A_{i-1}p_8 \tag{9-16a}$$

$$KA_i = A'_{i-1}p_8 \tag{9-16b}$$

p_9: Increment

This operation increments the contents of the A register by one; in other words, the register behaves like a synchronous binary counter with p_9 being the count command signal. In Sec. 6-5 it was shown that the flip-flop input functions for a synchronous binary counter are:

$$JA_1 = KA_1 = p_9$$

$$JA_i = KA_i = p_9 \cdot \prod_{j=1}^{i-1} A_j \qquad i = 2, 3, \ldots, n \tag{9-17}$$

where Π is a product sign designating the AND operation and the A_j's are the outputs of all less significant flip-flops. The Boolean function specified by Eq. 9-17 is implemented by one AND gate per stage. The number of inputs in each gate is a function of the flip-flop position in the register. If we number the position of the individual flip-flops from 1 to n starting from the right-most element, the number of inputs to an AND gate is equal to its position number. Thus, for stage number 2 the gate requires two inputs, one from A_1 and one from p_9. But the last stage requires an AND gate with n inputs, one from each of the previous flip-flops and one from p_9. It is interesting to note that the maximum propagation delay with this configuration is equal to the propagation time of only one AND gate. This arrangement is similar to the carry look-ahead scheme discussed in Sec. 9-3 and achieves the least waiting time for the carry to propagate through all the stages. It is possible to decrease the number of inputs to the AND gates if one can afford a longer propagation delay. The concept of partitioning into similar subunits with carry look-ahead is applicable to counting circuits as well, since a counter is a special adder that adds a constant equal to one. The implementation of Eq. 9-17 implies no partitioning; i.e., all n flip-flops form a single group. By partitioning into groups with a smaller number of flip-flops, the number of inputs to the AND gates is reduced. The accumulator circuit to be designed in this chapter is purposely partitioned into n identical groups with one flip-flop per group. Therefore a modification of Eq. 9-17 is needed to make each AND gate in each stage similar to any other stage. This can be achieved by defining an input signal E_i into a stage and an output signal E_{i+1} out of each stage as shown in Fig. 9-17. In this configuration, a total of n-1 AND gates are used, but each gate has only two inputs. The disadvantage is that the maximum propagation delay increases considerably. It is equal to the time required for the signal to propagate through n-1 gates. On the other hand, it standardizes the counting circuit needed for each stage into an AND gate with only two inputs. The flip-flop input functions for a typical stage are easily obtained from inspection of Fig. 9-17,

$$JA_i = E_i \tag{9-18a}$$

$$KA_i = E_i \tag{9-18b}$$

with the additional requirement that each stage generates an output E_{i+1} to be used as an input for the next stage on its left.

$$E_{i+1} = E_i A_i \quad i = 1, 2, \ldots, n-1$$
$$E_1 = p_9 \tag{9-19}$$

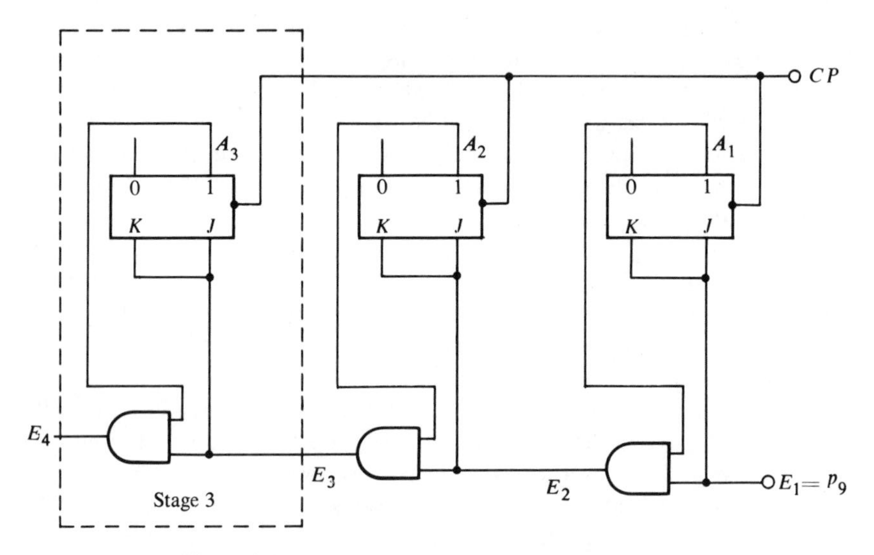

Figure 9-17 Counter circuit with carry propagation

z: Check if zero

This signal is an output from the accumulator and its function is different from the operations considered so far. It checks the content of the accumulator and produces an output z whose value is logic-1 when all the flip-flops of the register contain zeros. Obviously, this can be implemented with one AND gate having n inputs coming from the complement output of each flip-flop. The propagation delay of this circuit is the time required for the signal to propagate through this single AND gate. Again for the sake of standardization and partitioning into n identical stages we shall adopt the alternative implementation shown in Fig. 9-18. Here we are splitting the n-input AND gate into n AND gates of two inputs each and increasing the propagation delay to the time required for the signal to propagate through n gates. Each stage will then generate an output function z_{i+1} given by Eq. 9-20 to be used as an input for the next stage on its left.

$$z_{i+1} = z_i A_i' \quad i = 1, 2, \ldots, n$$
$$z_1 = 1 \tag{9-20}$$
$$z_{n+1} = z$$

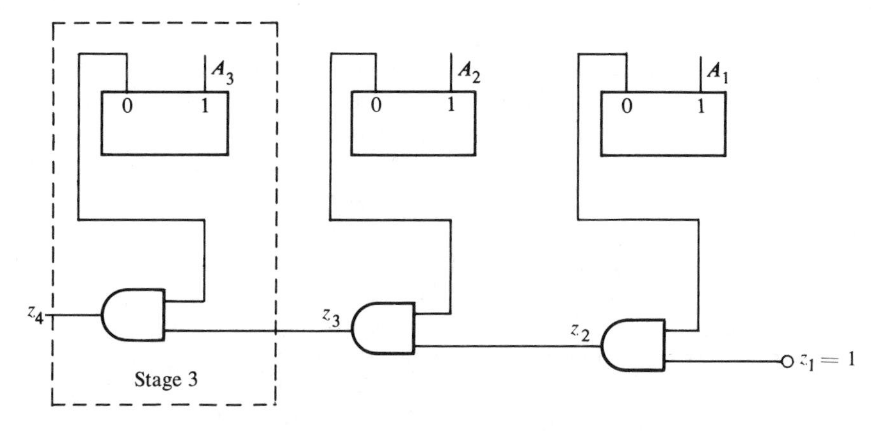

Figure 9-18 Chain of AND gates for checking zero content of register

9-8 COMPLETE ACCUMULATOR

The circuits for each individual operation derived in the last section must be integrated to form a complete accumulator unit. The reader is reminded that this digital unit is a multipurpose register capable of performing various elementary operations and although we have named it *accumulator*, any other name would serve as well. Now, the command signals p_1 to p_9 are mutually exclusive and therefore, their corresponding logic circuits can use the same flip-flop. To avoid interaction among the many outputs, an OR gate is inserted in each flip-flop unit. In other words, the logical sum of Eqs. 9-8, 9-10 to 9-16, and 9-18 is formed for each J and K input.

$$\begin{aligned} JA_i &= B_iC_i'p_1 + B_i'C_ip_1 + p_3 + B_ip_5 + B_ip_6 + A_{i+1}p_7 \\ &\quad + A_{i-1}p_8 + E_i \end{aligned} \tag{9-21a}$$

$$\begin{aligned} KA_i &= B_iC_i'p_1 + B_i'C_ip_1 + p_2 + p_3 + B_i'p_4 + B_ip_6 \\ &\quad + A_{i+1}'p_7A_{i-1}'p_8 + E_i \end{aligned} \tag{9-21b}$$

It is also necessary that in each stage we generate the output functions specified by Eqs. 9-9, 9-19, and 9-20.

$$C_{i+1} = A_iB_i + A_iC_i + B_iC_i \tag{9-22}$$

$$E_{i+1} = E_iA_i \tag{9-23}$$

$$z_{i+1} = z_iA_i' \tag{9-24}$$

These output functions are generated in stage i and applied as inputs to the next stage $i + 1$. The first stage placed in the least significant position has no neighbor on its right and must use as inputs logical values $C_1 = 0$, $E_1 = p_9$, and $z_1 = 1$. Boolean functions 9-21 to 9-24 give all the information necessary to obtain the logic diagram of each stage of the accumulator register.

The logic diagram of one typical stage of the accumulator register is shown in Fig. 9-19. Again, it should be emphasized that an n-bit register is comprised of n such stages, each interconnected with its neighbors. The single stage has one flip-flop A_i, an OR gate in each J and K input, and a variety of combinational circuits. There are six inputs to each stage: B_i is the input from the corresponding bit of register B, C_i is the carry from the previous stage on the right, A_{i-1} is the output of the flip-flop from the stage on the right, A_{i+1} is the output of the flip-flop from the stage on the left, E_i is the carry input from the increment operation, and z_i is used to form the output z. There are eight control inputs p_1 to p_8, one for each elementary operation listed in Table 9-1. There are four outputs: A_i is the output of the flip-flop, C_{i+1} is the carry for the next stage on the left, E_{i+1} is the increment carry for the next stage on the left, and z_{i+1} is applied to the next stage on the left to form the chain for the output z. There are three other inputs into each stage not directly related to the flow of information: clock pulses from the master-clock generator, at least one power supply input, and a ground terminal. This makes the total number of input and output leads in one stage equal to 21; the entire circuit could be incorporated into one IC chip with 21 terminals. The logic diagram is a direct implementation of the Boolean functions listed in Eqs. 9-21 to 24 and is self-explanatory.

The first stage $i = 1$ and the last stage $i = n$ need special consideration because they do not have neighbors on one side. The circuit shown in Fig. 9-19 can be used for these two stages and, although it may seem inefficient, it nevertheless facilitates the design to require the use of only standard modules. Some of the terminals in these two stages require inputs or outputs different from those specified for a typical stage. The input C_i in the first stage should be connected to 0, input E_i to p_9 and z_i to 1. The output z_{i+1} in the last stage is a logic-1 when the contents of the A register are zero. Output E_{n+1} is not of any use. Output C_{n+1} can be used as an overflow indicator. Inputs A_{i-1} of the first stage and A_{i+1} of the last stage need special consideration during shifting. Their connections determine the type of shift encountered.*

The three slowest functions of the accumulator are: the arithmetic addition, the incrementing process, and the chain that forms the z output. This is because the signals of these functions have to propagate through combinational gates before their outputs settle to the required values. Thus, enough time must be allowed for the carries to propagate through the combinational circuits ($2 \times n$ gates for addition and n-1 gates for incrementing) before the command signal p_1 or p_9 is enabled. Similarly, output z will not have the correct value until the signal propagates through n AND gates. Only in low-speed digital systems can the designer afford to use the elementary carry propagation circuits employed in Fig. 9-19. It was

*This subject is discussed further in the next section.

used in this chapter for simplicity and tutorial reasons. Improvement of the carry propagation time is required in many practical systems. This can be done by partitioning the accumulator into groups of more than one flip-

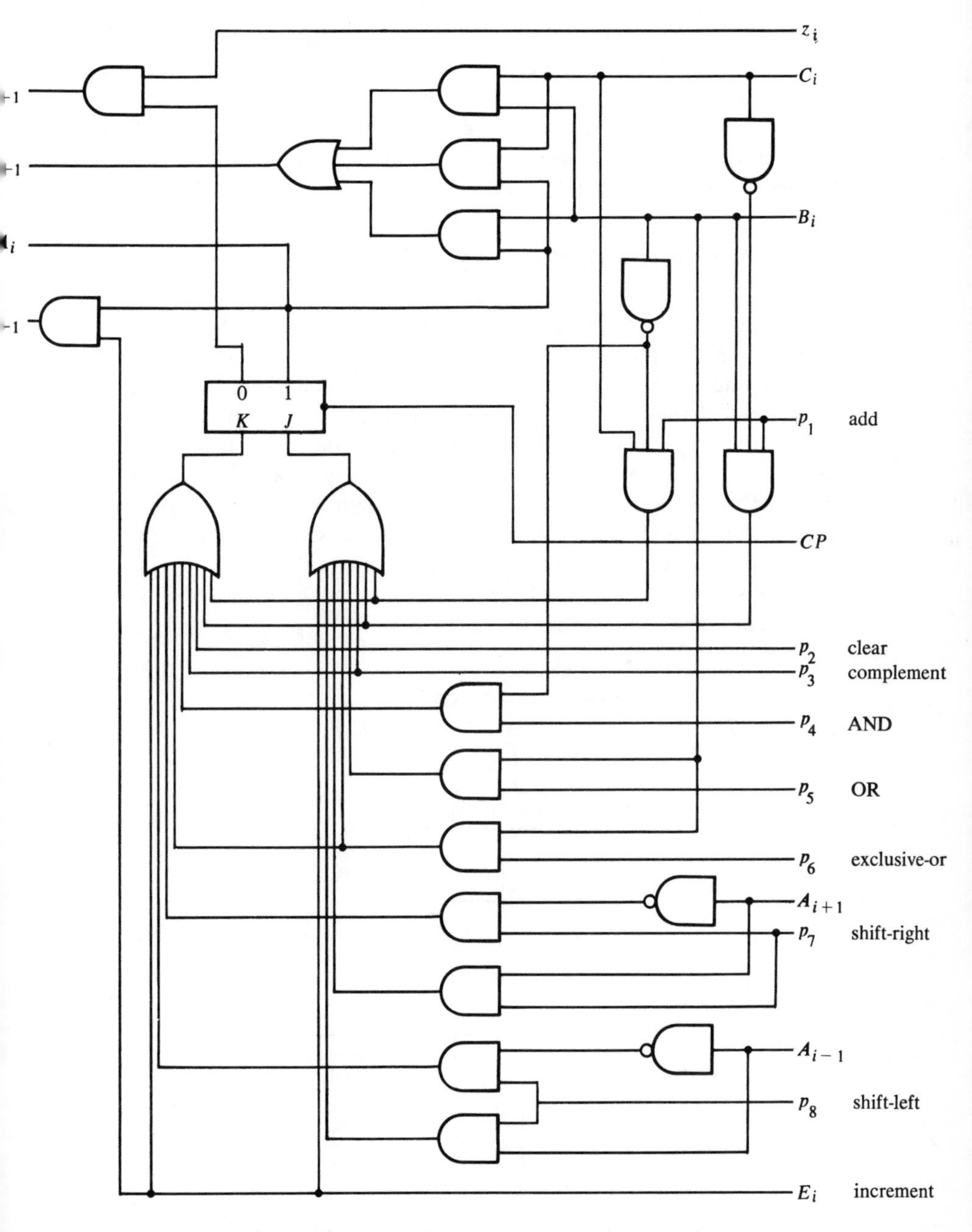

Figure 9-19 One typical stage of the accumulator register

flop, with each group generating a carry for the next group. Several other techniques can be used. In practice, one generally has an upper limit on the time allowed for the completion of an operation. Given this constraint, different schemes are examined to determine which will yield the speed required with the least amount of equipment. Then the simplest and most convenient technique is selected. It should also be pointed out that the development of IC technology facilitates the standardization of functions such as accumulators. General purpose registers that perform a variety of elementary operations are available in IC form as MSI or LSI devices (8, 9).

9-9 ACCUMULATOR OPERATIONS

The second operand needed for the binary operations in the accumulator is stored in the *B* register. This register, quite often, is the memory buffer register, whose function is to receive the word read from the memory unit or to hold the word to be stored into memory. One may think of the accumulator as a function register that accepts data read from memory and operates on it according to a programmed sequence of command signals generated in a control unit. Figure 9-20 helps clarify this concept. The accumulator environment consists of the *B* register and the control unit. The data for the various operations is transferred into the *B* register and the results from the accumulator are transferred back into the *B* register. The source and destination of data may be a memory unit, a set of

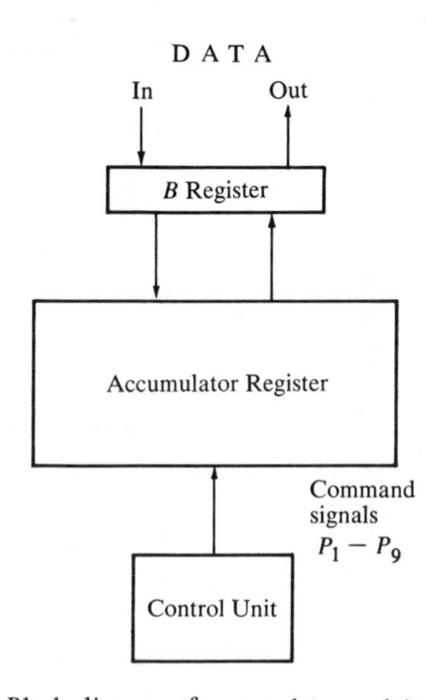

Figure 9-20 Block diagram of accumulator and its environment

terminals, or other registers. The sequence of operations is determined in the control unit, which generates command signals synchronized with the incoming data. The sequence of control signals may be either permanently wired or stored in memory as instructions. With this concept in mind, we shall proceed to describe some useful functions that can be processed in the accumulator. These processes may require either a single or a multiple sequence of elementary operations for their execution.

Simple Transfer

The simple transfer is not included in the list of elementary operations for the accumulator of Fig. 9-19. However, a simple transfer from register B into the accumulator can be implemented from the two elementary operations clear and add or clear and OR. Adding or ORing the contents of B to zero is equivalent to a straight transfer from B to A. The transfer from the accumulator register to B or any other register is easily implemented since the outputs of the A register are available as outputs from the accumulator. The transfer from any other register into the accumulator cannot be direct if the circuit of Fig. 9-19 is used since the inputs of the A register are not available as terminals. Some digital computers have a limited number of registers and such a transfer is not used. In larger systems with many other processing registers, such a transfer necessitates the availability of the input terminals of the A register.

Algebraic Addition and Subtraction

So far, in the discussion of binary addition, no mention was made of the sign of the operands. This was purposely avoided to decrease the complexity of the process. We shall now show that addition and subtraction of two signed binary numbers can be implemented in the accumulator.

The sign of a number is a binary quantity and can be stored in one flip-flop, with plus being represented by a 0 and minus by a 1. We shall use the left-most flip-flop of registers A and B to store the sign and use the complement representation for negative numbers as shown in Sec. 8-5; that is, negative numbers are represented either by 2's complement or by 1's complement.

The addition of two binary numbers with sign-complement representation is very simple and is stated by the following algorithms.*

1. *Addition with sign-2's complement representation.* The addition of two signed binary numbers with negative numbers represented by their 2's complement is obtained from the addition of the two numbers

*The proof of the two algorithms can be found in reference 3, Ch. 1. See also Prob. 12-19.

including their sign bits. A carry in the most significant (sign) bit is discarded.

2. *Addition with sign-1's complement representation.* The addition of two signed binary numbers with negative numbers represented by their 1's complement is obtained from the addition of the two numbers, including their sign bits. If there is a carry out of the most significant (sign) bit, the result is incremented by 1 and its carry discarded.

Addition and subtraction with sign-magnitude representation is somewhat more complicated and is covered in Sec. 12-5.

Numerical examples for addition with negative numbers represented by their 2's complement are shown below. Note that negative numbers must be initially in 2's complement representation, and that the sum obtained after addition is always in the required representation.

```
+ 6     0 000110          - 6    1 111010
               +                        +
+ 9     0 001001          + 9    0 001001
----    --------          ----   --------
+ 15    0 001111          + 3    0 000011


+ 6     0 000110          - 9    1 110111
               +                        +
- 9     1 110111          - 9    1 110111
----    --------          ----   --------
- 3     1 111101          - 18   1 101110
```

The two numbers in the four examples are added, including their sign bit. Any carry out of the sign bit is discarded and negative results are automatically in their 2's complement form.

The four examples are repeated below with negative numbers represented by their 1's complement. The carry out of the sign bit is returned and added to the least significant bit (end around carry).

```
+ 6     0 000110          - 6    1 111001
               +                        +
+ 9     0 001001          + 9    0 001001
----    --------                 --------
+ 15    0 001111                10 000010
                                        +
                                 └──────► 1
                          ----   --------
                          + 3    0 000011
```

```
+ 6     0 000110           - 9     1 110110
               +                          +
- 9     1 110110           - 9     1 110110
----    --------                  --------
- 3     1 111100               ┌─11 101100
                               │          +
                               └────────► 1
                           ----   --------
                           - 18   1 101101
```

The accumulator register can be used without modifications to add or subtract two signed binary numbers in 2's complement representation. For the 1's complement representation, an overflow flip-flop L is needed to store the end carry. One flip-flop in each register must be reserved to accomodate the sign of the two operands. By convention the two left-most flip-flops A_n and B_n are reserved for the signs, while the remaining registers store the magnitude of the operands in the chosen representation. If 2's complement representation is chosen, then an add-command signal p_1 will complete the addition (with carry C_{n+1} neglected). If 1's complement representation is used, the end-carry-out of C_{n+1} must be stored temporarily in the overflow flip-flop L to determine whether the result has to be incremented by 1. The value of C_{n+1} is transferred to flip-flop L with the add-command signal p_1. The procedure for adding two signed binary numbers in 1's complement representation is as follows: apply an add-command signal p_1, check the status of the overflow flip-flop; if it is set, increment the A register by 1 by applying a p_9 command signal.

Subtraction of two binary numbers can be accomplished by complementing the subtrahend and adding it to the minuend.* The 1's complement of a number stored in the accumulator is obtained from the application of a complement command signal p_3. The sign flip-flop A_n is also complemented, so that positive numbers change to negative and negative numbers to positive. To obtain the 2's complement of a number stored in the accumulator, two elementary operations are needed: the first complements the number (including the sign) and the second increments it by 1. A different procedure can be employed if the contents of register B are to be added to the 2's complement of the operand stored in the accumulator. The A register is first complemented, a signal equivalent to logic-1 is applied to the input carry C_1 of the first stage, and then an add-command signal is applied.

Subtraction of two signed numbers in complement form can be accomplished in one of two ways. In the first method, the subtrahend is first transferred into the accumulator, its complement formed, and then added to

*This procedure was explained in Sec. 1-5.

the minuend in the B register. This requires the following sequence of elementary operations:

For 2's complement representation:

Step	Elementary operation	Command signal	Remarks
1.	subtrahend $\Rightarrow B$		subtrahend transferred in
2.	$0 \Rightarrow A$	p_2	clear A
3.	$(A) + (B) \Rightarrow A$	p_1	transfer subtrahend to A
4.	$(\bar{A}) \Rightarrow A$	p_3	complement
5.	$(A) + 1 \Rightarrow A$	p_9	form 2's complement of subtrahend
6.	minuend $\Rightarrow B$		minuend transferred in
7.	$(A) + (B) \Rightarrow A$	p_1	forms the difference in A

For 1's complement representation:

Step	Elementary operation	Command signal	Remarks
1.	subtrahend $\Rightarrow B$		subtrahend transferred in
2.	$0 \Rightarrow A$	p_2	clear A
3.	$(A) + (B) \Rightarrow A$	p_1	transfer subtrahend to A
4.	$(\bar{A}) \Rightarrow A$	p_3	form 1's complement of subtrahend
5.	minuend $\Rightarrow B$		minuend transferred in
6.	$(A) + (B) \Rightarrow A$, $C_{n+1} \Rightarrow L$	p_1	add and set overflow flip-flop L if end-carry
7.	L: $(A) + 1 \Rightarrow A$	p_9	forms the difference in A

The difference formed in the accumulator may be transferred directly out or to register B and then to the external system, or may be left in the A register if needed for the next operation. Note that the only variation between the two sequences is in the step at which the incrementing command p_9 is applied. An extra overflow flip-flop L is required for the 1's complement and only if this flip-flop is set by an end-carry is the accumulator incremented.

A more efficient subtraction procedure results if the complement of the subtrahend is formed in register B instead of the accumulator. In this method the minuend is the first operand to be transferred to the accumulator; the subtrahend is left in register B. The subtrahend in register B is then complemented and added to the accumulator. For example, the arithmetic subtraction of two numbers from a third, X minus Y minus Z, is implemented with fewer steps by this procedure than by the previous method.

Comparison

An often-used function in digital data processing is the comparison of two numbers to determine their relative magnitude. The two numbers may be of the same or opposite signs. It is possible to design a digital circuit with three outputs to perform this function and produce a logic-1 in one and only one of three output terminals, depending on whether the first number is greater than, less than, or equal to the second number,* If an accumulator is available, the function can be implemented by the application of a sequence of elementary operations. The relative magnitude of two signed binary numbers X and Y may be found from the subtraction of Y from X and checking the sign bit of the difference in A_n. If it is a 1, the difference $X - Y$ is negative and $X < Y$; if it is a 0 either $X > Y$ or $X = Y$. A check of the z output distinguishes between these two possibilities, with $z = 1$ making it an equality. However, this last procedure is valid only in the 2's complement representation. In the 1's complement case, a zero may manifest itself in two forms; either by a number with all 0's or by a number with all 1's. Thus accumulators with 1's complement representation require an additional circuit for the z output in order to detect the second type of arithmetic zero.

Shifting Operations

The elementary operations of shift-right and shift-left as defined in Table 9-1 did not specify the data transferred to the left-most and right-most flip-flops of the register. The information transferred to input A_{i+1} of the nth flip-flop or the input A_{i-1} of the 1st determine the type of

*See Sec. 4-6.

shift implemented. A *logical shift* is one that inserts 0's into the extreme flip-flops. Therefore a logical shift-right command necessitates the application of a logic-0 to terminal A_{i+1} in the nth stage, while a logical shift-left requires the application of a logic-0 to terminal A_{i-1} of the 1st stage. A *circular shift* circulates the bits around the two ends. Therefore, for a circular left shift, one has to connect the output A_n of the nth stage to input A_{i-1} of the 1st stage. A circular right shift requires that output A_1 of the first flip-flop be connected to input A_{n+1} of the last. An *arithmetic shift* shifts only the magnitude of the number, without changing its sign. This type of shift is also called *scaling* or *shift with sign extension*. An arithmetic left shift multiplies a binary number by 2, and an arithmetic right shift divides it by 2. Remember that flip-flop n holds the sign of the number and that the magnitude of a negative number may be represented in one of three different ways. In sign-magnitude representation, the arithmetic shifts are logical shifts among the first $n - 1$ flip-flops. In sign-2's complement representation, a right shift leaves the sign bit unchanged and a left shift transfers 0's into stage 1. In sign-1's complement representation, a right shift leaves the sign bit unchanged and a left shift transfers the sign bit into the least significant bit position.*

Logical Operations

The logical operations are very useful for manipulating on individual bits or on part of a word stored in the accumulator. A common application of these operations is the manipulation of selected bits of the accumulator specified by the contents of the B register. Three such functions are the selective set, selective clear, and selective complement. A *selective-set* operation sets the bits of the A register only where there are corresponding 1's in the B register. It does not affect bit positions which have 0's in B. The following specific example may help to clarify this operation.

1100	B
1010	A before
1110	A after

The two left-most bits of B are 1's and so the corresponding bits of A are set. One of these two bits was already set and the other has been changed from 0 to 1. The above example may serve as a truth table from which one can clearly deduce that the selective-set operation is just an OR elementary operation. The *selective-complement* operation complements bits in A only where there are corresponding 1's in B. For example:

*The proofs of these algorithms are left for a problem at the end of the chapter.

1100	B
1010	A before
0110	A after

Again the two left-most bits of B are 1's and so the corresponding bits of A are complemented. This example again can serve as a truth table from which one can observe that the selective-complement operation is just an exclusive-or elementary operation. The *selective-clear* operation clears the bits of A only where there are corresponding 1's in B. For example:

1100	B
1010	A before
0010	A after

Again the two left-most bits of B are 1's and so the corresponding bits of A are cleared to 0. One can deduce that the logical operation performed on the individual bits is $B_i'A_i$. To implement the selective-clear operation, it is necessary to complement register B and then AND to A.

1100	B
0011	B'
1010	A before
0010	A after ANDing

If register B cannot be complemented, one can use De Morgan's theorem to express the operation as $(B_i + A_i')'$, which requires three elementary operations, complement A, OR B to A, and complement A again.

The logical operations are useful for packing and unpacking parts of words. For example, consider an accumulator register with 36 flip-flops receiving data from a punch card with 12 bits per column. The data from any three columns can be packed into one word in the accumulator. Other input systems transfer data in alphanumeric code consisting of six bits per character, so that six characters can be packed in the register. For such data, one may want to pack or unpack the individual items for various data processing applications. The packing of data may be accomplished by first clearing the A register and then performing an OR operation with a single item stored in B. If the individual items are placed in the least significant position of B, the next operation on A requires shifting the item to the left before ORing another item from B. To pack three columns of a card with 12 bits in each column, one requires three OR operations and three logical shift-left operations, with the shifting done on 12 bits at a time. The unpacking of words requires the isolation of items from a packed word. An individual item may be isolated from an entire word by an AND operation (sometimes called *masking*) with a word in B that has all 1's in the corresponding bits of the item in question and 0's everywhere else.

9-10 CONCLUDING REMARKS

In this chapter, a multipurpose accumulator register was specified and designed. The usefulness of such a register was demonstrated by several data processing examples. The most common operation encountered in an accumulator is arithmetic addition of two numbers. For this reason and because, relatively, this is the most involved function to design, a major portion of the chapter was devoted to this subject. To simplify the presentation, unsigned binary numbers were first considered. Only later in Sec. 9-9 was it shown that the same circuit can be used to add or subtract two signed binary numbers. Both serial and parallel adders were introduced and their differences discussed. The serial adder was found to require less equipment than its parallel counterpart, but, except for a few small digital systems, most other computers employ parallel adders because they are faster. The timing problems encountered in parallel adders as a result of the carry-propagation delay was explained and a method for decreasing this delay was briefly mentioned. The many possible techniques available for high-speed parallel addition are too numerous for inclusion in this text. The serious reader is advised to supplement this material with the references at the end of the chapter.

A set of elementary operations was specified in Sec. 9-6. This set is typical of operations of many accumulators found in digital computers. In some small computers the entire arithmetic unit or a large portion of it encompasses one such register. In larger systems, the accumulator is only a part of the central processing unit and sometimes more than one accumulator is available. In multiple accumulator systems it is customary to abbreviate the names of the registers by a letter and a number such as A1, A2, A3, or R1, R2, R3, etc. Although only nine elementary operations were specified for the accumulator register in this chapter, one must realize that many other operations could be included. Other possible operations are: transfer from and operations with other registers, shifting by more than one bit, add-and-shift combined operation (useful for multiplication), direct circuits for binary subtraction, and other logical operations. These operations are not chosen arbitrarily but emerge from the system requirements. The procedure for determining required register operations from the system specifications is clarified in Ch. 11, where a specific digital computer is specified. From these specifications, the register operations are determined.

The accumulator register specified in Sec. 9-6 was designed in Secs. 9-7 and 9-8. Two reduction procedures were utilized to simplify the design: the partition of the register into smaller and similar subregisters and the separation of the functions into mutually exclusive entities. The partitioning of the n-bit register into n similar stages, and the initial separation of the logic design for each elementary operation as outlined in Sec. 9-7, has reduced the design process to a very simple procedure. This methodology is basic in digital logic design and is the most important single lesson to be learned from this chapter. This procedure is employed again in Ch. 11. It is demonstrated there again that the partitioning and separation techniques simplifies logic design and facilitiates the derivation of the combinational circuits among the various registers.

It should be pointed out that digital computers may have processing units that operate on data different than the binary operands considered here. Such data may be floating-point operands, binary-coded decimal operands, or alphanumeric characters. An accumulator that operates on such data will be somewhat different and more involved than the one derived in this chapter. Floating-point data format was presented in Sec. 8-5 and decimal adders are discussed in Sec. 12-2. The reader interested in the arithmetic algorithms for these data representations is referred to Chu (3).

In Sec. 9-9 a few examples were presented to demonstrate the data processing capabilities of an accumulator register. It is apparent that if these processes are to be of any practical use, the register cannot remain isolated from its environment. A block diagram was presented in Fig. 9-20 to help visualize the source and destination of data and the source of command signals for the elementary operations. Admittedly, this environment was very loosely defined and needs further clarification. In the next chapter, this point will be taken again and the position of the accumulator among other registers and computer units will be clarified.

PROBLEMS

9-1. Register A in Fig. 9-1 holds the number 0101 and register B holds 0111.

(a) Determine the values of each S and C output in the four FA circuits.

(b) What signals must be enabled for the sum to be transferred to the A register?

(c) After the transfer, what is the content of A and what are the new values of S_1-S_4 and C_2-C_5?

9-2. Derive the logic diagram of a one-step and a two-step binary adder with a JK flip-flop.

9-3. Redraw Figs. 9-3 and 9-6 using only

(a) NAND gates

(b) NOR gates

9-4. Obtain the logic diagram of a one-step parallel adder with RS flip-flops. Show that the number of combinational gates is about the same as the FA in Fig. 9-1.

9-5. Design three different parallel binary subtractors to subtract the content of B from the content of A. Use each of the three methods discussed in Sec. 9-2.

(a) Full-subtractors and D flip-flops.
(b) One-step subtractor with T flip-flops.
(c) Two-step subtractor with T flip-flops.

9-6. Show the logic diagram of a one-stage parallel binary adder-subtractor circuit. Use the same combinational circuit to form the sum or the difference, and let command signals ADD or SUB determine whether the normal or complement output of A_i is used to form the carry or borrow for the next stage. Use D flip-flops.

9-7. A four-bit two-step adder, of which one stage is shown in Fig. 9-6, has the following binary numbers stored in the registers. Register A: 0101; Register B: 0111. Determine the values of A_1 through A_4 and C_2 through C_5

(a) before the $AD1$ command signal is applied,
(b) after the $AD1$ signal is applied, and
(c) after the $AD2$ signal is applied.

(Assume that $C_1 = 0$, and note that the carries have no meaning except for step b.)

9-8. The Boolean function of the carry look-ahead derived in Sec. 9-3 was for the input to the second group. Derive the carry C_7 for the third group as a function of C_4, A_4, A_5, A_6, and B_4, B_5, B_6. From this Boolean function, generalize the carry circuit needed in each group of three flip-flops.

9-9. What is the maximum carry propagation delay for a 16-bit accumulator that uses the circuit of Fig. 9-1? Assume the maximum propagation time of any gate to be 50 nsec.

9-10. In the serial adder of Fig. 9-8, the addend is 0111 and the accumulator holds the binary number 0101. Draw a timing diagram showing the clock pulses, the ADD command signal, the outputs A_1 through A_4, and the output of the carry flip-flop C during the process of addition.

9-11. Draw the logic diagram of a serial adder using D flip-flops for the shift register, a T flip-flop to store the carry, and NOR gates for the combinational circuit.

9-12. A 24-bit serial adder uses flip-flops and gates with maximum propagation delay of 200 and 100 nsec, respectively. The addend is transferred from a serial memory unit at a rate of 100,000 bits per second. What should the maximum clock-pulse frequency be and how long would it take to complete the addition?

9-13. Design a serial counter; in other words, determine the circuit needed for an increment elementary operation with a serial register. Note that it is necessary to have a carry flip-flop that can be set initially to 1 to provide the 1 bit in the least significant position.

9-14. Design a 40-bit register (using *T* flip-flops) with an increment elementary operation. Partition the register into 10 groups of four flip-flops each and show the logic diagram of one typical group with a carry look-ahead for the next group. Calculate the maximum propagation time, assuming that each AND gate has a maximum delay of 20 nsec.

9-15. An IC chip contains a four-bit accumulator with 15 operations q_1 to q_{15}. Only four terminals are available for specifying the 15 operations. Show the logic diagram of the decoder inside the chip and tabulate the code used to specify each operation. What should be the code when no operation is specified?

9-16. Derive the logic diagram of an accumulator with the same elementary operations as in Fig. 9-19 using

(a) *T* flip-flops
(b) *RS* flip-flops

9-17. Design one typical stage of an *A* register (using *JK* flip-flop) with the following elementary operations:

q_1	$(A) - 1 \Rightarrow A$	decrement
q_2	$(A) - (B) \Rightarrow A$	subtract
q_3	$(A) \wedge (\bar{B}) \Rightarrow A$	selective clear

9-18. (a) Perform the four different arithmetic computations in sign-2's complement representation of (±13) + (±7). (b) Repeat with sign-1's complement representation.

9-19. In the algorithm for addition with sign-2's complement representation stated in Sec. 9-9, the effect of overflow was neglected.

(a) Using the above mentioned algorithm and a seven-bit accumulator, perform the following additions: (+35) + (+40) and (-35) + (-40). Show that the answers are incorrect and explain why.
(b) State an algorithm for detecting the occurrence of an overflow by inspecting the sign bits of the addend, augend, and sum.
(c) Repeat parts (a) and (b) for sign-1's complement representation.

9-20. (a) Obtain the logic diagram of a combinational circuit that adds three bits and two input carries *A* and *B*. Show that the sum output is the exclusive-or of the input variables.
(b) Draw a block diagram of a parallel adder for adding three binary numbers $X + Y + Z$ stored in three registers. Use the three-bit adder obtained in part (a).
(c) Show the sum and carries generated from the three-bit adders during the addition of 15 + 15 + 15.

9-21. The binary numbers that follow consist of a sign bit in the left-most position followed by either a positive number after 0 or a negative number in 2's complement after a 1. Perform the subtraction indicated by taking the 2's complement of the subtrahend and adding it to the minuend. Verify your results.

(a) 010101 - 000011
(b) 001010 - 111001
(c) 111001 - 001010
(d) 101011 - 100110
(e) 001101 - 001101

9-22. Repeat Prob. 9-21, assuming that the negative numbers are in 1's complement representation.

9-23. List the sequence of elementary operations required to perform two subtractions; i.e., $X - Y - Z$. Assume that operands can be positive or negative and that all negative numbers are in their 1's complement form. The operand X enters the B register followed first by Y and then Z. Assume further that the B register can be complemented with a control signal q. The circuit model available is the one shown in Fig. 9-10 together with an overflow flip-flop L that accepts the value of C_{n+1} during the p_1 add-command.

9-24. State an algorithm for comparing the relative magnitudes of two signed numbers stored in two registers. Do not use subtraction. Assume that the binary numbers are represented in sign-magnitude.

9-25. Justify the algorithms stated in the text for arithmetic shift-right and shift-left of numbers stored in a register in

(a) sign-magnitude
(b) sign-2's complement
(c) sign-1's complement

9-26. List the sequence of elementary operations together with their symbolic notation for a selective-clear operation. Only the elementary operations of Table 9-1 are available.

9-27. It is necessary to design an accumulator that shifts by more than one bit. The number of shifts is specified by a binary number stored in a separate register. The shift command signals P_7 and P_8 are enabled only during one pulse period. Derive the logic diagram of the external register. (*Hint*: Use the register as a binary down-counter, enable the shift when the count is not zero, and disable when the count reaches zero.)

9-28. List the sequence of elementary operations required to pack six alphanumeric characters into one 36-bit word in the accumulator.

REFERENCES

1. Flores, I. *The Logic of Computer Arithmetic*. Englewood Cliffs, N. J.: Prentice-Hall, Inc., 1963.

2. MacSorley, O. L. "High-Speed Arithmetic in Binary Computers," *Proceedings of the IRE*, Vol. 49 (January 1961), 67-91.

3. Chu, Y. *Fundamentals of Digital Computer Design*. New York: McGraw-Hill Book Co., 1962.

4. Lehman, M., and N. Burla. "Skip Techniques for High-Speed Carry Propagation in Binary Arithmetic Units," *IRE Trans. on Electronic Computers*, Vol. EC-10 (December 1961), 691-98.

5. Gilchrist, B. J., B. J. H. Promerente, and S. Y. Wong. "Fast Carry Logic for Digital Computers," *IRE Trans. on Electronic Computers*, Vol. EC-4 (December 1955), 133-36.

6. Bartee, T. C., and D. J. Chapman. "Design of an Accumulator for a General Purpose Computer," *IEEE Trans. on Electronic Computers*, Vol. EC-14 (August 1965), 570-74.

7. Quatse, J. T., and R. A. Keir. "A Parallel Accumulator for a General Purpose Computer," *IEEE Trans. on Electronic Computers,* Vol. EC-16 (April 1967), 165-71.

8. Lucas, P. "An Accumulator Chip," *IEEE Trans. on Computers*, Vol. EC-18 (February 1969), 105-14.

9. Gordon, C. B., and J. Grason. "The Register Transfer Module Design Concept," *Computer Design*, Vol. 10, No. 5 (May 1971), 87-94.

10 COMPUTER ORGANIZATION

10-1 STORED PROGRAM CONCEPT

Digital systems may be classified as special or general purpose. A special purpose digital system performs a specific task with a fixed set of operational sequences. Once the system is built, its sequence of operations is not subject to alterations. Examples of special purpose digital systems can be found in numerous peripheral control units one of which is, for example, a magnetic-tape controller. Such a system controls the movement of the magnetic tape transport and the transfer of digital information between the tape and the computer. There exists a large class of special purpose systems that cannot be classified as digital computers. The name "digital computer" is customarily reserved for those digital systems that are general purpose. A general purpose digital computer can process a given set of operations and, in addition, can specify the sequence by which the operations are to be executed. The user of such a system can control the process by means of a *program,* i.e., a set of instructions that specify the operations, operands, and the sequence by which processing has to occur. The sequence of operations may be altered simply by storing a new program with different instructions.

A computer instruction is a binary code that specifies some register transfer operations. A program in machine code is a set of instructions that forms a logical sequence for processing a given problem. Programs may be written by the user in various programming languages, such as FORTRAN, COBOL, etc. However, a program to be executed by a computer must be in *machine language;* that is, in the specific binary code acceptable to the particular computer. There are specific machine-language programs known as *compilers* that make the necessary translation from the user programming language such as FORTRAN to the required machine code.

An instruction code is a group of bits that tell the computer to perform a specific operation. It is usually divided in parts, each having its own particular interpretation. The most basic part of an instruction code is its operation part. The operation code of an instruction is a group of bits that define an operation such as add, subtract, multiply, shift, complement, etc. The set of machine operations formulated for a computer depends on the processing it is intended to carry out. The total number of operations thus obtained determines the set of machine operations. The number of bits required for the operation part of the instruction code is a function of the total number of operations used. It must consist of at least n bits for a given 2^n (or less) distinct operations. The designer assigns a bit combination (a code) to each operation. The control unit of the computer is designed to accept this bit configuration at the proper time in a sequence and supply the proper command signals to the required destinations in order to execute the specified operation. As a specific example, consider a computer using 32 distinct operations, one of them being an ADD operation. The operation code may consist of five bits, with a bit configuration 10010 assigned to the ADD operation. When the operation code 10010 is detected by the control unit, a command signal is applied to an adder circuit to add two numbers.

The operation part of an instruction code specifies the operation to be performed. This operation must be executed on some data, usually stored in computer registers. An instruction code, therefore, must specify not only the operation but also the registers where the operands are to be found as well as the register where the result is to be stored. These registers may be specified in an instruction code in two ways. A register is said to be specified *explicitly* if the instruction code contains special bits for its identification. For example, an instruction may contain not only an operation part but also a memory address. We say that the memory address specifies explicitly a memory register. On the other hand, a register is said to be specified *implicitly* if it is included as part of the definition of the operation; in other words, if the register is implied by the operation part of the code.

Consider, for example, a digital computer with two accumulator registers $R1$ and $R2$ and 1024 words of memory. Assume that the operation part of an instruction consists of five bits and that an ADD operation has the code 10010. A precise definition of the ADD instruction can be formulated in a variety of ways, with each specific definition requiring a unique instruction code format and a different number of explicitly and implicitly specified registers or operands. We shall now proceed to consider four possible formats for an arithmetic addition instruction.

A zero-address instruction is an instruction code that contains only an operation part and no address part, as shown in Fig. 10-1(a). There are no bits to specify registers and, therefore, we say that all registers are implicitly specified. Since a binary operation such as addition requires two operands, both of which must be available in registers, the computer must have at least two accumulators. A possible interpretation of the instruction code may be: When the operation code 10010 is detected by the control unit, the contents of register R1 are added to the contents of register R2 and the sum is stored in register $R2$. Thus, the registers in a zero-address instruction are implicitly included in the definition of the operation.

A one-address instruction format is shown in Fig. 10-1(b). This format consists of a five-bit operation code and a ten-bit address. The number of bits chosen for the address part of an instruction is determined by the maximum capacity of the memory unit used. Since each word stored must have a specific address attached to it, there must be at least n bits available in an address that must specify 2^n distinct memory registers. In this particular example, the memory consists of 1024 words. Therefore, 10 bits are used for the address part of the instruction. If the operation part of the code is an arithmetic addition, the address part usually specifies a memory register where one operand is to be found. The register where the second operand is to be found, and the register where the sum is to be stored, must be implicitly specified. A possible interpretation of the instruction code whose

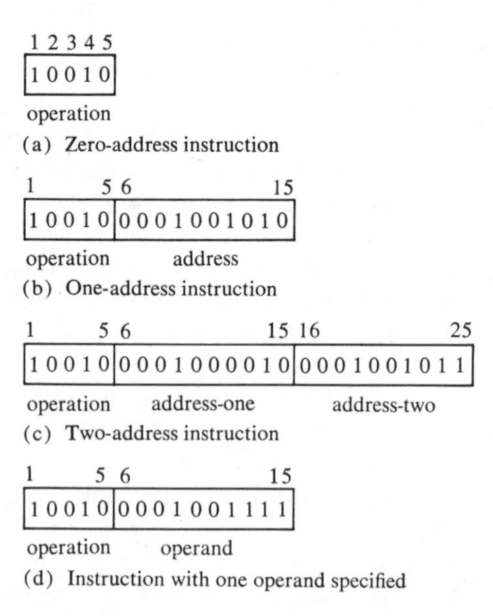

Figure 10-1 Instruction code formats

format is shown in Fig. 10-1(b) may be: Add the contents of the memory register specified by the address part of the instruction to the contents of register $R1$ and store the sum in register $R1$. The instruction code 100100001001010 shown in Fig. 10-1(b) has a five-bit operation part 10010 signifying an ADD operation and an address part 0001001010 equal to decimal 74. The control unit of the computer must accept this bit configuration at the proper timing sequence and send control signals to execute the instruction. This involves a transfer of the address part into the memory address register, reading of the contents of memory register 74, and arithmetically adding these contents to the accumulator register $R1$.

A two-address instruction format is shown in Fig. 10-1(c). The instruction consists of an operation part and two addresses. The first 5 bits signify an arithmetic addition operation and the last 20 bits give the address of two memory registers. These two addresses may either specify the location of the two operands or the location of one operand and the sum. In either case, one register remains to be implicitly specified. A possible interpretation of the instruction code with the format shown in Fig. 10-1(c) may be: Add the contents of the memory register specified by address-one to the contents of the accumulator register $R1$ and store the sum in the memory register specified by address-two. A three-address instruction is also possible. In this case, three registers can be explicitly specified to give the location of the two operands and the sum.

Another possible instruction format is shown in Fig. 10-1(d). The second part of this instruction code specifies, not the register in which to find the operand, but the operand value itself. A possible interpretation of this code format may be: Add the binary operand given in bits 6 to 15 to the contents of register $R1$ and store the result in $R1$. In this case, register $R1$ is implicitly specified to hold one operand and to be the register that stores the sum. The second operand is explicitly supplied in bits 6 to 15 of the instruction code.

The arithmetic addition operation, as well as any other binary operation, must specify two operands and the register where the result is stored. On the other hand, a unary operation such as shift-right must specify one register and possibly the number of bits to be shifted. Similarly, a simple transfer between two registers is an operation that must specify the two registers upon which the operation is to be performed. Thus, the number of registers to be specified in an instruction depends on the type of operation. The number of registers specified explicitly depends on the format chosen by the computer code designer. Most designers have found it efficient to use a one-address instruction format. Such computers include an accumulator register in the processing unit that is invariably implied implicitly by the operations. Computers with one-address instructions use the address part of the instruction to specify the address of a memory register where one operand is to be found. The other operand, as well as the sum, is always stored in the accumulator. The address part of a simple transfer operation will specify explicitly one register in the address part of the instruction code, while the accumulator is used as the second required register. Unary operations may specify in their address part either a memory register or the accumulator.

A flexibility for choosing the number of addresses for each operation is utilized in computers that employ variable-length instruction codes. In these computers, the variable length of the instruction code is determined by the operation encountered. Some operations have no address part, some have one address, and some have two addresses. This instruction format results in a more efficient code assignment.

Both instructions and operands are stored in the memory unit of a general purpose digital computer. Usually, an instruction code having the same number of bits as the operand word is chosen, although various other alternatives are possible. Instruction words are stored in consecutive location in a portion of the memory and constitute the *program*. They are read from memory in consecutive order, interpreted by the control unit, and then executed one at a time. The operand words are stored in some other location of memory and constitute the data. They are read from memory from the given address part of the instructions. Every general purpose computer has its own unique set of instructions repertoire. The stored program concept (the ability to store and execute instructions) is the most important property of a general purpose computer.

10-2 ORGANIZATION OF A SIMPLE DIGITAL COMPUTER

This section introduces the basic organization of digital computers and explains the internal processes by which instructions are stored and executed. The basic concepts are illustrated with the use of a simple computer as a model. Although commercial computers generally have a much more complicated logical structure than the one considered here, the chosen structure is sufficiently representative to demonstrate the basic organizational properties common to most digital computers. The organization and logical structure of the simple computer is described by first defining the physical layout of the various registers and functional units. A set of machine-code instructions is then arbitrarily defined. Finally, the logical structure is described by means of a set of register-transfer operations that specifies the execution of each instruction.

Physical Structure

A block diagram of the proposed simple computer is shown in Fig. 10-2. The system consists of a memory unit, a control section, and six registers.

The memory unit stores instructions and operands. The control section generates command signals to the various registers to perform the necessary register-transfer operations. All information processing is done in the six registers and their associated combinational circuits. The registers are listed in Table 10-1, together with a brief description of their function. The letter designation is used to represent the register in symbolic relations.

Table 10-1 List of Registers in the Simple Computer

Letter designation	*Name of register*	*Function*
A	Accumulator	general purpose processing register
B	Memory-Buffer	holds contents of memory word
C	Program-Control	holds address of next instruction
D	Memory-Address	holds address of memory word
I	Instruction	holds current instruction
P	Input-Output	holds input-output information

The *input-output P register* is a very simple way to communicate with the external environment. It will be assumed that all input information such as machine-code instructions and data is to be transferred into the P register by an external unit and that all output information is to be found in the same register. The *memory-address D register* holds the address of the memory register. It is loaded from the C register when an instruction is read from memory, and from the address part of the I register when an operand is read from memory. The *memory-buffer B register* holds the contents of the memory word read from, or written in, memory. An instruction word placed in the B register is transferred to the I register. A data word placed in the B register is accessible for operation with the A register or for transfer to the P register. A word to be stored in a memory register must first be transferred into the B register, from where it is written in memory.

The *program-control C register* holds the address of the next instruction to be executed. This register goes through a step-by-step counting sequence and causes the computer to read successive instructions previously stored in memory. It is assumed that instruction words are stored in consecutive memory locations and read and executed in sequence unless a branch instruction is encountered. A branch instruction is an operation that calls for a transfer to a nonconsecutive instruction. The address part of a branch instruction is transferred to the C register to become the address of the next instruction. To read an instruction, the contents of the C register are transferred to the D register and a memory read cycle is initiated. The instruction placed in the B register is then transferred into the I register.

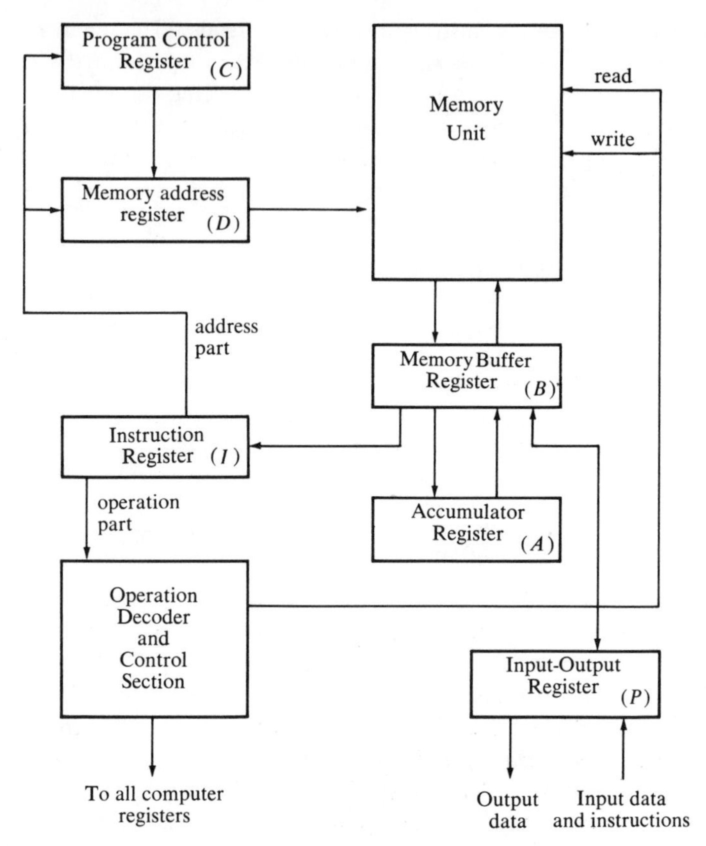

Figure 10-2 Block diagram of a simple computer

Since the C and D registers frequently contain the same number, it may seem as if one of them can be removed. Both registers are needed, however, because the C register must keep track of the address of the next instruction in case the current instruction calls for an operand stored in a memory register. The address of the operand loaded into the D register from the address part of the I register destroys the address of the current instruction from the D register.

The *accumulator A register* is a general purpose processing register that operates on data previously stored in memory. This register is used for the execution of most instructions. It is implicitly specified in binary, unary, and memory transfer operations. The *instruction I register* holds the instruction bits of the current instruction. The operation part of the code is decoded in the control section and used to generate command signals to the various registers. The address part of the instruction code is used to read an operand from memory or for some other function as given in the definition of the instruction.

Machine-Code Instructions

A list of instructions for the simple computer model is given in Table 10-2. This list is chosen to illustrate various types of instructions and does not represent a practical instruction repertoire. The reader may notice that the first nine instructions are identical to the list of elementary operations in Table 9-1 used for the accumulator register.

Table 10-2 List of Instructions for the Simple Computer

Operation Code	Name	*Address part*	*Function*
0001	Add	M	$(A) + (<M>) \Rightarrow A$
0010	OR	M	$(A) \vee (<M>) \Rightarrow A$
0011	AND	M	$(A) \wedge (<M>) \Rightarrow A$
0100	Exclusive-or	M	$(A) \oplus (<M>) \Rightarrow A$
0101	Clear	none	$0 \Rightarrow A$
0110	Complement	none	$(\bar{A}) \Rightarrow A$
0111	Increment	none	$(A) + 1 \Rightarrow A$
1000	Shift-right	none	$(A_{i+1}) \Rightarrow A_i$
1001	Shift-left	none	$(A_{i-1}) \Rightarrow A_i$
1010	Input	M	$(P) \Rightarrow <M>$
1011	Output	M	$(<M>) \Rightarrow P$
1100	Load	M	$(<M>) \Rightarrow A$
1101	Store	M	$(A) \Rightarrow <M>$
1110	Branch unconditional	M	$M \Rightarrow C$
1111	Branch-on-zero	M	if $(A) = 0$ then $M \Rightarrow C$ if $(A) \neq 0$ proceed to next instruction

The code format used for the instructions consists of an operation part and a one-address designated by the letter M. The first four instructions represent a binary operation; one is an arithmetic addition and three are logical operations. The address part M of the binary instructions specifies a memory register where one operand is stored. The accumulator holds the second operand and the sum. Instructions five through nine are unary operations for the accumulator register and do not use the address part. Instructions 10 through 13 consist of memory-transfer operations. The address part of the instruction specifies a memory register; the second register is implied implicitly. For the input and output operations, the implied register is P. For the load and store operations, the implied register is A.

The last two instructions are branch-type instructions. The branch-unconditional instruction causes the computer to continue with the instruction found in memory register specified by the address part M. The function of this instruction is to transfer the contents of the address part M

to the C register, since the latter always holds the address of the next instruction. The branch-on-zero instruction checks the contents of the A register. If they are equal to 0, the next instruction is taken from memory register whose address is given by the address part M. If the contents of the A register are not equal to 0, the computer continues with the next instruction in normal sequence.

Instructions and data are transferred into the memory unit of the simple computer through the P register. It is assumed for simplicity that the P register is loaded with a binary word from an external system and that the repeated execution of the input instruction (code 1010) transfers the words into consecutive memory locations. To start the execution of the stored program, the operator sets the C register to the first address and a "start" switch is activated. The first instruction is read from memory and executed. The rest of the instructions are read in sequence and executed in consecutive order unless a branch instruction is encountered. Instructions are read from memory and executed in registers by a sequence of elementary operations such as inter-register transfer, shift, increment, clear, add, complement, memory read, and memory write. The control section generates the sequence of command signals for the required elementary operations.

Logical Structure

Once the start switch is activated, the computer sequence follows a basic pattern—an instruction whose address is in the C register is read from memory and transferred to the I register. The operation part of the instruction is decoded in the control section. If it is one with an address part, the memory may be accessed again to read a required operand. Thus words read from memory into the B register can be either instructions or operands. When an instruction is read from memory, the computer is said to be in an instruction *fetch* cycle. When the word read from memory is an operand, the computer is said to be in a data *execute* cycle. It is the function of the control section to keep track of the various cycles.

An instruction is read from memory and transferred to the I register during a fetch cycle. The register-transfer elementary operations that describe this process are:

$(C) \Rightarrow D$		transfer instruction address
$(<D>) \Rightarrow B,$	$(C) + 1 \Rightarrow C$	memory read, increment C
$(B) \Rightarrow <D>,$	$(B) \Rightarrow I$	restore memory word, transfer instruction to I register

A magnetic-core destructive-read memory is assumed and therefore, a restoration of the word back to the memory register is required.* The C register is incremented by 1 to prepare it for the next instruction.

The fetch cycle is common to all instructions. The elementary operations that follow the fetch cycle are determined in the control section from the decoded operation part. The binary and memory-transfer instructions have to access the memory again during the execute cycle. The unary and branch instructions can be executed right after the fetch cycle. In all cases, control returns back to the fetch cycle to read another instruction. The register-transfer elementary operations during the execute cycle for the four binary operations are:

$(I[M]) \Rightarrow D$, $(I[\mathrm{Op}]) \Rightarrow$ Decoder		transfer address part
$(\langle D \rangle) \Rightarrow B$		read operand
$(B) \Rightarrow \langle D \rangle$		restore operand
$(A) + (B) \Rightarrow A$	if operation is add	
$(A) \vee (B) \Rightarrow A$	if operation is OR	
$(A) \wedge (B) \Rightarrow A$	if operation is AND	
$(A) \oplus (B) \Rightarrow A$	if operation is exclusive-or	
Go to fetch cycle.		

The symbol $(I[\mathrm{M}])$ designates the contents of the address part M of the I register. Only one of the binary operations listed is executed depending on the value of $(I[\mathrm{Op}])$; i.e., the operation part of the I register.

When the operation code is a memory-register transfer type, the execute cycle is described by the following elementary operations:

Input:

$(I[M]) \Rightarrow D,$ $(I[\mathrm{Op}]) \Rightarrow$ decoder	transfer address
$0 \Rightarrow \langle D \rangle,$ $(P) \Rightarrow B$	clear memory register, transfer word from P
$(B) \Rightarrow \langle D \rangle$	store word in memory register
Go to fetch cycle	

Output:

$(I[M]) \Rightarrow D,$ $(I[\mathrm{Op}]) \Rightarrow$ decoder	transfer address
$(\langle D \rangle) \Rightarrow B$	read word
$(B) \Rightarrow \langle D \rangle,$ $(B) \Rightarrow P$	restore and transfer word
Go to fetch cycle	

Load:

$(I[M]) \Rightarrow D,$ $(I[\mathrm{Op}]) \Rightarrow$ decoder	transfer address
$(\langle D \rangle) \Rightarrow B$	read word
$(B) \Rightarrow \langle D \rangle,$ $(B) \Rightarrow A$	restore and transfer word
Go to fetch cycle	

*See Sec. 8-3.

Store:

$(I[M]) \Rightarrow D,$	$(I[\text{Op}]) \Rightarrow$ decoder	transfer address
$0 \Rightarrow \langle D \rangle,$	$(A) \Rightarrow B$	clear memory register, transfer word
$(B) \Rightarrow \langle D \rangle$		store word in memory
Go to fetch cycle		

The input and store operations are similar, as are the output and load operations, the difference being only that the *P* register is used for the former and the *A* register for the latter.

When the operation code represents a unary or a branch-type instruction, the memory is not accessed for an operand and the execute cycle is skipped. The instruction is completed with the execution of one elementary operation, which can occur at the same time that the instruction word is restored back into memory during the conclusion of the fetch cycle. This elementary operation is listed in Table 10-2 under the column heading "function." The unary operations are executed in the *A* register. The branch-unconditional instruction causes a transfer of the address part into the *C* register. The branch-on-zero instruction causes the same transfer if $(A) = 0$; if $(A) \neq 0$, the address of the next instruction in the *C* register is left unchanged.

The computer sequence of operations is summarized in a flow chart in Fig. 10-3. Note that the completion of an instruction always follows a return to the fetch cycle to read the next instruction from memory.

An Example

We shall illustrate the use of the machine-code instructions of the simple computer by writing a program that will accept instructions or data words from the *P* register and store them in consecutive registers in memory starting from address 0. The program that will process this input information is to be stored in memory registers starting from address 750. The program in machine-code instructions is listed below. The mnemonic name of the operation is entered instead of its binary code.

Memory Location	*Instruction*		*Function*
	Operation	*Address*	
750	Input	000	$(P) \Rightarrow \langle M \rangle$
751	Load	750	$(\langle 750 \rangle) \Rightarrow A$
752	Increment		$(A) + 1 \Rightarrow A$
753	Store	750	$(A) \Rightarrow \langle 750 \rangle$
754	Branch-unconditional	750	Branch to location 750

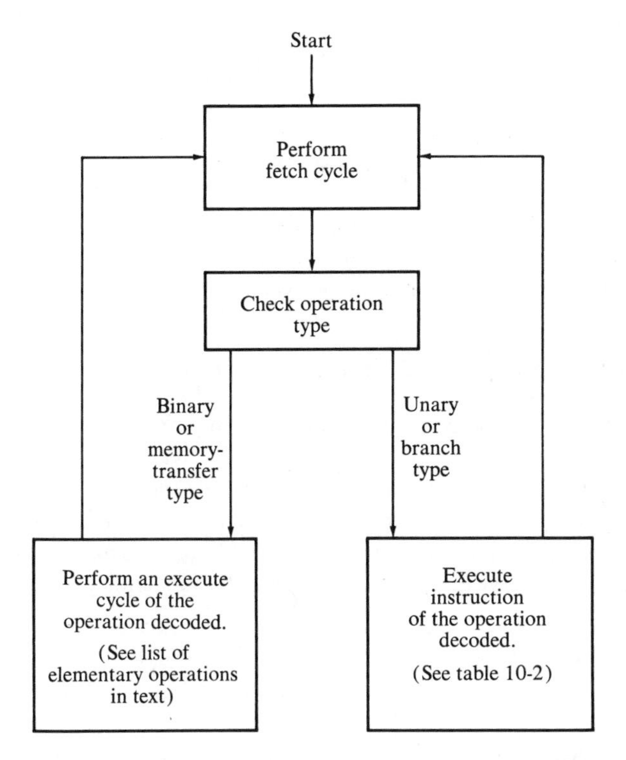

Figure 10-3 Flow diagram for the sequence of operations in the simple computer

The first word from the P register goes to the memory register in location 0. The instruction in location 750 is then loaded into the A register, incremented by 1, and stored back. Thus the new contents of location 750 become: "Input 001." The program then branches back to address 750. The execution of the modified instruction at location 750 causes a transfer of the second word from the P register to the memory register in location 001. The program loop is repeated again, causing consecutive input words to be stored in consecutive memory locations.

Although this example illustrates a very simple machine-code program, there are at least three difficulties associated with its execution. The first difficulty is the lack of provision for terminating the process. Obviously, if the input data goes beyond location 749, it will destroy its own program. One solution to this problem is to provide a "Stop" instruction and stop the computer after a programmed test. The details of such a test are left for an exercise. The second difficulty is concerned with the problem of initialization: if the program is used a second time, the initial address part of the instruction in location 750 will not necessarily be 0. The third difficulty may be stated as a question: how is the first program that enters other programs into the computer brought to memory? In other words, how is the above listed program accepted from the P register if it is not initially stored in memory? This problem is fundamental with general purpose computers and is concerned with the initial starting procedures referred to as "cold start." It is solved by a procedure called "bootstrapping," usually a manual operation done by the operator on the computer console that loads a basic initial program into memory, which in turn is capable of calling a more extensive program from an input device.

10-3 LARGE GENERAL PURPOSE COMPUTERS

A block diagram of a large general purpose computer is shown in Fig. 10-4. The memory unit is usually very large and divided into modules; each module is able to communicate with each of the three processors. The central processor is responsible for processing data and performing arithmetic, logical decisions, and other required processing operations. The system control supervises the flow of information among units, with each

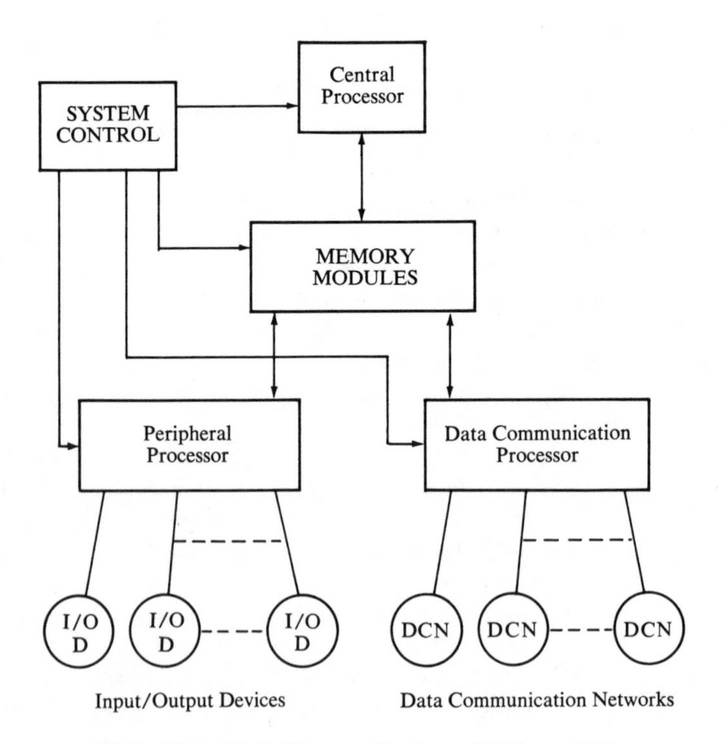

Figure 10-4 Block diagram of a large digital computer

unit usually having its own control section to take care of internal functions. The memory, processor, and control units are similar in small and large systems except that the large one has more capabilities and is much more complicated. Special equipment included in large computers, not normally found in small ones, are processors to control the information flow between the internal computer and external devices.

A large computer can use a range of equipment to perform input–output functions. These include card reader and punch, paper-tape reader and punch, teletype keyboard and printer, optical character reader, visual display, graph plotter, and data communication networks. Some input and output devices also provide additional storage to support the internal memory capacity. Such devices include magnetic tapes, magnetic disks, and magnetic drums. The purpose of the *peripheral processor* is to provide a pathway for the transfer and translation of data passing between input, output, external storage, and internal memory modules. The peripheral processor is sometimes called a *channel controller*, since it controls and regulates the flow of data to and from the internal and external parts of the computer.

Computers with time-sharing capabilities must provide communication with many remote users via communication lines. A separate *data communication processor* is sometimes provided to handle input and output data of many communication networks operating simultaneously. Information exchange with these devices is at so slow a rate that it is possible to service tens or even hundreds with a single unit. Its principles of organization and its relation to the rest of the computer are very similar to those of the peripheral processor. However, it is necessarily more complex since it communicates with more devices at the same time. The data communication processor provides a communication link between each remote user and the central processor.

Peripheral Processor

The input–output (I/O) register in the simple computer described in Sec. 10-2 connects directly to the memory-buffer register for direct transfer of data in and out of memory. This means that the entire computer is idle while waiting for data from a slow input device or for data to be accepted by a slow output device. The difference in information flow rate between the central processor (millions of operations or transfers per second) and that of the input–output devices (a few to few hundred-thousand transfers per second) makes it inefficient for a large computer to have direct communication between the memory modules and external equipment. The peripheral processor makes it possible for several external pieces of equipment to be operating essentially simultaneously, providing data to and taking data from internal storage.

The data format of input and output devices usually differs from the main computer format, so the peripheral processor must restructure data words from many different sources compatible with the computer. For example, it may be necessary to take four 8-bit characters from an input device and pack them into one 32-bit word before the transfer to memory. Each input or output device has a different information interchange rate. In the interest of efficiency, it is inadvisable to devote central processor time to waiting for the slower devices to transfer information. Instead, data is gathered in the peripheral processor while the central processor is running. After the input data is assembled, it can be transferred into memory using only one computer cycle. Similarly, an output word transferred from memory to the peripheral processor is transferred from the latter to the output device at the device transfer rate and bit capacity. Thus the peripheral processor acts as a transfer-rate buffer.

As each I/O device becomes ready to transfer information, it signals the peripheral processor by use of control bits called "flags." With many I/O devices operating concurrently, data becomes available from many sources at once. The peripheral processor must assign priorities to different devices. This assignment may be on a simple first-come, first-served basis. However, devices with higher transfer rate are usually given higher priority.

An I/O transfer instruction is initiated in the central processor and executed in the peripheral processor. For example, the instruction I/O M, where I/O is an input-output operation code and M is an address, is executed in the central processor by simply transferring the contents of memory register M to a peripheral processor register. The contents of memory register M specify a *control word* for the peripheral processor to tell it what to do.

A typical control word may have a binary code information as follows:

operation	device	address	count

The operation code specifies input, output, or control information such as start unit operating, rewind tape, etc. The device code is an identification number assigned to each I/O device. Each device reacts only to its own code number. The memory address part specifies the first memory register where the input information is to be stored or where the output information is available. The count number gives the number of words to be transferred.

To continue with a specific example, let us assume that the operation code specifies an input and that the device code specifies a card reader. The peripheral processor first activates the unit and goes to fulfill other I/O tasks while waiting for a signal from the card reader to indicate that it is ready to transfer information. When the ready signal is received, the peripheral processor starts receiving data read from cards. If each column punched in the card is transferred separately, the peripheral processor will accept 12 bits at a time. The 12-bit character from each card column can be immediately converted to an internal 6-bit alphanumeric code by the peripheral processor. Characters are then assembled into a word length suitable for memory storage. When a word is ready to be stored, the peripheral processor "steals" a memory cycle from the central processor; i.e., it receives access to memory for one cycle. The address part of the control word is transferred to the memory-address register and the data which was assembled goes to the memory-buffer register and the first word is stored.

The address and count numbers in the control word are decreased by 1, preparing for storage of the next word. If the count reaches 0, the peripheral processor stops receiving data. If the count is not 0, data transfer continues until the card reader has no more cards.

A peripheral processor is similar to a computer control section in that it sends signals to all registers to govern their operation. Among its duties are synchronizing data transfer with the main computer, keeping track of the number of transfers, checking flags, and making sure of data transfer to prescribed memory locations.

It is important to note that this method of data transfer solves the information transfer-rate-difference problem mentioned earlier. The central processor operates at high speed independently of the peripheral processor. The former's operation interferes with the latter's only when an occasional memory cycle is requested. The slow operations needed when communicating with input and output transfers are taken care of by the interface control between peripheral processor and the I/O devices.

Program Interrupt

The concept of program interrupt is used to handle a variety of problems which may arise out of normal program sequence. Program interrupt refers to the transfer of control from the *normal running program* to another *service program* as a result of an internally or externally generated signal. The interrupt scheme is handled by a *master program* which makes the control transfers. To better appreciate the concept, let us consider an example.

Suppose in the course of a calculation that the computer finds it must divide by zero (due, of course, to a careless programmer), what should be done? It would be easy enough to indicate the error and halt, but this would waste time. Suppose there is a flip-flop which would set whenever a divide by zero operation occurs. The output of this flip-flop could send a signal to interrupt the normal running program and load into the memory-address register the beginning address of a program designed to service the problem. This service program would take appropriate diagnostic and corrective steps and decide what to do next. The sensing of the flip-flop and the transfer to a service program is handled by the master program and is called a *program interrupt.*

There are many contingencies that demand the immediate attention of the computer and cause a program interrupt by setting a particular flip-flop in an interrupt register. The setting of any interrupt flip-flop results in transfer of control to the master program. This program would then check to see which flip-flop was set, steer control to the appropriate service program in memory and, upon completing the task, clear the flip-flop. The master program also insures that the contents of all processor registers remain unchanged during the interrupt so that normal processing could resume after the interrupt service has been completed.

Program interrupts are initiated when internal processing errors occur, when an external unit demands attention, or when various alarm conditions occur. Examples of interrupts caused by internal error conditions are register overflow, attempt to divide by zero, an invalid operation code, and an invalid address. These error conditions usually occur as a result of a premature termination of the instruction execution. The service program determines the corrective measures to be taken.

Examples of external request interrupts are I/O device not ready, I/O device requesting transfer of data, and I/O device finished transfer of data. These interrupt conditions inform the system of some change in the external environment. They normally result in a momentary interruption of the normal program process which is continued after servicing or recording the interrupt condition. Termination of I/O operations is handled by interrupts; this is the way the peripheral processor usually communicates with the central processor.

Interrupts caused by special alarm conditions inform the system of some detrimental change in environment. They normally result from either a programming error or hardware failure. Examples of alarm condition interrupts are running program is in an endless loop, running program tries to change the master program, or power failure. Alarm conditions due to programming error result in a rejection of the program that caused the error. Power failure might have as its service routine the storing of all the information from volatile registers into a magnetic-core memory in the few milliseconds before power ceases.

Data Communication

A data communication processor is an I/O processor that distributes and collects data from many remote terminals. It is a specialized I/O processor with the I/O devices replaced by data communication networks. A communication network may consist of any of a wide variety of devices such as teletypes, printers, display units, remote computing facilities, or remote I/O devices. With the use of a data communication processor, the computer can execute fragments of each user's program in an interspersed manner and thus have the apparent behavior of serving many users at once. In this way the computer is able to operate efficiently in a time-sharing environment.

The most common means of connecting remote terminals to a central computer is via telephone lines. The advantages of using already installed, comprehensive coverage lines are obvious. Since these lines are narrow-band voice channels (analog) and computers communicate in terms of signal levels (digital), some form of conversion must be used. This converter is the so-called Modem (from MODulator-DEModulator). The Modem converts the computer pulses into audio tones which may be transmitted over phone lines and also demodulates the tones for machine use. Various modulation schemes as well as different grades of phone lines and transmission speeds are used.

In a typical time-sharing system, the central computer is connected to many remote users. As in the case of the multitude of I/O units, the central computer must be able to single out terminals for singular communication. As before, each terminal is assigned a device code and is selected by matching the code broadcast by the computer. The binary code most commonly used for data transmission is the ANSCII seven-bit code (Table 10-3), with an eighth bit used for parity. This code consists of 95 *graphic characters* that include upper and lower case alphabetic characters, numerals zero to nine, punctuation marks and special symbols, and 33 *control characters*, 10 of which are used in communication control.

The control characters do the control operations necessary to route data properly and put it into the proper format. They are grouped into three functions: communication control, format effectors, and information separators. The 10 communication control functions listed in Table 10-4 will be useful in a subsequent description of terminal operation. Format effectors are functional characters that control the layout of printing or display devices. They include the familiar typewriter format controls such as backspace (BS), horizontal (HT) and vertical tabulation (VT), carriage return (CR), and line feed (LF). Information separators are used to separate blocks of data in a logical, hierarchical order such as sentences, paragraphs, pages, multiple pages, etc. These standard characters may be used in any manner the data communication designer wishes; although they are standard from a symbolic standpoint, their functions are by no means standard.

The information sent to or received from a remote device is called a *message*. Text (graphic characters) is usually prefaced by control characters.

Control information plus text constitute a message. The length and content of the text depends upon the application. Control information is needed to direct the data communications flow and report its status.

A data communication processor receives information either by individual bits in series or by a parallel transfer of all bits of a character called a *byte*. A byte is usually accumulated in an input register and then transferred as one character. The data communication processor accumulates bytes from many remote devices and builds messages by programmed procedures. It uses information tables describing the data network characteristics supplied by the particular installation. It interprets and translates codes for control characters but does not interpret the text transmitted.

Table 10-3 American National Standard Code for Information Interchange (ANSCII) (formerly USASCII and ASCII)

Bits		b_7 →	0	0	0	0	1	1	1	1
		b_6 →	0	0	1	1	0	0	1	1
		b_5 →	0	1	0	1	0	1	0	1
b_4 b_3 b_2 b_1 ↓	Row ↓	Column →	0	1	2	3	4	5	6	7
0 0 0 0	0		NUL	DLE	SP	0	@	P	`	p
0 0 0 1	1		SOH	DC1	!	1	A	Q	a	q
0 0 1 0	2		STX	EC2	"	2	B	R	b	r
0 0 1 1	3		ETX	DC3	#	3	C	S	c	s
0 1 0 0	4		EOT	DC4	$	4	D	T	d	t
0 1 0 1	5		ENQ	NAK	%	5	E	U	e	u
0 1 1 0	6		ACK	SYN	&	6	F	V	f	v
0 1 1 1	7		BEL	ETB	'	7	G	W	g	w
1 0 0 0	8		BS	CAN	(	8	H	X	h	x
1 0 0 1	9		HT	EM	)	9	I	Y	i	y
1 0 1 0	10		LF	SUB	*	:	J	Z	j	z
1 0 1 1	11		VT	ESC	+	;	K	[	k	{
1 1 0 0	12		FF	FS	,	<	L	\	l	¦
1 1 0 1	13		CR	GS	—	=	M	]	m	}
1 1 1 0	14		SO	RS	.	>	N	∧	n	~
1 1 1 1	15		SI	US	/	?	O	—	o	DEL

The process by which contact is established with the remote terminal is called Poll/Select. The data communication processor continuously polls all terminals by sending the code ENQ (0000101), which asks "Are you ready to send?". The remote terminal may indicate its readiness in a number of ways. If the polled unit is a simple teletypewriter, a send mode flag may be set. In more complex terminals, buffer memories may be included within the unit to transmit slowly gathered data at high speeds. In this case, the

Table 10-4 Communication Control Characters

Code	*Binary*	*Meaning*
SOH	0000001	Start of heading
STX	0000010	Start of text
ETX	0000011	End of text
EOT	0000100	End of transmission
ENQ	0000101	Inquiry
ACK	0000110	Acknowledge
DLE	0010000	Data link escape
NAK	0010101	Negative acknowledge
SYN	0010110	Synchronous idle
ETB	0010111	End of transmission block

unit assumes a transmit ready status only after its buffer memory has been filled with a complete message. It then waits for a poll from the data center to activate the message transmission. If the polling signal arrives and the terminal is not in a ready mode, it sends back a negative acknowledge (NAK) to indicate that it is not ready. The data center is then satisfied until the next poll. If the terminal is ready, it sends back an ACK, and communication proceeds.

A very simple format that may be used between the data communication processor and a remote terminal will now be given. The data communication processor establishes contact by sending the following characters, where Y is a station address and ENQ is a control character listed in Table 10-4:

Y Y ENQ

Assuming the remote device has information to be transmitted, it sends the following characters, where Y is its identification address and X is text:

SOH Y Y STX X X .... X ETX

If the data communication processor receives the data without errors, it sends back an acknowledge character and terminates:

Y Y ACK EOT

The speed of most remote terminals (especially teletype units) is extremely slow compared with computer speeds. This property allows multiplexing of many users to achieve greater efficiency in the time-sharing systems. Multiplexing is a systematic repeated selection of multiple inputs, which combines them into a single output. It is analogous to a multiple position rotary switch rotating at high speed, sampling one input at a time. This technique allows many users to operate at once and share a single communication line while being sampled at speeds comparable to the speed of the main computer.

Microprogramming

It was shown in Sec. 10-2 that a computer instruction is executed internally by a sequence of elementary operations. It is possible to derive combinational circuits to implement the elementary operations during appropriate control pulses. By wiring circuits permanently to do the elementary operations, however, the computer is limited to a fixed instruction repertoire prescribed by the original design.

A microprogrammed controlled computer has an independent control over the processes dictated by the instructions. It incorporates a control memory that holds *microprograms* for the interpretation of machine instructions. The microprogram specifies a sequence of elementary operations called *micro-operations*. The sequence of elementary operations needed to execute a machine instruction is not wired by unalterable circuits but instead is specified by a set of *microinstructions* in the microprogram. The usual procedure is to choose the micro-operation sequence for each computer instruction by specifying a set of microinstructions. The microprogrammed concept increases the flexibility of a computer in that it allows the user to define his own instructions to fit his needs.

A microprogrammed controlled computer needs a small computing section within the main computer to program and execute the microprograms. This inner computing unit replaces the control section of a conventional stored program computer. It is usually comprised of a control memory, input and output registers, and a control decoder. The control memory stores the individual microinstructions, which are interpreted by the control decoder. The most common storage media used for microprograms is the read-only memory. This is a memory unit whose binary cells have fixed, permanently wired values. Common algorithms are permanently stored in the read-only memory for the required micro-operation sequences. The control memory may also be a read–write memory, in which case any sequence of micro-operations the user wishes to define may be programmed in microinstructions.

A block diagram of a basic microprogrammed system is shown in Fig. 10-5. An instruction fetched from main memory specifies the first address of the microprogram that implements this instruction. The address for subsequent microinstructions may come from a number of sources. A simple way to specify the next micro instruction is to include its address with each microinstruction. This method is suitable for well-defined sequences of operations, where the algorithm for a certain machine instruction can be executed by consecutive micro-operations. It may be necessary, however, to make certain microinstruction executions dependent on the results of previous micro-operations. For this reason, the next address encoder receives information from either the present microinstruction, the main computer, or both.

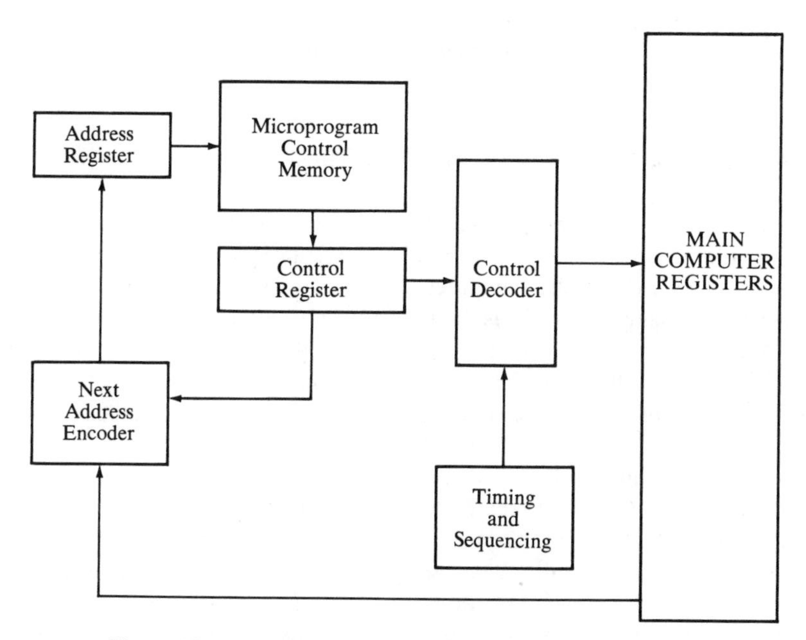

Figure 10-5 A generalized model of microprogram control

A microinstruction is a control word which is decoded by the control decoder. The control decoder generates signals for the main computer to perform micro-operations. The control word specified by the address register appears in the control register. This control word has many possible formats, depending on the specific microprograming scheme adopted. In specifying the format of a control word, the following questions must be considered:

1. How many control functions will be specified by a single memory word?

2. How will the control functions be executed? For example, it is possible to scan m bits of an n-bit control word sequentially or process all the n bits concurrently.

3. How will control words be decoded?

4. How will the execution sequence be established?

5. Where will the address of the next control word be derived?

Once these and other questions have been answered, design may proceed in exactly the same manner in which a stored program computer is designed (see Ch. 11).

Most microprograms are executed by calling the first address of the microprogram, sequencing through the micro-operations stored in the control memory, and then transferring back to main program control. A microprogram is thus nothing more than a subroutine at the elementary operation level.

10-4 CLASSES OF MACHINE INSTRUCTIONS

Instructions are represented inside the computer in binary-coded form since this is the most efficient method of representation in computer registers. However, the user of the computer would like to formulate his problem in a higher-level language such as FORTRAN or COBOL. The conversion from the user-oriented language to its binary-code representation inside the computer can itself be performed by a computer program. This is referred to as the compiling process; the program which performs the conversion is called a *compiler.*

It would be instructive to demonstrate the process of compiling with a simple example. Consider a FORTRAN assignment statement $X = Y + Z$ to be translated into the language of the simple computer described in Sec. 10-2. This statement, being one of many in the program, is punched on a card and entered into the computer. The FORTRAN statement is stored in memory with an alphanumeric binary code and occupies six consecutive character locations. The six characters are: $X = Y + Z$, and a symbol used for the end of statement. The translation from the set of characters to machine code is processed by the compiler program as follows. First the characters are scanned to determine if the statement is indeed a valid assignment statement. Then the characters X, Y, and Z are recognized as variables and each is assigned a memory address. The + character is interpreted to be an arithmetic operator and the = character as a replacement operator. The compiler program translates the statement into the following set of machine instructions:

Operation	Address	Binary code	Function
Load	17	1100000010001	Load value of Y into Ac
Add	18	0001000010010	Add value of Z to Ac
Store	19	1101000010011	Store result in X

It is assumed that the compiler chose addresses 17, 18, and 19 for the variables Y, Z, and X, respectively. A *value* entered for a variable through an input statement, or calculated from an assignment statement, is stored in the memory register whose address corresponds to the variable *name*. The compiler keeps an internal table that correlates variable names with memory addresses and refers to the corresponding address when the value of the variable is used or when the variable is given a new value. Thus, the *value* of the sum is stored in the memory register whose address corresponds to the *name* X.

The binary-code instructions are the ones actually stored in memory registers. The first four bits are derived from the operation code listed in Table 10-2. The last 10 bits represent the address part and are obtained from the conversion of the decimal number to its binary equivalent.

The number of instructions available in a computer may be greater than a hundred in a large system and a few dozen in a small one. Some large computers perform a given function with one machine instruction, while a smaller one may require a large number of machine instructions to perform that same function. As an example, consider the four arithmetic operations of addition, subtraction, multiplication, and division. It is possible to provide machine instructions for all four operations. It is also possible to provide only the *Add* operation as a machine instruction, with the other arithmetic operations being implemented by a sequence of many *Add* instructions. Functions executed internally with a single computer instruction are said to be implemented by *hardware*. Functions implemented with a subroutine program (i.e., a sequence of machine instructions), are said to be implemented by *software*. Large computers provide an extensive set of hardware instructions, smaller ones depend more on software implementation. Certain series of computer models possess similar external characteristics but are different in operating speed and computer costs. The high-speed expensive models have a large number of hardware-implemented instructions. The low-speed, less expensive models have only some basic hardware instructions, with many other functions being implemented by software subroutines.

The physical and logical structure of computers is normally described in a reference manual provided with the system. Such a manual explains the internal structure of the computer, including the processing registers available and their logical capabilities. It will list all hardware-implemented instructions, specify their binary-code format, and provide a precise definition of each instruction. The instructions for a given computer can be conveniently grouped into three major classes: (1) those that perform operations and change data value such as arithmetic and logical instructions, (2) those that move information between registers without transforming it, and (3) those concerned with testing of data and branching out of normal program sequence.

Operational Instructions

Operational instructions manipulate data in registers and produce intermediate and final results necessary for the solution of the problem. These instructions perform the needed calculations and are responsible for the bulk of activity involved in processing data in the computer. The most familiar example of operational instructions are the four arithmetic operations of addition, subtraction, multiplication, and division. These are the most basic functions from which scientific problems can be formulated using numerical analysis methods. Although implemented by hardware instructions in most computers, division and sometimes multiplication are implemented by software in some small computers. Arithmetic instructions require two operands as input and produce one operand as output. If the memory registers containing all three operands are specified explicitly, a three-address instruction is needed. This instruction may take the form:

Add M_1, M_2, M_3

where the mnemonic name Add represents the operation code and M_1, M_2, M_3 are the three addresses of the operands. The execution of this instruction requires three references to memory in addition to the one required during the fetch cycle. The elementary operations to implement the instructions are:

$(\langle M_1 \rangle) \Rightarrow B$ read 1st operand
$(\langle M_2 \rangle) \Rightarrow B, (B) \Rightarrow A$ read 2nd operand, transfer 1st to A
$(A) + (B) \Rightarrow A$ form the sum in the processor
$(A) \Rightarrow B$ transfer sum for storage
$(B) \Rightarrow \langle M_3 \rangle$ store sum in 3rd register

For simplicity, a nondestructive memory unit has been assumed, and the transfer of the address parts from the instruction register to the memory-address register have been omitted. Most computers use a reduced number of explicit memory registers in the instruction format and assume that some of the operands use implict registers as discussed in Sec. 10-1 and Prob. 10-2.

Hardware algorithms for processing machine multiplication and division require three registers. The B register is used for holding one operand; the A register, for executing repeated additions or subtractions; and a third register, for storing the second operand and/or the result. The third register is sometimes called the *MQ*-register because it holds the multiplier during multiplication and the quotient during division. These two arithmetic instructions require a large number of successive elementary operations of additions and shifts, or subtractions and shifts, for their execution. When the computer central control decodes the operation part of such a lengthy instruction, it is customary for it to transfer a special control signal to a local control section within the arithmetic unit. This local controller supervises the execution of the instruction among the arithmetic registers and then transfers back a signal to the central control upon completion of the instruction execution. (See Sec. 12-6.)

The four basic arithmetic operations are said to be *binary* because they use two operands. Unary and ternary arithmetic operations are also possible.

The square root of a number is a unary operation because it needs only one operand. A function such as $A + B \cdot C$ is a ternary operation because it uses three operands, A, B, and C, to produce a result equal to the product of the second and the third operand added to the value of the first. The square root function is well known to occur in many scientific problems, and the above-mentioned ternary function occurs repeatedly in matrix manipulation problems. Most computers perform these two function by software means. Some high-speed computers find it convenient to include such operations in hardware instruction. The disadvantage involved in the added equipment is offset by the gain obtained from the added speed provided by the direct implementation.

The data type assumed to be in the processing registers during the execution of arithmetic instructions is normally specified implicitly in the definition of the instruction. An arithmetic instruction may specify fixed-point or floating-point data, binary or decimal data, single-precision or double-precision data. It is not uncommon to find a computer with four types of *Add* instructions such as *Add binary integer, Add floating-point, Add decimal*, or *Add double-precision.*

Fixed-point binary numbers are assumed to be either integer or fractions; negative numbers are represented in registers by either sign-magnitude, 1's complement, or 2's complement. Floating-point arithmetic instructions require special hardware implementation to take care of alignment of fractions and scalling of exponents. Some computers prefer to use software subroutines for the floating-point operations to reduce the amount of circuits in the processor. Decimal data are represented by a binary code, with each decimal digit occupying four flip-flops of a register. Computers oriented towards business data processing applications find it more convenient to have arithmetic instructions that operate on decimal data. The number of bits in any computer register is of finite length and therefore, the results of arithmetic operations are of finite precision. Some computers provide double-precision arithmetic instructions where the length of each operand is taken to be the length corresponding to two computer registers.

There are many other operational-type instructions not classified as arithmetic operations. The common property of such instructions is that they transform data and produce results which conform with the rules of the given operation. The logical instructions AND, OR, etc., are operational instructions that require two input operands to produce a value for the output operand. Bit-manipulation instructions such as bit-set, bit-clear, and bit-complement transform data in registers according to the rules discussed in Sec. 9-9. Unary operations such as clear, complement, and increment are operational instructions that require no operand from memory and, therefore, do not use the address part of the instruction code. The shift instruction can utilize the part of the instruction code normally used to specify an address for storing an integer that specifies the number of places to be shifted. Other operational instructions are those concerned with format editing operations such as code translation, binary to decimal conversion, packing and unpacking of characters, and editing of data as, for example, during the preparation of output characters to a printer.

Inter-Register Transfer Instructions

Inter-register transfer instructions move data between registers without changing the information content. These instructions are essential for input and output movement of data and serve a necessary function before and after operational instructions. These instructions must specify a source and a destination register. The information at the source is invariably assumed to remain undisturbed. However, the previous information at the destination register must be destroyed after the transfer in order to accomodate the new information. An inter-register transfer instruction requires two addresses for its specification when the source and destination are both memory registers. However, the source or destination can be specified implicitly if they are chosen to be processor registers. A memory-to-memory inter-register transfer instruction is of the form:

Move M_1, M_2

where "Move" is a mnemonic name for the transfer operation code, M_1 is the address of the source register, and M_2 is the address of the destination register. This instruction can be implemented with a destructive-read type of memory as follows:

Read:

$(<M_1>) \Rightarrow B$ read source register
$(B) \Rightarrow <M_1>$ restore word
$(<M_1>)$ remain in B register

Write:

$0 \Rightarrow <M_2>$ clear destination register
$(<M_1>)$ are still in B register
$(B) \Rightarrow <M_2>$ store word in destination register

This two-address move instruction can be split into two one-address instructions. The first instruction loads the source information into a processor register; the second stores it in the destination register. These instructions were defined in the simple computer of Sec. 10-2 as

Load M_1
Store M_2

One-address computers move information from and to memory registers by load and store instructions. If the processor contains many registers, a load and a store instruction must be provided for each register, since processor registers are specified implicitly in the definition of the operation. An alternative method for specifying transfers between a memory register and one of a number of processor registers is an instruction format with a register-designate field. Thus, an instruction of the form:

Store M, R

where "Store" is an operation code, M is a memory address, and R is a binary number specifying one of the processor registers, will cause a transfer from the designated processor register to the designated memory register. The register-designate field R within the instruction format can be used to specify transfers between processor registers. An instruction of the form:

Transfer R_1, R_2

where "Transfer" is an operation code, R_1 a binary number that specifies the source register, and R_2 another binary number that specifies the destination register, will cause a transfer from the designated source register to the designated destination register.

Registers in the control unit and those associated with instruction sequencing are usually not available for transfer manipulation by programmable instructions. For example, the B, C, D, and I registers of the simple computer of Sec. 10-2 are used for internal control only. They are not available directly for manipulation with machine instructions. Only memory, processor, input, and output registers are normally directly specified in inter-register transfer instructions. However, it should be realized that computer instructions specify a function and in order to execute this function, it is necessary to move information in control registers. For example, the instruction

Add M

specifies the function

$$(A) + (\langle M \rangle) \Rightarrow A$$

Yet, to perform this function internally, the computer must perform the register transfers which are part of the fetch cycle and those necessary for reading the operand from memory. Only then is the specified function executed.

Test and Branch Instructions

Computer instructions are normally stored in consecutive memory registers and executed sequentially. This is accomplished, as explained in Sec. 10-2, by incrementing the address stored in the program-control register during the fetch cycle so that the next instruction is taken from the next memory register when execution of the current instruction is completed. This type of instruction sequencing uses a counter to calculate implicitly the address of the next instruction. Although this method is employed in practically all computers, it is possible also to have a computer with an explicit next-instruction scheme. By this method, the program-control register is eliminated. Instead, the address of the next instruction is explicitly specified in the present instruction. As an example, consider the two-address instruction:

Add M_1, M_2

with M_1 specifying an operand and M_2 specifying the address of the next instruction, rather than an operand. The function of the instruction may be defined to be $(<M_1>) + (A) \Rightarrow A$, which requires the following elementary operations for its execution:

$(I[M_1]) \Rightarrow D$	transfer address of operand
$(<D>) \Rightarrow B$	read operand
$(B) \Rightarrow <D>, (B) + (A) \Rightarrow A$	restore word, add to A

To fetch the next instruction from memory, the computer must proceed to use the second address M_2 as follows:

$(I[M_2]) \Rightarrow D$	transfer address of next instruction
$(<D>) \Rightarrow B$	read instruction
$(B) \Rightarrow <D>, (B) \Rightarrow I$	restore word, transfer instruction to I register

The new instruction, together with the address of the one following it, is now stored in the instruction register.

When the implicit counting instruction sequence method is used, the computer must include branch instructions in its repertoire to allow for the possibility of transfer out of normal sequence. The simplest branch instruction is the unconditional, which transfers control to the instruction whose address is given in the address part of the code. The unconditional, as well as one-conditional branch instruction, were illustrated in the simple computer of Sec. 10-2. Conditional branch instructions allow a choice between alternative courses of action, depending on whether certain test conditions are satisfied. The ability of branching to a different set of instructions as a result of a test condition on intermediate data is one of the most important characteristics of a stored program computer. This property allows the user to incorporate logical decisions in his data processing problem.

Test conditions that are usually incorporated with a computer instruction repertoire include comparisons between operands to determine the validity of certain relations such as equal, unequal, greater, smaller, etc. Other useful test conditions are associated with the relative magnitude of a number to check if it is positive, negative, or zero. Counting the number of times a program loop is executed is a useful test for branching out of the loop. Other typical test conditions encountered in digital computers are: input and output status bits, such as device ready or not ready; processor status error bits, such as overflow after an arithmetic addition; and status of console switches set by the operator. These status bits are stored in flip-flops and tested by conditional branch instructions.

The usual code format of a conditional branch instruction is of the form:

Branch-on-condition M

where "Branch-on-condition" is the mnemonic for the binary operation code which specifies the condition implicitly. When the condition being tested is satisfied, a branch to the instruction at address M is made. The program continues with the next consecutive instruction in sequence when the condition is not satisfied. Some typical conditional branch instructions are:

Branch to M if $(A) > 0$.

Branch to M_1 if $(<M_2>) < (A)$.

Branch to M if $(X) = 0$; where X is a processor register.

Branch to M if the overflow indicator is on.

These conditional branch instructions are executed by transferring the address part M (M_1 in example 2) from the instruction register to the program-control register if the test condition is satisfied. Otherwise, the address of the next instruction stored in the program-control register remains unaltered.

It is possible to have different definitions for the conditional branch instruction. For example, the branch to the instruction at address M may be made if the test condition is *not satisfied.* It is also possible to have a two-address branch instruction such as:

Branch on zero M_1, M_2	If $(A) = 0$ Branch to M_1
	If $(A) \neq 0$ Branch to M_2

It is also possible to have zero-address branch instructions. These instructions are called *skip*-type instructions. Their function is as follows: If the test condition is satisfied, the next instruction is skipped, otherwise the next instruction in sequence is executed. Consider the following three *consecutive* instructions:

Skip-on-zero accumulator	
Branch-Unconditional M_1	(This instruction is executed if $(A) \neq 0$)
Add M_2	(This instruction is executed if $(A) = 0$)

The first is a skip instruction which has no address part. If the condition of the skip instruction is satisfied, the next instruction is skipped and the contents of

$< M_2 >$ are added to the accumulator. If the condition is not satisfied, program control is transferred to the instruction at address M_1.

A very important branch instruction that any computer must provide is associated with entering and returning from subroutines. A subroutine is an independent set of consecutive instructions that performs a given function. During normal execution of a program, a subroutine may be called to perform its function with a given set of data. The subroutine may be called many times at various points of the program. Every time it is called, it executes its set of instructions and then transfers control back to the main program. The place of return to the main program is the instruction whose address follows the address of the instruction that called the subroutine. This address was in the program-control register just prior to the subroutine transfer. These concepts can be clarified with the following example. Consider a main program stored in memory starting from location 200, and a subroutine located in locations 500 to 565. "Enter-Subroutine" is a mnemonic name for an operation code that calls a subroutine whose first instruction is found in the address part of the calling instruction.

Main Program

	instruction	
location	operation	address
200	Enter-subroutine	500
201	Add	351
.	.	.
.	.	.
.	.	.
253	Enter-subroutine	500
254	OR	365
etc.		

Subroutine

location	instruction
500	first instruction of subroutine
.	.
.	.
.	.
.	.
.	.
565	Exit from subroutine (last instruction)

The instruction in location 200 calls a subroutine located at address 500. The execution of this instruction causes a branch to location 500 and then control continues sequentially to execute the instructions of the subroutine. When the last instruction in location 565 is decoded, the control unit interprets the operation code whose mnemonic name is "Exit from Subroutine" as a branch instruction back to the main program at location 201, the location following the instruction that called the subroutine. When the main program reaches location 253, the process of entering and returning from the same subroutine is repeated again, except that now the return address is to location 254. The "Exit from Subroutine" instruction must know to branch back to location 201 the first time and to location 254 the second time. The simplest method for providing the return address is to store the contents of the program-control register (which holds the address of the next instruction) in some temporary register before leaving the main program and entering the subroutine. This temporary register may be a reserved memory register, a reserved processor register, or the memory register prior to the subroutine program (location 499 in this example). One possible way to execute the two subroutine branch instructions with elementary operations is:

Enter-subroutine M:

$(I\ [M]) \Rightarrow C,\ (C) \Rightarrow X$

Exit from subroutine: (no address part)

$(X) \Rightarrow C$

where I is the instruction register, C is the program-control register, and X is a reserved processor register used implicitly for subroutine transfers.

10-5 CHARACTER MODE

A *character* is a binary code that represents a letter, a decimal digit, or a special symbol. A computer is said to operate in the *character mode* when characters are the units of information stored in memory registers or used during manipulation of information in the processor. This is in contrast with computers that operate in the *word mode* where the unit of information stored in registers represents data types used during the solution of numerical problems. The processing of scientific problems requires data types such as integers or floating-point operands, which require large word lengths, normally in excess of 24 bits. The processing of business data or other symbol-manipulation applications requires the use of character codes whose unit of information consists of six or eight bits.

The type of instructions useful in word mode operation is discussed in Sec. 10-4. Character mode instructions are similar, except that the unit of information is a character instead of a word. The arithmetic operations are the ones most often used when operating in the word mode. The instructions most commonly used in character mode are editing operations and data movement of character strings. The following are some applications that require character mode representation and operation.

1. The input and output data employed by the human user is invariably in decimal form and in a special format such as integer, floating-point, etc. The user-oriented data is represented in computer registers with a character code. It has to be translated into the corresponding data type used internally in computer registers.
2. A user-oriented programming language such as FORTRAN consists of letters, decimal digits, and special symbols. The computer stores the input program in memory registers as string of characters.
3. The translation from the user's program to machine-language code requires many character mode operations on the input character string.
4. Business and commercial applications deal with names of people, names of items, identification numbers, and similar information, represented in computer registers with a character code. The processing of such information requires a variety of character mode operations.

The differences between the needs of scientific computation and business data processing brought about the design and production of specialized computers oriented toward either word or character mode representation and operation. Later computer models, recognizing some common characteristics and needs of the two modes, have incorporated both types in one machine. Computers may be grouped in four categories according to their representation and use of basic units of information such as words and characters.

1. *Word machine.* This type of computer can operate only in the word mode. Memory and processor register lengths are fixed; an instruction always specifies an entire word. This type of machine is usually called a "scientific computer." However, character mode operations are needed during input, output, and translation of user-oriented programs. These operations are usually manipulated by indirect and inefficient means using the existing word-oriented instructions.
2. *Character machine.* This computer type uses a length of memory and processor registers equal to the number of bits used to represent a character. Data is stored as a string of characters and is referenced from memory by specifying the address of the first character in the string and some indication as to the last character. Last character

indication may be the address of the last character in the string, a character count, or, for variable-length data, a mark in one bit of the character code that designates end of data. This type of machine uses decimal arithmetic with a four-bit code to represent a decimal digit (the remaining bits of the character code are either removed or used for other purposes). Instruction codes have their operation part in one character; addresses are placed in succeeding character locations in memory. Data execution is usually accomplished through manipulation of one character at a time. This type of machine is usually called a "business-oriented computer."

3. *Word machine with character mode.* This type of computer is basically a word machine, except that character mode representation and operation is also provided. The memory and processor registers are of fixed word length but may, for character mode, be considered as storing a fixed number of distinguishable characters. For example, a computer with a word length of 48 bits and a character code of 6 bits can store eight characters in a single word. Character mode instructions have in their address part, in addition to the bits specifying the address of a word, a three-bit character designation field that specifies a character within the word.

4. *Byte machine.* A *byte* is a unit of information eight bits long. It can represent a variety of different information items depending on the particular operating mode of the computer. The various units of information are chosen to represent multiples of bytes, their length is recognized by the type of instruction used. For example, data may be represented as follows:

decimal digit	half byte or four bits
alphanumeric character	one byte or eight bits
binary integers	two bytes for single precision
	four bytes for double precision
binary floating-point numbers	eight bytes

Instructions may be of variable length, with the operation code occupying one byte. The number of addresses belonging to the instruction is determined from the type of operation code and may be from no-address to possibly two or three. Word-oriented instructions will normally use fixed-length binary numerical data. Character mode instructions may use variable-length decimal data or character strings.

10-6 ADDRESSING TECHNIQUES

The sequence of operations that implement a machine instruction has been divided in Sec. 10-2 into an instruction fetch cycle and a data execute cycle. The instruction is read from memory during the fetch cycle and placed in the instruction register. During the execute cycle, the operand (if needed) is read from memory and the instruction is then executed. If no operand is needed, the instruction is executed without a second reference to memory. To describe some additional features used to determine the address of operands, the sequence of operations needed during the execute cycle is more conveniently subdivided into two phases: the data-fetch and the instruction-execution. The instruction-execution phase consists of those operations that execute the instruction after the operands are available. The *data-fetch* phase is responsible for the selection of memory registers used for data when the instruction is executed. It is skipped for those instructions that do not use memory registers for data, but may otherwise consist of a fairly involved sequence of elementary operations. The data-fetch phase considered so far consists of one reference to memory for reading or storing an operand. This phase becomes somewhat more complicated when the computer uses certain addressing techniques such as indirect and relative addressing.

Indirect Address

Added to the operation part of an instruction may be a second part designated as the address part M. However, the binary field M may sometimes designate an operand or the address of an operand. When the second part of the instruction code is the actual operand, as in Fig. 10-1(d), the instruction is said to have an *immediate* address, or a *literal* designation. When the address part M specifies the address of an operand, the instruction is said to have a *direct* address. This is in contrast with a third possibility called *indirect* address, where the field M designates an address of a memory register in which the *address* of the operand is found. It is customary to use one bit in the instruction code, sometimes called a *tag* or a *mode* bit, to distinguish between a direct and an indirect address. Consider for example the instruction code format shown in Fig. 10-6. It consists of a four-bit operation code, a five-bit address part M, and an addressing mode bit. The mode bit is designated by d and is equal to 0 for a direct address and to 1 for an indirect address. The operation code of the

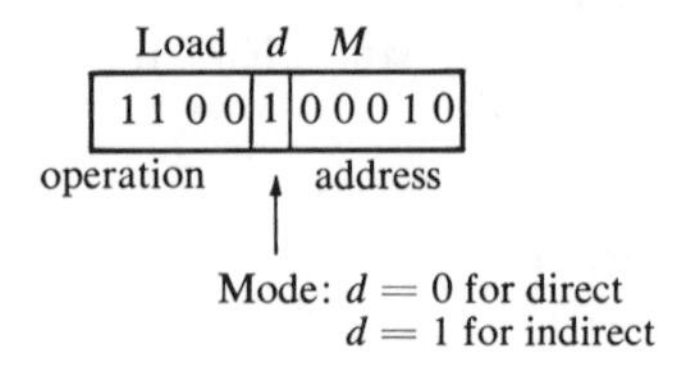

Figure 10-6 Instruction code format with mode bit

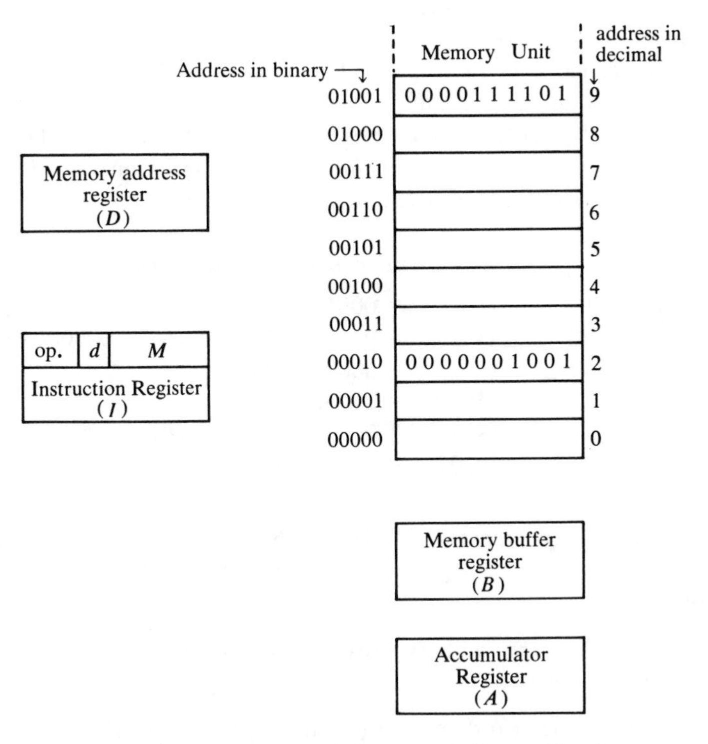

Figure 10-7 Example to demonstrate indirect addressing

instruction specifies a load function. The mnemonic designation which has been used for such an instruction with the mode bit equal to 0 is:

Load M

When the mode bit is a 1, it is customary to write the instruction as

Load $* M$

where $*$ stands for indirect. The function of this instruction can be specified in symbolic notation as

$(<M>) \Rightarrow A$ for direct address
$(<(<M>)>) \Rightarrow A$ for indirect address

$(<M>)$ is a symbol for the contents of the memory register whose address is M. It follows that the symbol shown for the indirect address is the contents of the memory register whose address is $(<M>)$.

The indirect address instruction requires two references to memory to fetch an operand. The first reference is needed to read the address of the operand; the second is for the operand itself. This can be demonstrated with an example using the numerical values of Fig. 10-7 to implement the load instruction specified in Fig. 10-6. For the direct mode, $d = 0$ and the A register receives the contents of memory register 2, which is 0000001001. For the indirect mode as shown, $d = 1$ and the content of memory register 2 is the address of the operand. This address is 9 and therefore the A register receives the contents of memory register 9, which is 0000111101.

The indirect load instruction requires the following sequence of elementary operations after the fetch cycle:

	Content of registers when a change occurs	
1. $(I[M]) \Rightarrow D$	$(D) = 00010 = (2)_{10}$	
2. $(\langle D \rangle) \Rightarrow B$	$(\langle 2 \rangle) = 0$	$(B) = 0000001001$
3. $(B) \Rightarrow \langle D \rangle$	$(\langle 2 \rangle) = 0000001001$	
4. $(B[M]) \Rightarrow D$	$(D) = 01001$	
5. $(\langle D \rangle) \Rightarrow B$	$(\langle 9 \rangle) = 0$	$(B) = 0000111101$
6. $(B) \Rightarrow \langle D \rangle$	$(\langle 9 \rangle) = 0000111101$	
7. $(B) \Rightarrow A$		$(A) = 0000111101$

Steps 1 to 6 encompass the data-fetch phase; step 7 executes the operation. The above sequence performs the following elementary operations:

Step 1 transfers the address part of the instruction register to the memory-address register.

Step 2 reads the word into the memory-buffer register. This word contains the address of the operand.

Step 3 restores the word to the memory register.

Step 4 transfers the address part of the memory-buffer register (address of operand) to the memory-address register.

Step 5 reads operand into the memory-buffer register.

Step 6 restores the word to the memory register.

Step 7 transfers operand to accumulator register.

Relative Address

A relative address does not indicate a storage location itself, but a location relative to a reference address. The reference address may be a special processor register or a special memory register. When the address field M of the instruction code is relative address, the actual address of the operand, called the *effective* address, is calculated during the data-fetch phase as follows:

$$\text{effective address} = \text{contents of a special register} + M$$

The effective address may be considered to consist of the sum of a *base address* and a *modifier address*. It is convenient to distinguish between two types of relative addressing schemes. The first type contains a *base-address register* within the control unit and allows the field M to designate the modifier address. The second type uses the field M as the base address and a special register, called an *index register*, acts as the modifier. We shall proceed to explain in more detail the second type and then return to the first.

An index register is a processor or a special memory register whose contents are added to the address part M to obtain the effective address of the operand. A digital computer may have several index registers. It is customary to include an index-register field in the instruction code to specify the particular one selected. This is demonstrated in the instruction-code format shown in Fig. 10-8. The index-register field designated by X may specify register $X1$, $X2$, or $X3$ with the binary code 01, 10, or 11, respectively. The code $X = 00$ specifies no index register. The effective address is obtained during the data-fetch phase with the elementary operation

$$(I[M]) + (Xn) \Rightarrow D \qquad n = 1, 2, 3$$

When $X = 0$, no index register is used and the effective address is M:

$$(I[M]) \Rightarrow D$$

Fig. 10-9 shows a numerical example for implementing the load instruction whose format is specified in Fig. 10-8. The relative address is $M = 2$, and the index-register field specifies $X2$. The contents of $X2$, equal to 7, are added to M to obtain the effective address 9. The contents of memory register 9 are read from memory and transferred to the A register.

Computers with index registers contain a variety of instructions which manipulate with these registers. The instructions are similar to the ones considered in Sec. 10-4 and can be divided into the usual three classes:

1. Inter-register transfer instructions load and store contents of index registers. Some examples are:

Load-index M, X	$(<M>) \Rightarrow Xn$
Load-index immediate M, X	$M \Rightarrow Xn$
transfer A to index register X	$(A) \Rightarrow Xn$

   where X designates the index-register field and M is the address part of the instruction.

2. Operational-type instructions involve the usual unary and some binary operations. Some examples are:

Clear X	$0 \Rightarrow Xn$
Increment X	$(Xn) + 1 \Rightarrow Xn$

Decrement X	$(Xn) - 1 \Rightarrow Xn$
Increment X, D	$(Xn) + D \Rightarrow Xn$
Decrement X, D	$(Xn) - D \Rightarrow Xn$

where the field D contains explicitly the amount of incrementing or decrementing requested. The increment or decrement instructions without the D field specify the value of 1 implicitly.

3. Test and branch instructions usually test for a specific value of Xn to cause a branch. Some examples are:

Branch on index zero M, X	if $(Xn) = 0$, then $M \Rightarrow C$
Decrement and branch on zero M, X	$(Xn) - 1 \Rightarrow Xn$
	if $(Xn) = 0$, then $M \Rightarrow C$

The convenience of indexing can be demonstrated by a simple example. The addition of 100 numbers stored in consecutive memory registers starting from location 50 can be implemented with 101 instructions as follows:

Clear A
Add 50
Add 51
Add 52
⋮
Add 149

The use of index registers reduces the number of instructions considerably. The following program uses one index register to increment the effective address and another to count the number of additions.

Location	Instruction		Function
200	load immediate	100, $X1$	$100 \Rightarrow X1$
201	load immediate	50, $X2$	$50 \Rightarrow X2$
202	clear A		$0 \Rightarrow A$
203	add	0, $X2$	$(<X2>) + (A) \Rightarrow A$
204	increment	$X2$	$(X2) + 1 \Rightarrow X2$
205	decrement and		$(X1) - 1 \Rightarrow X1$
	branch on zero	207, $X1$	if $(X1) = 0$, then $207 \Rightarrow C$
206	branch	203	$203 \Rightarrow C$
207	next instruction		

The second type of relative addressing scheme uses a base-address register in the control unit and allows M to designate an address relative to it. As an example, consider an instruction format having seven bits for the address part. Suppose that the memory unit consists of 16,384 words, each of which require an address of 14 bits. The seven least significant bits of the address can be specified by M, while the most significant seven bits can be

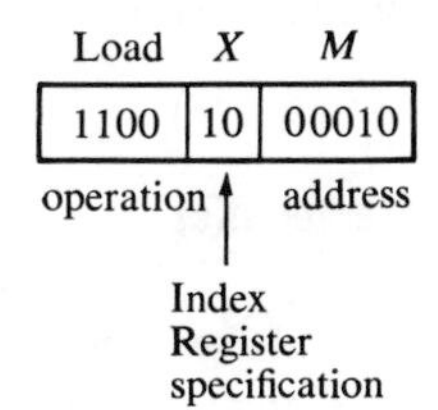

Figure 10-8 Instruction code format with index register

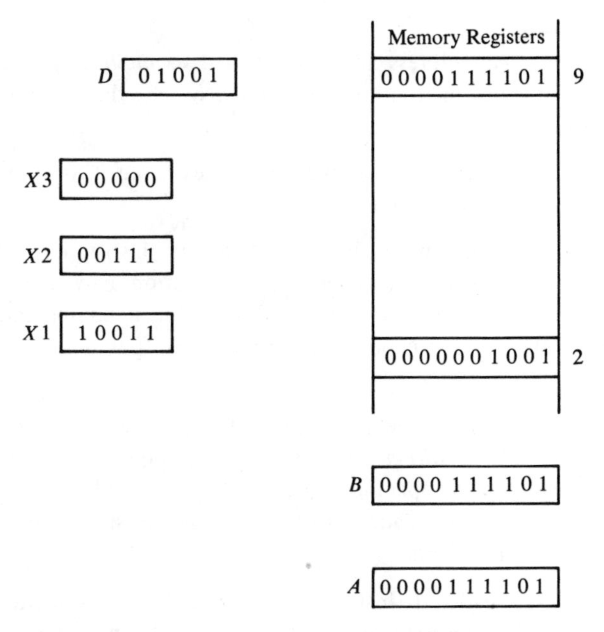

Figure 10-9 Example to demonstrate the use of index registers

stored in the base register. During the data-fetch phase, the effective address is obtained by combining the two parts. The value of the base register can be changed by the program with an instruction like:

Load base register M $\qquad M \Rightarrow$ Base register

And this value remains unchanged as long as instructions reference operands relative to the present value of the base register.

Relative addressing with base registers is useful in small computers for increasing the range of memory when only a limited number of bits are available for the address part of the instruction. Some large computers use base registers for the ease of relocation of programs and data. When programs and data are moved from one sector of memory to another, as required in multiprogramming and time-sharing systems, the address field of instructions do not have to change. Only the value of the program base register requires updating.

Summary of Addressing Characteristics

Various addressing techniques have been introduced throughout this chapter. The addressing characteristics that have been mentioned are now summarized in order to place the various possibilities in their proper perspective.

1. *Implication*: An address may be implied *explicitly* by a binary code as part of the instruction or *implicitly* in the definition of the operation.
2. *Number*: The number of explicit addresses in an instruction may vary from zero to three or more.
3. *Type*: In general, the address field of an instruction refers to an address of a memory register. An instruction may also have other address fields for the purpose of designating a unique index or processor register, or for choosing an input or output unit.
4. *Mode*: An *immediate* address is in reality not an address as such because the address field is taken to be the actual operand. An instruction with a *direct* address mode considers the address field as the address of an operand. The address field of an *indirect* mode instruction specifies an address of a memory register where the address of the operand is found.
5. *Relative*: An *indexed* instruction uses its address field as a base address to be modified by the contents of an index register. An address field relative to a *base-address register* specifies a certain number of lower significant bits of the effective address, while the base-address register holds the remaining higher significant bits.
6. *Resolution*: The address field of an instruction may specify a memory register consisting of a number of bits referred to as a *word*. Word sizes may range from 12 to 64 bits. The address field may specify a memory register that stores one alphanumeric *character*. A character length may be six or eight bits. The address field could, if necessary, have a resolution to each *bit* of memory.
7. *Length*: The length of the memory register specified by the address field or fields may be *fixed*, as is usually the case in word-oriented instructions. It may be made *variable* by specifying the first and last memory register or the first memory register and a register count.

PROBLEMS

10-1. A digital computer has a memory unit with a capacity of 4096 words, each 36 bits long. The instruction set consists of 55 operations.

(a) What is the minimum number of bits required for the operation code?

(b) How many bits are needed for the address part of an instruction:

(c) How many instructions can be packed in one word using zero-address, one-address, or two-address instruction formats?

(d) What happens to the fetch-cycle when two instructions are packed in one word?

10-2. Using the simple computer structure of Fig. 10-2 with two accumulator registers R1 and R2, list the register transfer elementary operations needed to execute the instructions given in Fig. 10-1 and defined in the text. Assume a magnetic-core memory.

10-3. Given: the simple computer structure of Fig. 10-2. For each instruction defined below:

(a) give a possible one-address format and a definition of the instruction in your own words, and

(b) list the sequence of elementary operations required to execute the instruction. Assume a nondestructive read-out memory.

(1) Add input and store: $(A) + (P) \Rightarrow <M>$

(2) Three-way test: If $(A) > 0$, branch to $<M>$; if $(A) < 0$, skip next instruction; if $(A) = 0$, proceed to next instruction in sequence.

(3) Increment memory register: $(<M>) + 1 \Rightarrow <M>$

(4) Replace: $(A) \Rightarrow <M>$ and $(<M>) \Rightarrow A$

(5) Skip; if not, increment: If $(<M>) = 0$, skip next instruction; if $(<M>) \neq 0$, then $(<M>) + 1 \Rightarrow <M>$.

10-4. Give the binary code of each instruction in location 750-754 of the program listed in Sec. 10-2. Use the operation code of Table 10-2 and a memory unit with a capacity of 2048 words of 15 bits each.

10-5. Assume that subtraction can be done in the simple computer by using the 2's complement representation (see Sec. 9-9). Modify the machine-code program of Sec. 10-2 to include a test for the number of input words entered from the P register and branch to a "Stop" instruction after this number reaches 750.

10-6. A memory unit has a capacity of 65,536 words of 25 bits each. It is used in conjunction with a general purpose computer.

(a) What is the length of the memory-address and buffer registers?

(b) How many bits are available for the operation part of a one-address instruction word?

(c) What is the maximum number of operations that may be used?

(d) If one bit of a binary integer word is used to store the sign of a number, what are the largest and smallest integers that may be accommodated in one word?

10-7. Explain the difference between a central processor, a peripheral processor, and a data communications processor.

10-8. Specify a format of a control word for the peripheral processor that requests input of 256 blocks from a magnetic-tape unit starting from block number 150. A block of storage contains 128 characters of six bits each.

10-9. List the sequence of operations necessary to service a program interrupt.

10-10. Give a character format between a data communication processor and a data communication network for a transmission of text from the computer to the remote terminal.

10-11. What is the difference between:

(a) a computer program and a microprogram?

(b) a computer instruction and a microinstruction?

(c) an operation and a micro-operation?

(d) an elementary operation and a micro-operation?

10-12. Give the mnemonic and binary instructions generated by a compiler from the following FORTRAN program. (Assume integer variables.)

```
SUM = 0
SUM = SUM + A + B
DIF = DIF + C
SUM = SUM + DIF
```

10-13. Repeat Prob. 10-12 using the following FORTRAN program:

```
IF (ALPHA + BETA) 10,20,10
10 X = A .AND. B .OR. C .AND. D
20 X = .NOT. X
```

10-14. List the sequence of machine instructions needed to multiply two integers. Use a repeated additional method. For example, to multiply 5 × 4, the computer performs the repeated addition 5 + 5 + 5 + 5.

10-15. Write a program; that is, list the sequence of machine instructions to perform an arithmetic addition of two floating-point numbers. Assume that each floating-point number occupies two consecutive memory registers; the first stores the exponent and the second stores the coefficient. Assume to have a computer with the instructions listed in Table 10-2 in addition to the following two instructions:

Branch on positive accumulator M.
Subtract M; i.e., $(A) - (\langle M \rangle) \Rightarrow A$

10-16. List the sequence of elementary operations needed to execute the single machine instruction described by the function:

$$(\langle M_1 \rangle) \oplus (\langle M_2 \rangle) \Rightarrow \langle M_3 \rangle$$

10-17. List the sequence of elementary operations needed to execute the single machine instruction whose mnemonic notation is

Move M_1, M_2, M_3

The instruction performs the following functions:

$(\langle M_1 \rangle) \Rightarrow \langle M_2 \rangle$
$(\langle M_1 + 1 \rangle) \Rightarrow \langle M_2 + 1 \rangle$
$(\langle M_1 + 2 \rangle) \Rightarrow \langle M_2 + 2 \rangle$ etc., up to
$(\langle M_1 + M_3 \rangle) \Rightarrow \langle M_2 + M_3 \rangle$

10-18. Let a "Branch to Subroutine" instruction with an address part M be executed as follows:

$(C) \Rightarrow \langle M \rangle$
$M + 1 \Rightarrow C$

Explain in your own words how the branching is done and where the beginning of the subroutine is. Suggest an instruction for returning to main program.

10-19. Give the sequence of elementary operations when both an index register and indirect address are specified in one machine instruction. The instruction is first modified by an index register (if specified) and then checked for an indirect address.

REFERENCES

1. Burks, A, W., H. H. Goldstein, and J. von Neuman. "Preliminary Discussion of the Logical Design of an Electronic Computing Instrument (1946)," *John von Neuman Collected Works.* New York: The Macmillan Co., 1963. Vol. V, 34-79.
2. Buchholz, W. (Editor). *Planning a Computer System.* New York: McGraw-Hill Book Co., 1962.
3. Chu, Y. *Introduction to Computer Organization.* Englewood Cliffs, N.J.: Prentice-Hall, Inc., 1970.
4. Flores, I. *Computer Organization.* Englewood Cliffs, N.J.: Prentice-Hall, Inc., 1969.
5. Martin, J. *Teleprocessing Network Organization.* Englewood Cliffs, N.J.: Prentice-Hall, Inc., 1970.

6. Vandling, G. C., and D. E. Waldecker. "The Microprogram Control Technique for Digital Logic Design," *Computer Design*, Vol. 8 No. 8. (August 1969), 44-51.

7. Husson, S. S. *Microprogramming: Principles and Practice.* Englewood Cliffs, N.J.: Prentice-Hall, Inc., 1970.

8. Hellerman, H. *Digital Computer System Principles.* New York: McGraw-Hill Book Co., 1967.

11 COMPUTER DESIGN

11-1 INTRODUCTION

This chapter presents the design of a small general purpose digital computer starting from its functional specifications and culminating in a set of Boolean functions. Though this computer is simple, it is far from useless. Its scope is quite limited when compared with commercial electronic data processing systems yet it encompasses enough functional capabilities to demonstrate the design process. It is suitable for construction in the laboratory with ICs, and the finished product can be a useful system capable of processing digital data.

The computer consists of a central processor unit, a memory unit, a teletype input unit, and a teleprinter output unit. The logic design of the central processor unit is given in detail. The other units are assumed to be available as finished products with known external characteristics.

The design of a digital computer may be divided into three interrelated phases: system design, logic design, and circuit design. System design concerns the specifications and the general properties of the system. This task includes the establishment of design objectives, design philosophy, computer architecture, required operations, speed of operation, and economic feasibility. The specifications of the computer structure are translated into Boolean functions by the logic design. The circuit design specifies the components and circuits for the various logic circuits, memory circuits, electromechanical equipment, and power supplies. The computer hardware design is greatly influenced by the software system, which is normally developed concurrently and which constitutes an integral part of the total system. The design of a digital computer is a formidable task that requires thousands of man-hours of effort. One cannot expect to cover all aspects of the design in one chapter. Here we are concerned with the system and logic design phases of a small digital computer whose specifications are formulated somewhat arbitrarily in order to establish a minimum configuration for a very small, yet practical machine. The procedure outlined in this chapter can be useful in the logic design of more complicated systems.

The design process is divided into six phases:

(1) the decomposition of the digital computer into registers which specify the general configuration of the system,

(2) the specifications of machine instructions,

(3) the formulation of a timing and control network,

(4) the listing of the sequence of elementary operations needed to execute each machine instruction,

(5) the listing of the elementary operations to be executed on each register under the influence of control signals, and

(6) the derivation of the input Boolean functions for each flip-flop in the system and the Boolean functions necessary to implement the control network.

The step-by-step execution of the operations specified in (4) and (5) are described by register transfer relations. The Boolean functions obtained in step (6) culminate the logic design. The implementation of these functions with logic circuits and flip-flops has been covered extensively in previous chapters.

11-2 SYSTEM CONFIGURATION

The configuration of the computer is shown in Fig. 11-1. Each block represents a register except for the memory unit, the console, the master-clock generator, and the control logic. This configuration is assumed to be unalterable. In practice, however, the designer starts with a tentative system configuration and constantly modifies it during the design process. The name of each register is written inside the block, together with a letter in parentheses. This letter represents the register in symbolic register transfer relations and in Boolean functions. The number in the lower right corner of each block is the number of flip-flops in the register. The configuration shown in Fig. 11-1 is very similar to the one used for the simple computer introduced in Sec. 10-2.

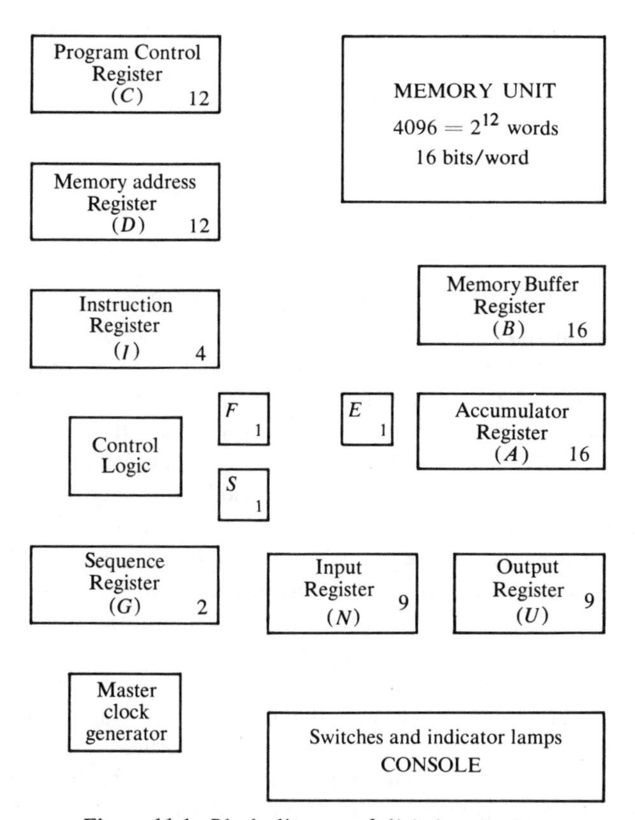

Figure 11-1 Block diagram of digital computer

We shall restrict the design of the computer to the parts that can be described by Boolean functions. Certain parts do not fall in this category and are assumed to be available as finished products with known external characteristics. These parts are the master-clock generator, the memory unit, the console and its associated circuits, and the I/O devices. The rest of the computer can be constructed with flip-flops and combinational circuits.

Master-Clock Generator

The master-clock generator is a common clock-pulse source, usually an oscillator, which generates a periodic train of pulses. These pulses are fanned-out by means of amplifiers and distributed over the space the system occupies. Each pulse must reach every flip-flop at the same instant. Phasing delays are needed intermittently so that the difference in transmission delays is uniform throughout. The frequency of the pulses is a function of the speed by which the system operates. We shall assume a frequency of 1 megahertz (one pulse every microsecond) for this computer. This frequency is lower than what is normally used in commercial systems but was chosen here for the sake of having a round number and to avoid problems of circuit propagation delays.

The Memory Unit

The memory unit is a magnetic-core random-access type with a capacity of 4096 words of 16 bits each. The capacity was chosen to be large enough for meaningful processing. A smaller size may be used if the computer is to be constructed in the laboratory under economic restrictions. Twelve bits of an instruction word are needed to specify the address of an operand, which leaves four bits for the operation part of the instruction.

The memory unit communicates with the address register and the buffer register in the usual manner. The memory cycle control signal, designated by m, initiates one read-write cycle for a period of 4 microseconds. During the first 2 microseconds, the word located at the specified address given by the address register is accessed, read into the buffer register, and erased from memory (destructive reading). During the following 2 microseconds, the word in the buffer register is stored in the location specified by the address register. Thus, a word can be read during the first half of the cycle and restored during the second half if the contents of the address and buffer registers are undisturbed. On the other hand, a new word can be stored in the addressed location if the address register is undisturbed but the contents of the buffer register are changed prior to the beginning of the second half of the memory cycle. To simplify the logic design, we shall not distinguish between a read and a write cycle. The memory cycle control signal (m) will automatically initiate a *read* followed by a *write* as described above. A timing diagram for the memory unit is shown in Fig. 11-4.

The Console

The computer console consists of switches and indicator lamps. The lamps indicate the status of registers in the computer. The normal output of a flip-flop applied to a lamp (with amplification if needed) causes the light to turn on when the flip-flop is set and to turn off when it is cleared. The purpose of the switches is to start, stop, and manually communicate with the computer. The function of the switches and the circuits they drive is explained later; their purpose can be better understood after some aspects of the computer are better known. The design of the computer is carried out without regard to these switches. In Sec. 11-8 we shall enumerate the console switches, explain their function, and specify the circuits associated with their operation.

The Input-Output Device

The I/O device is not shown in Fig. 11-1. It is assumed to be an electromechanical unit with a typewriter keyboard, a printer, a paper tape reader, and a paper tape punch. The input device consists of either the typewriter or the paper tape reader, with a manual switch available for selecting the one used. The output device consists of the printer or the paper tape punch, with another switch available for selecting either device. The unit uses an eight-bit alphanumeric code and has available two control signals indicating the status of the I/O devices. These control signals activate flip-flops, which indicate whether the device is busy or ready to transfer information. Registers are physically mounted in the unit and are used to store the input and output eight-bit information and control signals. The specifications for the I/O registers are given in more detail below.

The Registers

The part of the digital computer to be designed subsequently is decomposed into register subunits which specify the configuration of the system. Any digital system that can be decomposed in this manner can be designed by the procedure outlined in this chapter. Very briefly, this procedure consists of first obtaining a list of all the elementary operations to be executed on the registers, together with the *conditions* under which these operations are to be executed. The list of conditions is then used to design the control circuits, while the input Boolean functions to registers are derived from the list of elementary operations.

Using the frame of reference from Ch. 10, the register configuration of Fig. 11-1 is an absolute necessity. The following paragraphs explain why the registers are needed and what they do. A list of the registers and a brief description of their function can be found in Table 11-1. Registers that hold memory words are 16 bits long. Those that hold an address are 12 bits long. Other registers have different numbers of bits depending on their function.

Memory-Address and Memory-Buffer Registers

The memory-address register (D) is used to address specific memory locations. The D register is loaded from the C register when an *instruction* is to be read from memory and from the 12 least significant bits of the B register when *data* is to be read from the memory. The memory-buffer register (B) holds the word read from or written into memory. The operation part of an instruction word placed in the B register is transferred to the I register and the address part is left in the register for further manipulation. A data word placed in the B register is accessible for operation with the A register. A word to be stored in memory must be loaded into the B register at the beginning of the second half of a memory cycle.

Table 11-1 List of Registers

Letter Designation	*Name*	*Number of Bits*	*Function*
A	Accumulator Register	16	multipurpose register
B	Memory-Buffer Register	16	holds contents of memory word
C	Program Control Register	12	holds address of next instruction
D	Memory-Address Register	12	holds address of memory word
E	Extended flip-flop	1	extra flip-flop for accumulator
F	Fetch flip-flop	1	determines fetch or execute cycle
G	Sequence Register	2	sequence counter
I	Instruction Register	4	holds current operation
S	Start-Stop flip-flop	1	determines if computer is running or stopped
N	Input Register	9	holds information from input device
U	Output Register	9	holds information for output device

Program Control Register (C)

The C register is the computer's program counter. This means that this register goes through a step-by-step counting sequence and causes the computer to read successive instructions previously stored in memory. When the program calls for a transfer to another location or for skipping the next instruction, the C register is modified accordingly, thus causing the program to continue from a different location in memory. To read an instruction, the contents of the C register are transferred to the D register and a memory cycle is initiated. The C register is always incremented by 1 right after the initiation of a memory cycle that reads an instruction. Therefore, the address of the next instruction after the one presently being executed is always available in the C register.

Accumulator Register (A)

The A register is a multipurpose register that operates on data previously stored in memory. This register is used to execute most instructions and for accepting data from the input device or transferring data to the output device. This register, together with the B register, make up the so-called "arithmetic" unit of the computer. Although most data processing systems include more registers for this unit, we have chosen to include only two in order not to complicate the computer. With two registers in the arithmetic unit, only addition and/or subtraction can be implemented directly. Other operations, such as multiplication and division, are implemented by a sequence of instructions that form a subroutine.

Instruction Register (*I*)

The *I* register holds the operation bits of the current instruction. This register has only four flip-flops, since the operation part of the instruction is four bits long. The operation part of an instruction is transferred to the *I* register from the *B* register, while the address part of the instruction is left in the *B* register. The operation part must be taken out of the *B* register because an operand read from memory into the *B* register will destroy the previously stored instruction. (The operation part of the instruction is needed to determine what is to be done to the operand just read.)

Sequence Register (*G*)

This register is a sequence counter that produces timing signals for the entire computer. It is a two-bit counter-decoder excited by clock pulses with a period of 1 μsec. The four outputs of the decoder supply four timing signals each with a period of 4 μsec and a phase delay between two adjacent signals of 1 μsec. The relation between the timing signals and the memory unit is clarified in Sec. 11-4.

E, *F*, and *S* Flip-Flops

Each of these flip-flops is considered a one-bit register. The *E* flip-flop is an extension of the *A* register. It is used during shifting operations, receives the end-carry during addition, and otherwise is a useful flip-flop that can simplify the data processing capabilities of the computer. The *F* flip-flop distinguishes between the fetch and execute cycles. When *F* is 0, the word read from memory is treated as an instruction. When *F* is 1, the word is treated as an operand. *S* is a start-stop flip-flop that can be cleared by program control and manipulated manually from the console. When *S* is 1, the computer runs according to a sequence determined by the program stored in memory. When *S* is 0, the computer stops its operation.

Input and Output Registers

The input register (*N*) consists of nine bits. Bits 1 to 8 hold alphanumeric input information; bit 9 is a control flip-flop. The control bit is set when new information is available in the input device and cleared when the information is accepted by the computer. The control flip-flop is needed to synchronize the timing rate by which the input device operates compared to the computer. The normal process of information transfer is as follows: Initially, the control flip-flop N_9 is cleared by a "Start" pulse from the console. When the teletype key in the input device is struck, an eight-bit code character is loaded into the N register and N_9 is set. As long as N_9 is set, the information in the N register cannot be changed by striking another key. The computer checks the control bit; if it is a 1, the information from the N register is transferred into the accumulator register and N_9 is cleared. Once N_9 is cleared, new information can be loaded into the N register by striking a key again.

The output register (U) works similarly but the direction of information flow is reversed. A "Start" pulse from the console clears the control bit U_9. The computer checks the control bit; if it is a 0, the information from the accumulator register is transferred into U_1 to U_8 and the control bit U_9 is set. The output device accepts the coded information, prints the corresponding character, and clears the control bit U_9. The computer does not load a new character into the output device when U_9 is set because this condition indicates that the output device is in the process of printing the character.

11-3 MACHINE INSTRUCTIONS

The number of instructions available in a computer and their efficiency in solving the problem at hand is a good indication of how well the system designer foresaw the intended application of the machine. Medium- to large-scale computing systems may have hundreds of instructions, while most small computers limit the list to 40 or 50. The instructions must be chosen carefully to supply sufficient capabilities to the system for solving a wide range of data processing problems. A minimum requirement of such a list should include a capability for storing and loading words from memory, a sufficient set of arithmetic and logical operations, some address modification capabilities, unconditional branching and branching under test conditions, register manipulation capabilities, and I/O instructions. The instruction list chosen for our computer is believed to be close to the absolute minimum required for a restricted but practical data processor.

The formulation of a set of instructions for the computer goes hand in hand with the formulation of the formats for data and instruction words. A memory word consists of 16 bits. A word may represents either a unit of data or an instruction. The formats of data words are shown in Fig. 11-2. Data for arithmetic operations are represented by a 15-bit binary number with the sign in the 16th bit position. Negative numbers are assumed to be in their 2's complement equivalent. Logical operations are performed on individual bits of the word, with bit 16 treated as any other bit. When the

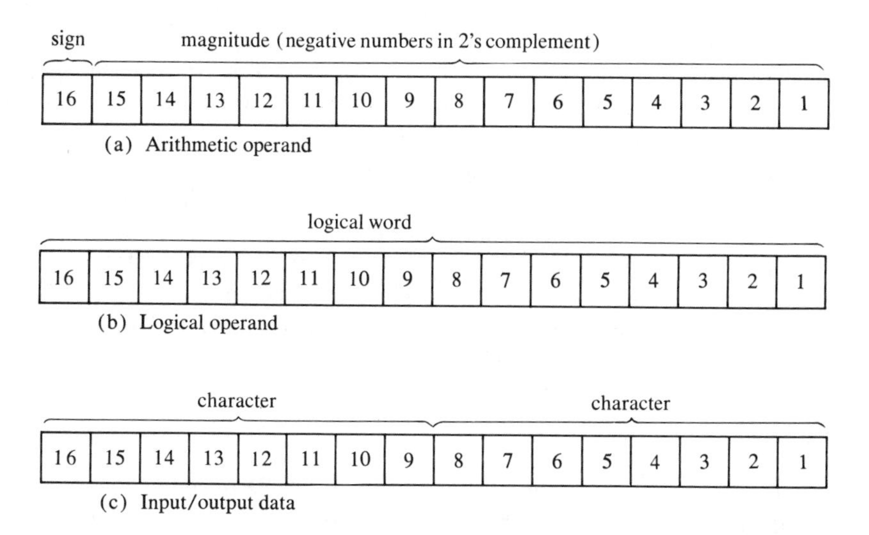

Figure 11-2 Data formats

computer communicates with the I/O device, the information transferred is considered to be eight-bit alphanumeric characters. Two such characters can be accommodated in one computer word.

The formats of instruction words are shown in Fig. 11-3. The operation part of the instruction contains four bits; the meaning of the remaining twelve bits depends on the operation code encountered. A *memory-reference* instruction uses the remaining 12 bits to specify an address. A *register-reference* instruction implies a unary operation on, or a test of, the A or E register. An operand from memory is not needed; therefore, the 12 least significant bits are used to specify the operation or test to be executed. A register-reference instruction is recognized by the code 0110 in the operation part. Similarly, an *input-output* instruction does not need a reference to memory and is recognized by the operation code 0111. The remaining 12 bits are used to specify the particular device and the type of operation or test performed.

Only four bits of the instruction are available for the operation part. It would seem, then, that the computer is restricted to a maximum of 16 distinct operations. However, since register-reference and input-output instructions use the remaining 12 bits as part of the operation code, the total number of instructions can exceed 16. In fact, the total number of instructions chosen for the computer is 23.

Out of the 16 distinct operations that can be formulated with four bits, only 8 have been utilized by the computer. This is because the left-most bit of all instructions (bit 16) is always a 0. This leaves open the possibility of adding new instructions and extending the computer capabilities if desired.

The 23 computer instructions are listed in Tables 11-2, 11-3, and 11-4. The mnemonic designation is a three-letter word and represents an abbreviation intended for use by programmers and users. The hexadecimal code listed is the equivalent hexadecimal number of the binary code used for the instruction. A memory-reference instruction uses one hexadecimal digit (four bits) for the operation part; the remaining three hexadecimal digits (twelve bits) of the instruction represent an address designated by the letter M. The register-reference and input-output instructions use all four hexadecimal digits (16 bits) for the operation. The function of each instruction is specified by symbolic notation that denotes the machine operation that it executes. A further clarification of each instruction is given below, together with an explanation of its utility.

AND to A

This is a logical operation that performs the AND operation on corresponding pairs of bits in A and the memory word and leaves the result in A. This instruction, together with CMA instruction (see Table 11-3) that performs the NOT logical operation, supply a sufficient set of logical operations to perform all the binary operations listed in Table 2-5.

ADD to A

This operation adds the contents of A to the contents of the memory word and transfers the sum into A. The sign bit is treated as any other bit

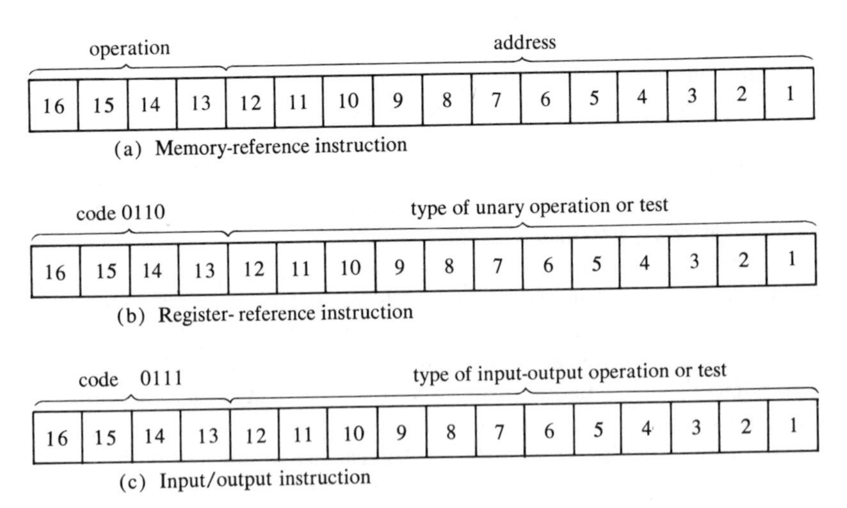

Figure 11-3 Instruction formats

Table 11-2 Memory Reference Instructions

Mnemonic	*Hexa-decimal Code*	*Description*	*Function*
AND M	0	AND to A	$(<M>) \wedge (A) \Rightarrow A$
ADD M	1	add to A	$(<M>) + (A) \Rightarrow A$, carry $\Rightarrow E$
ISZ M	2	increment and skip next instruction if zero	$(<M>) + 1 \Rightarrow <M>$; if $(<M>) + 1 = 0$ then $(C) + 1 \Rightarrow C$
STO M	3	store A	$(A) \Rightarrow <M>$
BSB M	4	branch to subroutine	$5000 + (C) \Rightarrow <M>$; $M + 1 \Rightarrow C$
BUN M	5	branch unconditionally	$M \Rightarrow C$

Table 11-3 Register Reference Instructions

Mnemonic	*Hexa-decimal Code*	*Description*	*Function*
NOP	6000	no operation	
CLA	6800	clear A	$0 \Rightarrow A$
CLE	6400	clear E	$0 \Rightarrow E$
CMA	6200	complement A	$(\bar{A}) \Rightarrow A$
CME	6100	complement E	$(\bar{E}) \Rightarrow E$
SHR	6080	shift-right	$(A_1) \Rightarrow E$; $(E) \Rightarrow A_{16}$; $(A_{i+1}) \Rightarrow A_i$ $i = 1, \ldots, 15$
SHL	6040	shift-left	$(A_{16}) \Rightarrow E$; $(E) \Rightarrow A_1$; $(A_{i-1}) \Rightarrow A_i$ $i = 2, \ldots, 16$
ICA	6020	increment A	$(A) + 1 \Rightarrow A$
SPA	6010	skip next instruction if A is positive	if $(A_{16}) = 0$ then $(C) + 1 \Rightarrow C$
SNA	6008	skip next instruction if A is negative	if $(A_{16}) = 1$ then $(C) + 1 \Rightarrow C$
SZA	6004	skip next instruction if A is zero	if $(A) = 0$ then $(C) + 1 \Rightarrow C$
SZE	6002	skip next instruction if E is zero	if $(E) = 0$ then $(C) + 1 \Rightarrow C$
HLT	6001	halt	$0 \Rightarrow S$

according to the 2's complement addition algorithm stated in Sec. 9-9. The end-carry out of the sign bit is transferred to the E flip-flop. This instruction, together with register-reference instructions, is sufficient for programming any other arithmetic operation such as subtraction, multiplication, or division.* This instruction can be used for loading a word from memory

*The sequence of operations needed to perform subtraction is enumerated in Sec. 9-9.

Table 11-4 Input-Output Instructions

Mnemonic	*Hexa-decimal Code*	*Description*	*Function*
SIN	7800	skip next instruction if input control is a 1	if $N_9 = 1$ then $(C) + 1 \Rightarrow C$
INP	7400	transfer from input device	$(N_{1-8}) \Rightarrow A_{1-8}$; $0 \Rightarrow N_9$
SOT	7200	skip next instruction if output control is a 0	if $U_9 = 0$ then $(C) + 1 \Rightarrow C$
OUT	7100	transfer to output device	$(A_{1-8}) \Rightarrow U_{1-8}$; $1 \Rightarrow U_9$

into the accumulator by first clearing the A register with CLA (see Table 11-3). The required word is then loaded from memory by adding it to the cleared accumulator.

ISZ: Increment and Skip if Zero

The increment and skip instruction is useful for address modification and for counting the number of times a program loop is executed. A negative number previously stored in memory at address M is read by the ISZ instruction. This number is incremented by 1 and stored back into memory. If, after it is incremented, the number reaches zero, the next instruction is skipped. Thus at the end of a program loop one inserts an ISZ instruction followed by a branch unconditionally (BUN) instruction to the beginning of the program loop. If the stored number does not reach zero, the program returns to execute the loop again. If it reaches zero, the next instruction (BUN) is skipped and the program continues to execute instructions after the program loop.

STO: Store A

This instruction stores the contents of the A register in the memory word at address M. This instruction and the combination of CLA and ADD are used for transferring words to and from memory and the A register.

BSB: Branch to Subroutine

This instruction is useful for transferring program control to another portion of the program (a subroutine) that starts at location $M + 1$. When executed, this instruction stores the address of the next instruction (of the main program) held in register C into bits 1 to 12 of the memory word at location M. It also stores the operation code 0101 (BUN) into bits 16-13 of the same word in location M. The contents of the address part M plus 1 are transferred into the C register to serve as the next instruction to be executed (the beginning of the subroutine). The return to the main program from the subroutine is accomplished by means of a BUN M instruction placed at the end of the subroutine, which transfers control to location M and, through it, back to the main program.

BUN: Branch Unconditionally

This instruction transfers control unconditionally to the instruction at the location specified by the address M. This instruction is listed with the memory-reference instructions because it needs an address part M. However, it does not need a reference to memory to read an operand, as is required by the other memory-reference instructions.

Register-Reference Instructions

The register-reference instructions are listed in Table 11-3. Of the 13 instructions, most are self-explanatory. Each register-transfer instruction has an operation code of 0110 (hexadecimal 6) and contains a 1 in only one of the remaining 12 bits. The skip instructions are used for program control conditioned on the sign of the A register or the value of the E flip-flop. To skip the next instruction, the C register (which holds the address of the next instruction) is increased by 1 so that the next instruction read from memory is two locations down from the location of the present (skip) instruction. The halt instruction is usually placed at the end of a program and its execution stops the computer by clearing the start-stop S flip-flop.

Input-Output Instructions

These instructions are listed in Table 11-4. They have an operation code of 0111 (hexadecimal 7) and each contains a 1 in only one of the remaining 12 bits. Two of these instructions (SIN and SOT) check the status of the control flip-flop to determine if the next consecutive instruction is executed. This next consecutive instruction will normally be a Branch (BUN) back to the previous SIN or SOT instruction so the computer remains in a two-instruction loop until the control flip-flop becomes a 1 in the input register or 0 in the output register. The control flip-flop is changed by the external device when it is ready to send or receive new information. So when the SIN instruction detects that the input-register has a character available ($N_9 = 1$) or when the SOT instruction detects that the output register is empty ($U_9 = 0$), the next instruction in sequence (BUN) is skipped and an INP or an OUT instruction (placed after the BUN instruction) is executed.

Instructions and data are transferred into the memory unit either manually by means of console switches or from the input paper tape reader. A 16-bit instruction or data word can be prepared on paper tape in two columns of eight bits each. The first punched column represents half of the word and the second column supplies the remaining bits of the word. The two parts are packed into one computer word and stored in memory. Normally, instructions and data occupy different parts of memory, with a halt instruction terminating the instruction sequence. To start the computer, the operator loads the first address of the program from console switches into the C register and activates a "start" switch. The first instruction is read from memory and executed. The rest of the instructions are executed in consecutive order unless a branch or a skip instruction is encountered.

11-4 TIMING AND CONTROL

The timing for the entire computer is controlled by the master-clock generator, whose clock pulses are applied to all flip-flops in the computer. The clock pulse period is 1 μsec but the memory cycle time is 4 μsec. While the memory is busy during its cycle, four clock pulses are available to activate registers and execute required elementary operations. The purpose of the G register is to supply four distinct signals during a memory cycle. This register is a two-bit counter-decoder circuit* that generates the timing signals for the control unit. These timing signals are designated by t_0, t_1, t_2, and t_3 and are shown in Fig. 11-4. Each timing signal is 1 μsec long and occurs once every 4 μsec. We are assuming that triggering of flip-flops occurs during the trailing edge of the clock pulse and that the memory cycle control signal initiates a memory cycle also on the trailing edge of a pulse. The relation between the timing signals and the memory cycle is demonstrated in the timing diagram of Fig. 11-4. A memory cycle is initiated at the trailing edge of a memory cycle control signal designated by m. During the first 2 μsec of a memory cycle, reading takes place; i.e., the word whose address is specified by the D register is read into the B register and erased from memory. The word in the B register settles to its final value at least 0.2 μsec prior to the termination of signal t_2. During the second half of the memory cycle, writing takes place; i.e., the contents of the B register are stored in the word whose address is specified by the D register. While the memory is busy reading and writing a word, timing signals t_0-t_3 are used to initiate elementary operations on computer registers.

Certain timing relations must be specified for the memory unit in order to synchronize it with the timing signals used for registers. As mentioned before, the memory cycle control signal m initiates a read cycle which is *always* followed by a write cycle. To be able to transfer information into

*The design of counter-decoder circuits is covered in Sec. 7-6.

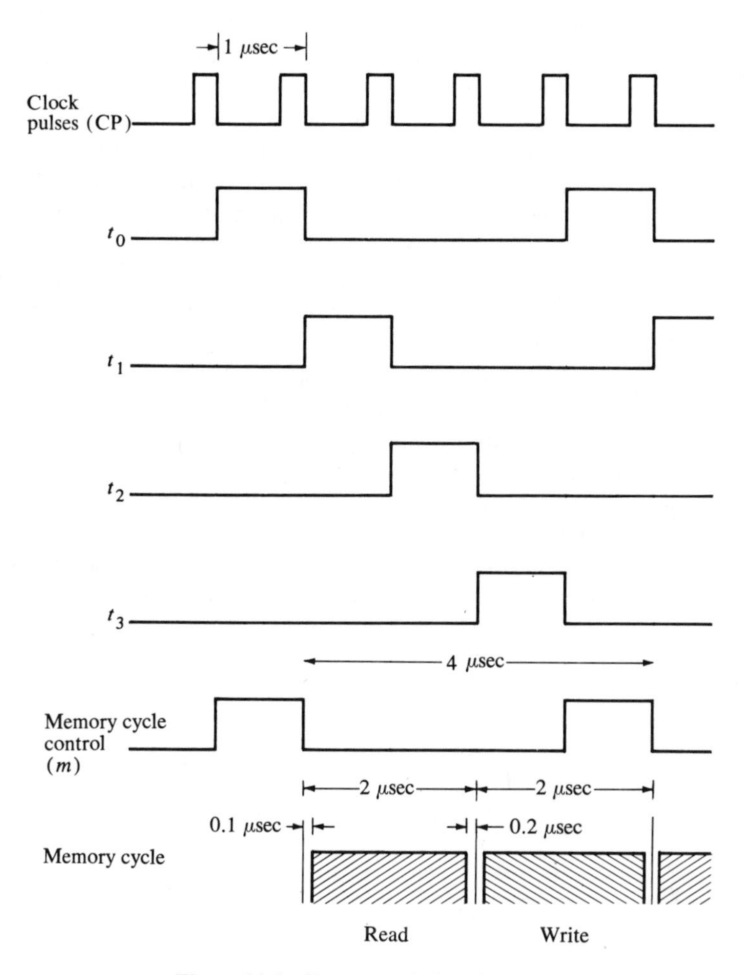

Figure 11-4 Computer timing signals

the address and buffer registers simultaneously with signals t_0 and t_2, we specify that the memory unit has to wait at least 100 nsec before using the outputs of these registers at the beginning of the read cycle, as well as at the beginning of the write cycle. This implies that the propagation delay of flip-flops can be at most 100 nsec. It is further assumed that the word read from memory during the read cycle settles to its final value in the buffer register not later than 1.8 μsec after the memory cycle control signal is applied.

The digital computer operates in discrete steps and elementary operations are performed during each step. An instruction is read from memory and executed in registers by a sequence of elementary operations. When the control receives an instruction, the operation code is transferred into the I

register. Logic circuits are required to generate appropriate control signals for the required elementary operations. A block diagram of the control logic that generates the signals for register operations is shown in Fig. 11-5. The operation part of the instruction in the I register is decoded into eight outputs q_0-q_7; the subscript number being equal to the hexadecimal code of the operation. The outputs of the G register are decoded into the four timing signals t_0-t_3, which determine the timing sequence of the various control functions. The status of various flip-flops and registers is sometimes needed to determine the sequence of control. The block diagram of Fig. 11-5 is helpful in visualizing the control unit of the computer when the register operations are derived during the design process.

The start-stop S flip-flop is shown going into the G register as well as the timing decoder. All elementary operations are conditioned on the timing signals. The timing signals are generated only when $S = 1$. When $S = 0$, the control sequence stops and the computer halts. The S flip-flop can be set or cleared from switches in the computer console, or cleared by the HLT (halt) instruction. We shall assume that when S is set by a "start" switch, signal t_0 is the first to occur; and when S is cleared by a "stop" switch, the current instruction is executed before the computer halts.

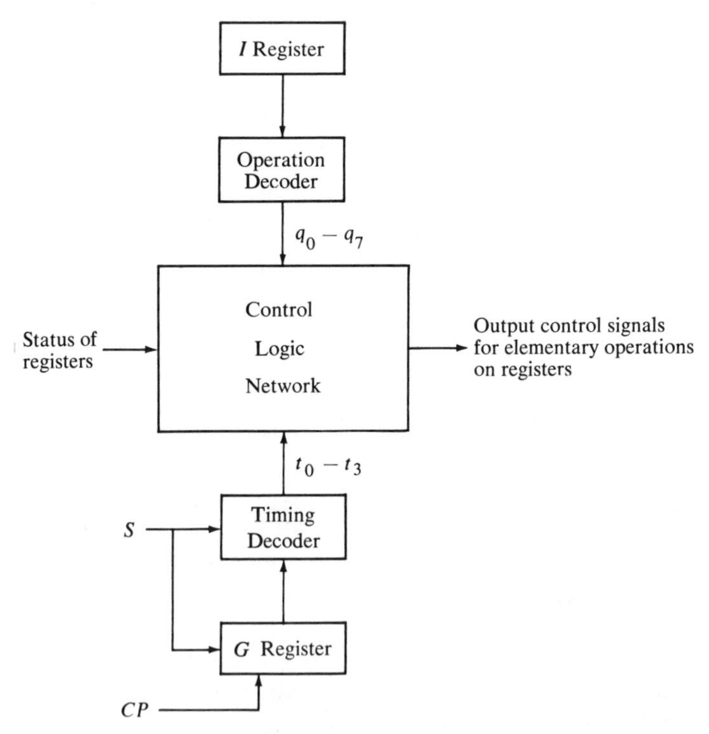

Figure 11-5 Block diagram of control logic

11-5 EXECUTION OF INSTRUCTIONS

Up to now, the system design of the computer has been considered. We have specified the register configuration, the set of computer instructions, a timing sequence, and the configuration of the control unit. In this section, we start with the logic design phase of the computer. The first step is to specify the elementary operations, together with the control signals, needed to execute each machine instruction. The list of elementary operations needed for each register will be derived in Sec. 11-6. In Sec. 11-7 we shall translate the elementary operations into Boolean functions.

The register operation sequence will describe precisely the process of information transfer within the registers of the computer. Each line of the sequence will consist of a control function, followed by a colon, followed by one or more elementary operations in symbolic notation. The control function is always a *Boolean function* whose variables are the timing signals t_0-t_3, the decoded operation q_0-q_7, and various outputs of flip-flops. The elementary operations are specified in accordance with the symbolic notation defined in Table 8-1.

Once a "start" switch is activated, the computer sequence follows a basic pattern. An instruction whose address is in the C register is read from memory. Its operation part transferred to the I register, and the C register is incremented by 1 to prepare it for the address of the next instruction. When the instruction is a memory-reference type (excluding BUN), the memory is accessed again to read the operand. Thus, words read from memory into the B register can be either instructions or data. The F flip-flop is used to distinguish between the two. When $F = 0$, the word read from memory is interpreted by the control logic to be an instruction, and the computer is said to be in an instruction *fetch* cycle. When $F = 1$, the word read from memory is taken as an operand, the computer is said to be in a data *execute* cycle.

An instruction is read from memory during the fetch cycle. The register-transfer relations that specify this process are:

$$F't_0: \quad (C) \Rightarrow D, m$$
$$F't_1: \quad (C) + 1 \Rightarrow C$$
$$F't_2: \quad (B_{13\text{-}16}) \Rightarrow I$$

When $F = 0$, the timing signals t_0, t_1, t_2 initiate a sequence of operations that transfer the contents of the C register into the D register, initiate a memory cycle, increment the C register to prepare it for the next instruction, and transfer the operation part to the I register. All elementary operations are executed when the control function (as specified on the left of the semi-colon) becomes logic-1 *and* when a clock pulse occurs. Thus the elementary operations in register flip-flops as well as the memory initiate cycle, are executed at the trailing edge of a clock pulse which corresponds to the trailing edge of the timing signals (See Fig. 11-4).

The next step depends on the value of q_i, which is decoded from the I register. If it is a memory-reference instruction, an operand is needed. If not, the instruction is executed during time t_3.

We shall find below that the BUN instruction does not need an operand from memory, although it is listed as a memory-reference instruction. When the operation code 0, 1, 2, 3, or 4 is encountered, the computer has to go to an execute cycle and read the operand. This condition is detected from the decoder associated with the I register and causes a transfer to the execute cycle by setting flip-flop F, thus:

$$F' (q_0 + q_1 + q_2 + q_3 + q_4)t_3: \quad 1 \Rightarrow F$$

The register operations common to all instructions during the fetch cycle are listed in Table 11-5.

When the computer is in an execute cycle, i.e., when $F = 1$, the basic cycle is as follows: read the operand from memory, execute the instruction, and return to the fetch cycle to read the next instruction. The operand is read during time $t_1 + t_2$ and is available in the B register at least 0.2 μsec prior to the termination of t_2. While the operand (or another word) is stored in memory during time $t_3 + t_0$, the instruction can be executed. The elementary operations needed to execute the instruction depend on the q_i signal received from the control logic. However, during the execute cycle, there are common operations independent of the value of q_i. These are:

$$Ft_0: \quad (B_{1\text{-}12}) \Rightarrow D, m$$
$$Ft_3: \quad 0 \Rightarrow F$$

Table 11-5 Instruction Fetch Cycle

Control function	*Elementary operations*	*Comments*
$F't_0$:	$(C) \Rightarrow D, m$	Read instruction
$F't_1$:	$(C) + 1 \Rightarrow C$	Increment C-register
$F't_2$:	$(B_{13-16}) \Rightarrow I$	Transfer operation
$F'(q_0 + q_1 + q_2 + q_3 + q_4)\, t_3$:	$1 \Rightarrow F$	Go to execute cycle
$(q_5 + q_6 + q_7)\, t_3$	(see Tables 11-7, 11-8 and 11-9)	Execute appropriate instruction

A memory cycle is initiated at t_0 to read the operand. The address of the operand is in bits 1-12 of the instruction, which is in the B register. Therefore, a transfer is needed from $B_{1\text{-}12}$ to the D register. At the completion of the execute cycle, the computer always returns to the fetch

Table 11-6 Common Operations for Execute Cycle

Control function	*Elementary operations*	*Comments*
Ft_0:	$(B_{1-12}) \Rightarrow D, m$	Transfer address part of instruction and read operand from memory.
$F(t_2 + t_3)$	(see Table 11-7)	Execute memory-reference instruction
Ft_3:	$0 \Rightarrow F$	Return to fetch cycle

Table 11-7 Execution of Memory-Reference Instructions

Instruction	*Control functions*	*Elementary operations*	*Comments*
AND	Fq_0t_3:	$(A) \wedge (B) \Rightarrow A$	Logical AND
ADD	Fq_1t_3:	$(A) + (B) \Rightarrow A$, carry $\Rightarrow E$	Arithmetic addition
ISZ	Fq_2t_2:	$(B) + 1 \Rightarrow B$	Increment B-register
	$Fq_2z_Bt_3$:	$(C) + 1 \Rightarrow C$	Skip next instruction if $(B) = 0$
STO	Fq_3t_2:	$(A) \Rightarrow B$	Transfer (A) to B-register
BSB	Fq_4t_2:	$(C) \Rightarrow B_{1-12}$, $0101 \Rightarrow B_{16-13}$, $(D) \Rightarrow C$	Store (C) in memory together with BUN operation. Next instruction taken from $M + 1$
	Fq_4t_3:	$(C) + 1 \Rightarrow C$	
BUN	q_5t_3:	$(B_{1-12}) \Rightarrow C$	Next instruction taken from M

cycle with the t_3 signal. The common operations performed during the execute cycle are listed in Table 11-6.

There remains now to specify the execution of instructions. The memory-reference instructions (excluding BUN) are executed when $F = 1$ and with timing signals t_2 and/or t_3. All other instructions are executed with timing signal t_3 without changing the value of F. Table 11-7 summarizes the operations required to execute each memory-reference instruction. The decoded operation q_i determines which operation is executed.

The AND instruction is executed with timing signal t_3 although it could be executed with t_2 as well. The ADD instruction is executed with timing signal t_3, which leaves 1 μsec for the carry to propagate in the adder circuits. The ISZ instruction is executed with the following two elementary operations:

$$Fq_2t_2: \quad (B) + 1 \Rightarrow B$$
$$Fq_2z_Bt_3: \quad (C) + 1 \Rightarrow C \qquad (z_B = 1 \text{ if } (B) = 0)$$

The instruction requires the incrementing of the word read from location M prior to its restoration during the memory write period. Since the word is available in the B register 0.2 μsec prior to the termination of timing signal t_2, there is available 0.2 μsec for the carry to propagate in the combinational gates that perform the counting. The B register is then incremented with timing signal t_2 and the new value is stored in location M. While the incremented number is being stored in memory its value, which is still available in the B register, is checked; if it is equal to 0, the C register is incremented by 1, causing a skip of one instruction. The symbol z_B is a signal from the B register. Its value is logic-1 when the contents of the register are equal to zero.

The store instruction specifies a transfer of contents from the A register to the memory word at location M. The word in location M is erased (and transferred to the B register) during the memory read period. By transferring the contents of A into B with timing signal t_2, the word stored during the memory write period is a new word equal to the contents of A.

The branch to subroutine (BSB) instruction is the most complicated instruction available in the computer. One possible way to execute this instruction is as follows:

$$Fq_4t_2: \quad (C) \Rightarrow B_{1-12}, \quad 0101 = B_{16-13}, \quad (D) \Rightarrow C$$
$$Fq_4t_3: \quad (C) + 1 \Rightarrow C$$

The address of the next instruction available in the C register is transferred to the address part (bits 1-12) of the B register and the code 0101 (BUN) is transferred to the operation part (bits 13-16). This is done with timing signal t_2 so that this value is stored in memory at location M. Remember that the address register D contains the address part M (see Table 11-6 at time Ft_0) of the original BSB instruction, so that a transfer of contents from the D register to the C register results in transferring M into C. Register C is incremented with timing signal t_3 so that the next instruction will be read from memory at location $M + 1$.

The branch unconditional (BUN) instruction does not need an operand. It merely specifies that the next instruction be taken from location M. Therefore, with signal t_3, the address part of the instruction in $B_{1\text{-}12}$ is transferred to register C to become the address of the next instruction to be executed. This can be done during the fetch cycle without having to set flip-flop F.

The execution of all register-reference instructions is listed in Table 11-8. These instructions are executed with timing signal t_3 during the termination of the fetch cycle. This is possible because the instructions need only one elementary operation for their execution, which can be performed while the instruction is stored back into memory. The control function is partly determined by one of the 12 least significant bits of the B register, where the instruction is available during time t_3. For example, the instruction

Table 11-8 Execution of Register-Reference Instructions

Instruction	*Hexadecimal Code*	*Control Functions*	*Elementary Operations*
NOP	6000	q_6:	Nothing
CLA	6800	$q_6B_{12}t_3$:	$0 \Rightarrow A$
CLE	6400	$q_6B_{11}t_3$:	$0 \Rightarrow E$
CMA	6200	$q_6B_{10}t_3$:	$(\bar{A}) \Rightarrow A$
CME	6100	$q_6B_9t_3$:	$(\bar{E}) \Rightarrow E$
SHR	6080	$q_6B_8t_3$:	$(A_1) \Rightarrow E, (E) \Rightarrow A_{16}, (A_{i+1}) \Rightarrow A_i$ $i = 1, 2, \ldots, 15$
SHL	6040	$q_6B_7t_3$:	$(A_{16}) \Rightarrow E, (E) \Rightarrow A_1, (A_{i-1}) \Rightarrow A_i$ $i = 2, 3, \ldots, 16$
ICA	6020	$q_6B_6t_3$:	$(A) + 1 \Rightarrow A$
SPA	6010	$q_6B_5A'_{16}t_3$:	$(C) + 1 \Rightarrow C$
SNA	6008	$q_6B_4A_{16}t_3$:	$(C) + 1 \Rightarrow C$
SZA	6004	$q_6B_3z_At_3$:	$(C) + 1 \Rightarrow C$
SZE	6002	$q_6B_2E't_3$:	$(C) + 1 \Rightarrow C$
HLT	6001	$q_6B_1t_3$:	$0 \Rightarrow S$

CLA has the hexadecimal code 6800, which gives a binary equivalent 0110 1000 0000 0000. The operation code is decoded from the I register and is equal to q_6. Bit 12 in the B register is a 1 so that the control function that executes this instruction is $q_6B_{12}t_3$.

The first seven register-reference instructions perform a clear, complement, shift, and increment operation on the A register or E flip-flop. The next four instructions are skip instructions executed only if the stated condition is satisfied. The skipping of the instruction is achieved by incrementing register C once again (in addition to the incrementing during $F't_1$, See Table 11-5). The required condition for skipping becomes part of the control function. Thus, the accumulator is positive if $A_{16} = 0$ and negative if $A_{16} = 1$. The symbol z_A designates an output from the A register which is logic-1 when its contents equal zero. The halt instruction clears the start-stop flip-flop S and stops the timing sequence.

The input-output instructions are executed with timing signal t_3 during the fetch cycle. The control functions for these instructions contain the operation q_7 and the corresponding bit of the B register. The I/O instructions consist of two transfer instructions and two skip instructions, which depend on the status of control flip-flop 9 in the input or output register.

11-6 REGISTER OPERATIONS

The sequence of register operations for the entire computer is listed in Tables 11-5 to 11-9. The difference between these tables and Tables 11-2

Table 11-9 Execution of Input-Output Instructions

Instruction	*Hexa-decimal Code*	*Control Functions*	*Elementary Operations*
SIN	7800	$q_7 B_{12} N_9 t_3$:	$(C) + 1 \Rightarrow C$
INP	7400	$q_7 B_{11} t_3$:	$(N_{1-8}) \Rightarrow A_{1-8}$ $0 \Rightarrow N_9$
SOT	7200	$q_7 B_{10} U_9' t_3$:	$(C) + 1 \Rightarrow C$
OUT	7100	$q_7 B_9 t_3$:	$(A_{1-8}) \Rightarrow U_{1-8}$, $1 \Rightarrow U_9$

to 11-4 is that the latter constitute a set of specifications for the computer, while the former specify in detail how the instructions are to be executed internally.

A digital system is completely specified by its sequence of elementary operations and control functions. The procedure for designing any synchronous digital system is the same; that is, from knowledge of the system requirements one formulates a control network and obtains the sequence of elementary operations for the system. Once this sequence is available, the rest of the design is straightforward. In many installations it is completed by design automation techniques.

To obtain the flip-flop functions to each register of the computer, we must obtain the list of elementary operations for each register separately. The elementary operations for each register can be retrieved from Tables 11-5 to 11-9 and arranged in a convenient order to facilitate the next step of the design. This is done in Table 11-10, where each register is listed together with its own required elementary operations and control functions. The information contained in Table 11-10 is identical to that available in the previous tables. The only difference is in the order in which it is arranged.

The control functions listed in Table 11-10 are assigned a Boolean variable name. This will help shorten the algebraic representation of input functions for the registers. The control function variables are assigned a lower-case letter identical to the capital letter reserved to symbolize the corresponding register. The control functions within a register are distinguished by a numerical subscript.

Table 11-10 is easily derived from Tables 11-5 to 11-9. The register to which an elementary operation belongs is recognized from the letter symbol found on the right of the double-line arrow. For example, to recognize the elementary operations belonging to register A we scan the operations listed in Tables 11-5 to 11-9 and retrieve all those that have an A to the right of the double-line arrow. The elementary operations for the other registers are obtained in a similar manner.

The G register was mentioned in Sec. 11-4 and shown in Fig. 11-5 as a two-bit counter that, together with a decoder, supplies the four basic timing

Table 11-10 Elementary Operations for Registers

Memory-Control		
t_0	m	Memory Cycle Control
A-Register		
$a_1 = Fq_0 t_3$:	$(A) \wedge (B) \Rightarrow A$	AND
$a_2 = Fq_1 t_3$:	$(A) + (B) \Rightarrow A$	add
$a_3 = q_6 B_{12} t_3$:	$0 \Rightarrow A$	clear
$a_4 = q_6 B_{10} t_3$:	$(\bar{A}) \Rightarrow A$	complement
$a_5 = q_6 B_8 t_3$:	$(E) \Rightarrow A_{16}, (A_{i+1}) \Rightarrow A_i$ $i = 1, 2, \ldots, 15$	shift-right
$a_6 = q_6 B_7 t_3$:	$(E) \Rightarrow A_1, (A_{i-1}) \Rightarrow A_i$ $1 = 2, 3, \ldots, 16$	shift-left
$a_7 = q_6 B_6 t_3$:	$(A) + 1 \Rightarrow A$	increment
$a_8 = q_7 B_{11} t_3$:	$(N_{1-8}) \Rightarrow A_{1-8}$	transfer
	$z_A = 1$ if $(A) = 0$	check (A) if 0
B-Register		
$b_1 = Fq_2 t_2$:	$(B) + 1 \Rightarrow B$	increment
$b_2 = Fq_3 t_2$:	$(A) \Rightarrow B$	transfer
$b_3 = Fq_4 t_2$:	$(C) \Rightarrow B_{1-12}$, $0101 \Rightarrow B_{16-13}$	transfer
	$z_B = 1$ if $(B) = 0$	check (B) if 0
C-Register		
$c_1 = F't_1 +$ $(q_2 z_B + q_4)Ft_3 +$ $(B_5 A'_{16} + B_4 A_{16} +$ $B_3 z_A + B_2 E')q_6 t_3 +$ $(B_{12} N_9 + B_{10} U'_9)q_7 t_3$:	$(C) + 1 \Rightarrow C$	increment
$c_2 = Fq_4 t_2$:	$(D) \Rightarrow C$	transfer
$c_3 = q_5 t_3$:	$(B_{1-12}) \Rightarrow C$	transfer
D-Register		
$d_1 = F't_0$:	$(C) \Rightarrow D$	transfer
$d_2 = Ft_0$:	$(B_{1-12}) \Rightarrow D$	transfer
I-Register		
$i_1 = F't_2$:	$(B_{13-16}) \Rightarrow I$	transfer
E-Flip-flop		
$e_1 = q_6 B_{11} t_3$:	$0 \Rightarrow E$	clear
$e_2 = q_6 B_9 t_3$:	$(\bar{E}) \Rightarrow E$	complement
$a_2 = Fq_1 t_3$:	carry $\Rightarrow E$	transfer carry
$a_5 = q_6 B_8 t_3$:	$(A_1) \Rightarrow E$	shift-right
$a_6 = q_6 B_7 t_3$:	$(A_{16}) \Rightarrow E$	shift-left

Table 11-10 Elementary Operations for Registers (Continued)

F-Flip-flop		
$f_1 = F'(q_0 + q_1 + q_2 + q_3 + q_4)t_3$:	$1 \Rightarrow F$	set
$f_2 = Ft_3$:	$0 \Rightarrow F$	clear
S-Flip-flop		
$s_1 = q_6B_1t_3$:	$0 \Rightarrow S$	clear
G-Register		
S:	$(G) + 1 \Rightarrow G$	count
U-Register		
$u_1 = q_7B_9t_3$:	$(A_{1-8}) \Rightarrow U_{1-8}$, $1 \Rightarrow U_9$	transfer and set flag
N-Register		
$n_1 = q_7B_{11}t_3$:	$0 \Rightarrow N_9$	clear flag

signals. This register is not listed in Tables 11-5 to 11-9 but must be included in Table 11-10. The G register counts clock pulses as long as the start-stop flip-flop S is equal to 1. All operations are conditional on the timing signals, and the timing signals are generated only when flip-flop $S = 1$. When $S = 0$, the control sequence stops and the computer halts.

11-7 DERIVATION OF BOOLEAN FUNCTIONS

Table 11-10 has all the necessary information needed to derive the Boolean functions for the control network and for the inputs to all registers. In addition, we need the Boolean functions for the two decoders shown in the block diagram of the control unit in Fig. 11-5. Signals q_0-q_7 are generated by the operation decoder and t_0-t_3 by the timing decoder. The Boolean functions for the two decoders are listed in Table 11-11. Note that the timing signals are generated only when the start-stop flip-flop $S = 1$.

The control logic network shown in the block diagram of Fig. 11-5 can be implemented with combinational gates. The Boolean functions for these gates are specified by the control functions listed in Table 11-10. Table 11-11 lists all the Boolean functions needed to implement the control unit of the computer.

The input Boolean functions for each register are derived from the elementary operations listed in Table 11-10. These functions together with other combinational circuits for each register are listed in Table 11-12. The input Boolean functions for the accumulator and all other registers are derived by the procedure outlined in Sec. 9-7. In fact, all the operations listed for the accumulator in Table 11-10 (except the transfer from the input register) are listed in Table 9-1. The accumulator designed in Ch. 9 is almost identical to the one employed by the computer. Note that the carry of the accumulator-adder has been designated by the letter K. This is because the letter C (used in Sec. 9-7 for the carry) is reserved here for the C register. Also note that the end-carry K_{17} (out of the last stage) is transferred to the E flip-flop.

The procedure for deriving the input Boolean functions for the registers from the list of elementary operations will not be carried out in detail here. The reader is referred to Sec. 9-7 for a thorough coverage of this procedure. The capital letters in the input functions of Table 11-12 represent registers; the subscripts denote individual flip-flops. Lower case letters represent control signals; their numerical subscripts have no specific meaning other than the definition given in Table 11-10.

It is necessary to check the propagation delay of critical operations to insure proper timing. The registers use JK flip-flops with trailing-edge triggering and a propagation delay of 100 nsec. All combinational gates are assumed to have a maximum propagation delay of 30 nsec. Some of the control unit circuits may require up to four levels of gates, but no propagation delay problems are to be anticipated since the elapsed time between two elementary operations is 1 μsec. The maximum carry propagation delay in the accumulator-addition circuits is $2 \times 30 \times 16 = 960$ nsec, but the word in the B register settles to its final value 200 nsec before the termination of t_2 (see Fig. 11-4). The add control signal propagates during time t_3. So the total time available for signal propagation is 1000 nsec. The maximum propagation delay for the increment operation in the accumulator

Table 11-11 Boolean Functions for Control Logic

Operation Decoder	*Timing Decoder*	*Control Logic Network (consists of the following control functions which are specified in Table 11-10)*		
$q_0 = I_4'I_3'I_2'I_1'$	$t_0 = G_2'G_1'S$	a_1	b_1	i_1
$q_1 = I_4'I_3'I_2'I_1$	$t_1 = G_2'G_1S$	a_2	b_2	e_1
$q_2 = I_4'I_3'I_2I_1'$	$t_2 = G_2G_1'S$	a_3	b_3	e_2
$q_3 = I_4'I_3'I_2I_1$	$t_3 = G_2G_1S$	a_4	c_1	f_1
$q_4 = I_4'I_3I_2'I_1'$		a_5	c_2	f_2
$q_5 = I_4'I_3I_2'I_1$		a_6	c_3	s_1
$q_6 = I_4'I_3I_2I_1'$		a_7	d_1	u_1
$q_7 = I_4'I_3I_2I_1$		a_8	d_2	n_1

Table 11-12 Boolean Input Functions to Registers

Accumulator Register

Input Functions:

$JA_i = B_i'K_ia_2 + B_iK_i'a_2 + a_4 + A_{i+1}a_5 + A_{i-1}a_6 + V_i + N_ja_8$

$i = 2, 3, \ldots, 15$

$j = i = 2, 3, \ldots, 8$

$KA_i = B_i'a_1 + B_i'K_ia_2 + B_iK_i'a_2 + a_3 + a_4 + A_{i+1}'a_5 + A_{i-1}'a_6 + V_i + N_j'a_8$

$JA_1 = B_1a_2 + a_4 + A_2a_5 + Ea_6 + a_7 + N_1a_8$

$KA_1 = B_1'a_1 + B_1a_2 + a_3 + a_4 + A_2'a_5 + E'a_6 + a_7 + N_1'a_8$

$JA_{16} = B_{16}'K_{16}a_2 + B_{16}K_{16}'a_2 + a_4 + Ea_5 + A_{15}a_6 + V_{16}$

$KA_{15} = B_{16}'a_1 + B_{16}'K_{16}a_2 + B_{16}K_{16}'a_2 + a_3 + a_4 + E'a_5 + A_{15}'a_6 + V_{16}$

Combinational circuits needed for input functions:

$K_{i+1} = A_iB_i + A_iK_i + B_iK_i$ — $i = 1, 2, \ldots, 16$ (carry)

$K_1 = 0$

$V_{i+1} = V_iA_i$ — $i = 1, 2, \ldots, 15$ (increment)

$V_1 = a_7$

also needed for detecting $(A) = 0$

$z_{i+1} = z_iA_i'$ — $i = 1, 2, \ldots, 16$

$z_1 = 1$

$z_A = z_{17}$ — (zero content)

Memory Buffer Register

Input Functions:

$JB_i = Y_i + A_ib_2 + Z_ib_3$ — $i = 1, 2, \ldots, 16$

$KB_i = Y_i + A_i'b_2 + Z_i'b_3$

Combinational circuits needed for input functions:

$Y_{i+1} = Y_iB_i$ — $i = 1, 2, 3, 4, 5, 7, 8, 9, 10, 12, 13, 14, 15$

$Y_1 = b_1$

$Y_7 = B_1B_2B_3B_4B_5B_6b_1$ — (increment)

$Y_{12} = B_1B_2B_3B_4B_5B_6B_7B_8B_9B_{10}B_{11}b_1$

$Z_i = C_i$ — $i = 1, 2, \ldots, 12$ (transfer from C register)

$Z_i = 1$ — $i = 13, 15$ (transfer BUN

$Z_i = 0$ — $i = 14, 16$ operation)

Table 11-12 Boolean Input Functions to Registers (Continued)

also need for detecting $(B) = 0$

$y_{i+1} = y_i B'_i$ $\quad i = 1, 2, \ldots, 16$

$y_1 = 1$

$z_B = y_{17}$ $\quad$ (zero content)

Program Control Register

$JC_i = X_i + D_i c_2 + B_i c_3$ $\quad i = 1, 2, \ldots, 12$

$KC_i = X_i + D'_i c_2 + B'_i c_3$

with:

$X_{i+1} = X_i C_i$ $\quad i = 1, 2, \ldots, 11$ (increment)

$X_1 = c_1$

Memory Address Register

$JD_i = C_i d_1 + B_i d_2$ $\quad i = 1, 2, \ldots, 12$

$KD_i = C'_i d_1 + B'_i d_2$

Instruction Register

$JI_i = B_{i+12} i_1$ $\quad i = 1, 2, 3, 4$

$KI_i = B'_{i+12} i_1$

Sequence Register

$JG_1 = S$ $\quad JG_2 = G_1 S$

$KG_1 = S$ $\quad KG_2 = G_1 S$

Output and Input Registers

$JU_i = A_i u_1$ $\quad i = 1, 2, \ldots, 8$ $\quad JU_9 = u_1$

$KU_i = A'_i u_1$ $\quad KN_9 = n_1$

Single flip-flops

$JE = e_2 + K_{17} a_2 + A_1 a_5 + A_{16} a_6$

$KE = e_1 + e_2 + K'_{17} a_2 + A'_1 a_5 + A'_{16} a_6$

$JF = f_1$

$KF = f_2$

$KS = s_1$

must be less than 1000 nsec. But the same operation in the B register must be completed in 200 nsec or less (during the ISZ instruction). The increment operation implemented in the accumulator has a maximum propagation delay of 450 nsec, but the same operation for the B register is implemented with carry look-ahead to the 12th position and is therefore reduced to a maximum delay of 180 nsec. (See Prob. 11-6.)

The set of Boolean functions for the digital computer given by Table 11-12 culminates the logic design phase of the system. The implementation of the Boolean functions requires some knowledge of the type of combinational gates used, their fan-in, fan-out, and any other restrictions. To complete the design of the computer, a wiring diagram is needed and a prototype should be built to check any wiring and logical errors. The final evaluation of the machine rests with the users whose opinion must be solicited after the computer is in operation.

11-8 COMPUTER CONSOLE

Any computer has a control panel or console with switches and lamps to allow manual and visual communication between an operator and the computer. This communication is needed for starting the operation of the computer (bootstrapping) and for maintenance purposes. For the sake of completeness we shall enumerate a set of useful console functions for the computer, although the circuits required to implement these functions will not be shown.

Lamps indicate to the operator the status of registers in the computer. The normal output of a flip-flop when connected to an indicator lamp will cause it to light when the flip-flop is set and to turn off when the flip-flop is cleared. The registers whose outputs are to be observed in the computer console are: A, B, C, D, I, E, F, and S. When a count is made of the total number of flip-flops involved, we find that 63 indicator lamps are needed.

A set of switches and their functions for the console may include the following:

1. Sixteen "word" switches to set manually the bits of one word.
2. A "start" switch to set the S flip-flop. The signal from this switch also clears flip-flop F, N_9, and U_9 and register G.
3. A "stop" switch to clear the S flip-flop. To insure the completion of the current instruction, the signal from the switch is ANDed with the Boolean function $(F + q_5 + q_6 + q_7)t_3$ before it is applied to clear S.
4. A "load address" switch to transfer an address to the C register. When this switch is activated, the contents of 12 "word" switches are transferred to register C.
5. A "deposit" switch to manually store words into memory. When this switch is activated, the contents of register C are transferred to register D and a memory cycle is initiated. After 2 μsec, the contents of the 16 "word" switches are transferred into the B register and register C is incremented by 1.
6. A "display" switch to examine the contents of a word in memory. When this switch is activated, the contents of register C are transferred to register D, a memory cycle is initiated, and register C is incremented by 1. The contents of the memory word specified by the

address in register C is in register B and can be seen in the corresponding indicator lamps.

To insure that the computer is not running when the power is turned on, the S flip-flop must have a special circuit that forces it to turn always to the clear position right after the application of power to the machine.

11-9 CONCLUDING REMARKS

The register configuration established in Fig. 11-1 is almost an absolute minimum for even the smallest computer. Registers A and B are used for arithmetic processing and only one arithmetic instruction (ADD) is included. Most other computers use more and possibly longer registers for arithmetic operation. With two registers in the "arithmetic" unit; only addition and/or subtraction can be implemented directly. However, other operations such as multiplication and division can be programmed by a sequence of instructions that forms a subroutine. Accuracy greater than 16 bits is also possible by using subroutines that perform double precision arithmetic.

In designing the computer, we have provided transfer paths directly from each source register to each destination register. This form of information transfer is efficient only if the number of registers available is small. When a large number of registers is involved, direct transfer paths are not economical because they require an excessive number of interconnections. The most common technique employed for inter-register transfer in computers is a bus system, as discussed in Sec. 8-1.

Many commercial computers provide more than one accumulator register in the processor unit. For example, a computer may have 16 accumulator and index registers and the particular register(s) used may be specified by the instruction. This implies that the augend and addend values for the arithmetic addition circuits may come from many sources and that the sum may go to any one of many possible destinations. Again, the best way to provide all possible data paths is through a bus system. In fact, many digital computers are built with an independent arithmetic adder circuit whose inputs and outputs are connected to a bus system. Control signals are used to specify the registers to be connected to the adder bus during a particular ADD elementary operation.

A peripheral processor is not included in the over-all system configuration since only one I/O device is used. Peripheral processors are an integral part of modern digital data processing systems. Their function is to supervise the communication from and to many input and output peripheral units and the central processor. The logic design of a peripheral processor can be carried out by the procedure outlined in this chapter once the specific functional characteristics of the unit are formulated.

Two features not included with the computer but found in almost any commercial machine are: indirect addressing in the instruction format and index registers with an appropriate set of instructions. These two features were not included here because they would have complicated the design process. Indirect addressing would require another memory cycle in addition to the fetch and execute cycles. The inclusion of index registers would increase the number of registers, as well as the number of instructions for the computer, without adding any new significant design features. It is worth mentioning that index-register capabilities can be simulated for this computer by means of the ISZ instruction.

The basic timing cycle shown in Fig. 11-4 consists of four timing signals during one memory cycle. By increasing the frequency of the clock-generator, more timing signals could be generated to increase the number of operations performed on registers during a memory cycle. The four timing signals employed for this computer were sufficient for the execution of any instruction during one or two memory cycles. Certain instructions such as multiplication and division require a longer sequence of elementary operations and a computer having such instructions would be operating efficiently if more timing signals occur during one memory cycle. The execution of long arithmetic instructions usually take computer time equivalent to several memory cycles.

The control portion of the computer as shown in Fig. 11-5 demonstrates a simple approach to control logic design. Another method that may be used to generate control sequences is by means of asynchronous delay elements and time-sequencing circuits. The asynchronous time-sequencing signals detect the completion of one operation and initiate the next. Another configuration for generating control signals is by means of a sequential circuit. The state of the memory elements in the sequential circuit specifies the present operation. The present state, with possible other conditions, determines the next state and the next operation. Another way to organize the control unit of a digital computer is with a read-only memory (ROM) and a microprogrammed control unit as discussed in Sec. 10-3.

The symbolic relations used throughout this design are sometimes called a "register-transfer language." The word "register-transfer" implies the possibility of performing an operation and transferring the result to the same or another register. The word *language* is borrowed from computer users who employ this term to programming systems. More details about this computer design language can be found in references (1-5). Duley and Dietmeyer (6) extended this language to include in the set of symbols the register configuration, the memory unit, and timing and control information. They have called this extended symbology a Digital System Design Language (DDL). In a second paper (7), they describe a computer program that accepts as input the complete specifications of the digital computer in DDL and translates it to the required set of Boolean functions. G. B. Gerace (8) gives examples of a transfer language for various arithmetic operations and describes a procedure for minimizing required Boolean functions. Iverson (9, 10) developed a common symbolic language for hardware and software systems that is useful for listing the sequence of elementary operations in a computer. Others (11-15) used an existing programming language called ALGOL and adapted it for a design language that describes the logic behavior of digital systems. The purpose of all these languages is to express a digital system in a precise, concise manner and then automate the design procedure by using a computer program to translate the specifications of the system into logic equations and other pertinent design information.

PROBLEMS

11-1. Show that with the three instructions AND M, CMA, and CLA, it is possible to perform each of the 16 logical operations listed in Table 2-5. Assume that one logical word (x in Table 2-5) is in memory location M and the other logical word (y in Table 2-5) is in the accumulator. The result of the operation is to remain in the accumulator. (Note that ADD M and STO M instructions are sometimes needed for transfer between memory and accumulator.)

11-2. Write a program to accept characters from the input device, pack two characters in one 16-bit word, and store the words in memory starting with location 100. Repeat using the instructions defined in Prob. 11-17.

11-3. The following program is a list of instructions in hexadecimal code. The computer executes the instructions starting from location 100. What is the value of the A register and the contents of memory register number 103 when the computer halts?

Location	Hexadecimal Code
100	4103
101	6020
102	6001
103	0000
104	6800
105	6200
106	5103

11-4. Write a program to add two numbers and check for overflow. The E flip-flop should be set to 1 if overflow occurs and cleared to 0 otherwise. Use the algorithm developed in Prob. 9-19(b) for detecting an overflow.

11-5. What is executed in the computer if an instruction is encountered and: (a) bit 16 is equal to 1 or (b) a register-reference instruction has two or more bits of B_{1-12} set to 1?

11-6. Verify that the maximum signal propagation delay for the circuit that increments the B register is equal to 180 nsec. Assume that the maximum propagation delay of a single AND gate is 30 nsec.

11-7. Design one typical stage of registers A, B, and C that perform the following functions. Use JK flip-flops.

P_1:	$(\bar{A}) \Rightarrow B$, $(\bar{B}) \Rightarrow C$, $(\bar{C}) \Rightarrow A$	
P_2:	$(A) \vee (B) \vee (C) \Rightarrow A$	logical OR
P_3:	$(A) \wedge (B) \wedge (C) \Rightarrow B$	logic AND
P_4:	$(A) \oplus (B) \oplus (C) \Rightarrow C$	exclusive-or

11-8. Derive the flip-flop input functions to all registers of the computer and verify the equations listed in Table 11-12. Start from the elementary operations given in Table 11-10.

11-9. Design the computer whose specifications were given in Sec. 10-2. Use the register configuration from Fig. 10-2 and the machine instructions listed in Table 10-2. Use a memory unit of 1024 words with 14 bits per word.

11-10. The ADD instruction assumes that negative numbers are in 2's complement representation. It is required to change the hardware execution of this instruction (operation code 1) so it will add numbers in sign-1's complement representation.

(a) Define the execution of the instruction with elementary operations and control functions.

(b) Change the input Boolean functions of the accumulator register so that the instruction is executed in one memory cycle (after the fetch cycle).

(c) Modify the execution of the instruction SZA to take into consideration this new representation of negative numbers.

11-11. Let bit 16 of a memory-reference instruction be a mode bit d to designate an indirect address (see Sec. 10-6).

(a) Modify the operation decoder.

(b) Modify Tables 11-6 and 11-17 to take into consideration the fact that if $d = 1$, then M (bits 1-12 of the instruction) designates an indirect address.

11-12. Include with the computer a 12-bit index register designated by X. Let bit 16 of a memory-reference instruction designate the index register field (see Sec. 10-6). When bit 16 is 1, the effective address (the address of the operand) is $M + (X)$.

(a) Modify the operation decoder.

(b) Modify the fetch cycle to obtain the correct effective address.

11-13. Add the following three instructions to the computer:

Operation Code	Description	Function
1000 M	2's complement subtract	$(A) - (<M>) \Rightarrow A$
1001 M	branch on zero	if $(A) = 0$ then $M \Rightarrow C$
1010 M	add to memory	$(<M>) + (A) \Rightarrow <M>$
		$(<M>) + (A) \Rightarrow A$

Show the sequence of elementary operations together with the control functions for the execution of each of the above instructions (include the fetch and execute cycle operations). *Note*: An instruction may require three or more memory cycles for execution. Add a flip-flop R to the computer and use it in conjunction with the F flip-flop to distinguish between four different memory cycles, thus:

$RF = 00$	for the fetch cycle
$RF = 01$	for reading an operand
$RF = 10$ or 11	available for other functions

11-14. List all changes and additions in the Boolean functions of Tables 11-10, 11-11, and 11-12 necessary to implement the computer with the addition of the three instructions of Prob. 11-13.

11-15. Show the sequence of elementary operations together with the control functions for the execution of the following instructions (see note in Prob. 11-13):

Operation code: 1011 M
Description: Add to memory, leave accumulator unchanged.
Function: $(<M>) + (A) \Rightarrow <M>; (A)$ remain the same.

11-16. Show the sequence of elementary operations together with the control functions for the execution of the following instruction.

operation: 1100 M
description: selective clear
function: $(<\bar{M}>) \wedge (A) \Rightarrow A$.

11-17. Show the sequence of elementary operations together with the control functions for the execution of the following instructions:

Hexadecimal Code	Description
$D800$	Swap left and right characters; $(A_{1-8}) \Rightarrow A_{9-16}$, $(A_{9-16}) \Rightarrow A_{1-8}$.
$D40x$	arithmetic shift accumulator eight positions to the right $(15 \geqslant x \geqslant 0)$.
$D20x$	circular shift accumulator x positions to the right $(15 \geqslant x \geqslant 0)$.
$D10x$	circular shift accumulator x positions to the left $(15 \geqslant x \geqslant 0)$.

Arithmetic and circular shifts were defined in Sec. 9-9.

11-18. Show the sequence of elementary operations, together with the control functions, for the execution of the following instruction:

operation code:	1110 M
description:	load to accumulator immediate
function:	$M \Rightarrow A_{1-12}$, sign extended, i.e., if $M_{12} = 0$ (positive), then $0 \Rightarrow A_{13-16}$. If $M_{12} = 1$ (negative), then $1 \Rightarrow A_{13-16}$.

11-19. Show the sequence of elementary operations, together with the control functions, for the execution of the following instruction:

operation:	1111 M
description:	compare ($< M >$) and (A) and skip 3 ways; leave (A) unchanged
function:	If $(< M >) > (A)$, then $(C) + 1 \Rightarrow C$ (skip next instruction) If $(< M >) < (A)$, then $(C) + 2 \Rightarrow C$ (skip next two instructions) If $(<M>) = (A)$, then proceed to next instruction.

11-20. Derive the input Boolean functions to registers for the implementation of the instructions in Prob. 11-15 to 11-19.

11-21. Design a digital computer with 16 operations and 34 instructions specified by Tables 11-2 to 11-4 and Probs. 11-13 to 11-19.

REFERENCES

1. Chu, Y. *Fundamentals of Digital Computer Design*, Ch. 11. New York: McGraw-Hill Book Co., 1962.
2. Bartee, T. C., I. L. Lebow, and I. S. Reed. *Theory and Design of Digital Machines*. New York: McGraw-Hill Book Co., 1962.
3. Wood, P. E., Jr. *Switching Theory*, Ch. 8. New York: McGraw-Hill Book Co., 1968.
4. Reed, I. S. "Symbolic Design Techniques Applied to a Generalized Computer," M.I.T. Lincoln Lab. Tech. Rept. 141, January 1957.
5. Schorr, H. A. "Computer-Aided Digital System Design and Analysis Using a Register Transfer Language," *IEEE Trans. on Electronic Computers*, Vol. EC-13, 730-737, December 1964.
6. Duley, J. R., and D. L. Dietmeyer. "A Digital System Design Language (DDL)," *IEEE Trans. on Computers*, Vol. C-18, 850-861, September 1968.
7. Duley, J. R., and D. L. Dietmeyer. "Translation of DDL Digital System Specifications to Boolean Equations," *IEEE Trans. on Computers*, Vol. C-18, 305-313, April 1969.

8. Gerace, G. B. "Digital System Design Automation–A Method for Describing a Digital System as a Sequential Network System," *IEEE Trans. on Computers*, Vol. C-17, 1044-1061, November 1968.
9. Iverson, K. E. *A Programming Language*. New York: John Wiley & Sons, Inc., 1962.
10. Iverson, K. E. "A Common Language for Hardware, Software, and Applications," 1962 Fall Joint Computer Conference, *AFIPS Proc.*, Vol. 22, 121-129, Washington,. D. C.: Spartan.
11. Schlaeppi, H. P. "A Formal Language for Describing Machine Logic, Timing and Sequencing (LOTIS)," *IEEE Trans. on Electronic Computers*, Vol. EC-13, 439-448, August 1964.
12. Gorman, D. F., and J. P. Anderson. "A Logic Design Translator," 1962 Fall Joint Computer Conference, *AFIPS Proc.*, Vol. 22, 251-261, Washington, D. C.: Spartan.
13. Proctor, R. M. "A Logic Design Translator Experiment Demonstrating Relationship of Language to Systems and Logic Design," *IEEE Trans. on Electronic Computers*, Vol. EC-13, 442-430, August 1964.
14. Chu, Y. "An ALGOL-like Computer Design Language," *Comm. of the ACM*, Vol. 8, 607-615, October 1965.
15. Parnas, D. L. "A Language for Describing the Functions of Synchronous Systems," *Comm. of the ACM*, Vol. 9, 72-76, February 1966.

12 LOGIC DESIGN

12-1 INTRODUCTION

The over-all objective of logic design is the analysis and synthesis of interconnected digital circuits that accomplish specific information processing tasks. Digital systems vary in size and complexity from a simple combinational or sequential circuit to a complex of interconnected and interacting digital computers and peripheral units. A large system is usually divided into subsystems, each of which is designed separately. The various subsystems are then interconnected with common data transfer and control signal paths.

The process of logic design and analysis is a complex undertaking, especially since the development of IC technology. Many installations develop special digital computer automated design and analysis techniques to facilitate the design process. Design-automation programs are extensively used to perform such tasks as:

(1) logic minimization and simplification of Boolean functions,

(2) simulation of the system prior to construction to check for logical error,

(3) derivation of Boolean functions from a list of register-transfer operations,

(4) assignment of logic circuits and their interconnections from the given Boolean functions,

(5) generation of wiring lists, and

(6) generation of documents for manufacturing and maintenance.

However, the specifications for a system and the development of algorithmic procedures for achieving the various data processing tasks cannot be automated and require the mental reasoning of a human designer.

The most challenging and creative part of the design is the establishment of design objectives and the formulation of algorithms and procedures for achieving the stated objectives. This task requires a considerable amount of experience and ingenuity on the part of the designer. The knowledge acquired in previous chapters coupled with the process of logical reasoning will be used later to demonstrate the derivation of design algorithms.

Throughout the book, we have adopted various symbolic and tabular procedures for describing the interrelation between digital circuits and systems. Combinational circuits are described by Boolean algebra or truth tables. Clocked sequential circuits are expressed by means of flip-flop input functions or state tables. Control sequences are specified with timing diagrams or state diagrams. Register-transfer relations express operations among registers and modules. During the design process we may use one or all of these design tools, depending on the system and its complexity.

The binary information found in a digital system is either data or control. Data are discrete elements of information that are manipulated by combinational circuits. Control information provides command signals for specifying the needed manipulation sequences. The logic design of a digital system is a process for deriving the digital circuits that perform data processing and the digital circuits that provide control signals. It should be noted that the control logic network is the most difficult part of the system to design. We have shown in Ch. 11 the configuration of a simple timing and control network. Other approaches to control logic design are introduced later in this chapter.

The design examples in this chapter demonstrate some important features of logic design. The first example is a combinational circuit with nine input variables. The second example is a sequential circuit that requires a control sequence. The next two examples deal with arithmetic operations and demonstrate a process for deriving design algorithms.

12-2 DECIMAL ADDER

A combinational circuit with a large number of input variables is difficult to design by paper and pencil techniques. A computer program based on the tabulation or similar method of Boolean function simplification would be helpful in such a case. However, it is sometimes possible to divide a large combinational circuit into smaller portions that are easier to manage and then, by interconnecting the pieces together, form the required circuit. This technique is demonstrated in the design of the decimal adder undertaken in this section.

Computers that perform arithmetic operations with the decimal system represent decimal numbers by decimal codes.* An adder for such a computer must employ arithmetic circuits that accept coded decimal numbers and present results in the accepted code. For binary addition, it was sufficient to consider a pair of significant bits at a time, together with a previous carry. A decimal adder requires a minimum of nine input lines and five output lines, since a minimum of four bits are required to represent each decimal digit in a binary code. Of course, there is a wide variety of possible adder circuits dependent upon the code used to represent decimal digits.

Consider the addition of two decimal digits in BCD, together with a possible carry from a previous stage. Each decimal digit is applied through four separate lines to be designated by A_8, A_4, A_2, A_1 and B_8, B_4, B_2, B_1, where the subscript number denotes the weight of the bit in the code. A ninth input line must be included for the carry from the lower significant stage. Note that the carry is never greater than 1, regardless of the base being used, and therefore a single line is sufficient for its transmission. The sum digit contains four output lines designated S_8, S_4, S_2, S_1, and an output carry on a single line is needed for the next higher stage. If the code used for the decimal digits contains more than four bits, the number of input and output lines of the adder must be increased accordingly.

The design of a nine-input, five-output combinational circuit requires a truth table with $2^9 = 512$ entries. Many of the input combinations are don't-care conditions, since each digit has six combinations that never occur. The simplified Boolean functions for the circuit may be obtained by the tabulation method. However, doing it with paper and pencil requires a considerable amount of time. A computer program would do the job easier and with greater accuracy.

Another way of solving this problem is to use full-adder circuits together with some modification that will compensate for the fact that six combinations are not used. Consider for example a four-bit adder; i.e., four full-adders in cascade that add any two four-bit binary numbers. When the inputs to these adders are decimal digits represented in BCD, the inputs may be considered binary numbers restricted to a range from 0000 to 1001. This is shown schematically in the upper portion of Fig. 12-1. The uncorrected digit output lines in the diagram range in value from binary 0 to binary 19 (10011), since the numbers added form the maximum sum $9 + 9 + 1 = 19$; the 1 in the sum being an input carry. Table 12-1 lists the possible outputs of the uncorrected digit. However, the output sum of the decimal adder must be represented in BCD and appear in the form listed in the second column of the table. Since the corresponding numbers in each row of the table should be the same, the problem is to find a simple set of rules by which the uncorrected digit sum can be corrected.

*See Sec. 8-5 for the representation of decimal numbers in registers.

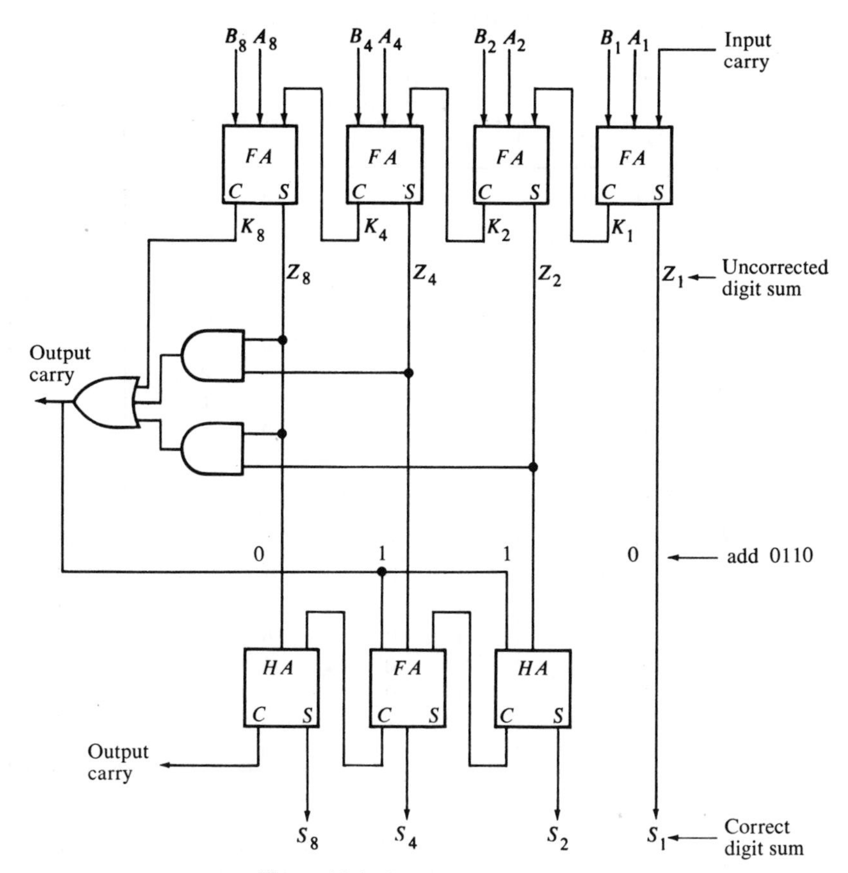

Figure 12-1 Decimal BCD adder

In examining the contents of the table, it is apparent that when the uncorrected digit is equal to or less than 01001, the correct digit sum is the same and therefore, no correction is needed. When the uncorrected digit is greater than 1001, the addition of binary 6 (0110) to the uncorrected digit makes the digit sum correct and the output carry appears.

The logic circuit that detects the necessary correction can be obtained from the table entries. It is obvious that a correction is necessary when $K_8 = 1$ or when $Z_8Z_4 = 1$ or when $Z_8Z_2 = 1$. Moreover, a necessary correction also implies that an output carry equal to 1 must be generated. The output carry is generated by the Boolean function

$$C = K_8 + Z_8Z_4 + Z_8Z_2$$

This output is used to form the needed correction.

To add binary 6 (0110) to the uncorrected digit, we use one full-adder and two half-adders (*HA*), as shown in the lower portion of Fig. 12-1.

Table 12-1 Derivation of BDC Adder

Possible outputs of the uncorrected digit sum					*Corresponding correct digit sum*					*Decimal*
K_8	Z_8	Z_4	Z_2	Z_1	C	S_8	S_4	S_2	S_1	
0	0	0	0	0	0	0	0	0	0	0
0	0	0	0	1	0	0	0	0	1	1
0	0	0	1	0	0	0	0	1	0	2
0	0	0	1	1	0	0	0	1	1	3
0	0	1	0	0	0	0	1	0	0	4
0	0	1	0	1	0	0	1	0	1	5
0	0	1	1	0	0	0	1	1	0	6
0	0	1	1	1	0	0	1	1	1	7
0	1	0	0	0	0	1	0	0	0	8
0	1	0	0	1	0	1	0	0	1	9
0	1	0	1	0	1	0	0	0	0	10
0	1	0	1	1	1	0	0	0	1	11
0	1	1	0	0	1	0	0	1	0	12
0	1	1	0	1	1	0	0	1	1	13
0	1	1	1	0	1	0	1	0	0	14
0	1	1	1	1	1	0	1	0	1	15
1	0	0	0	0	1	0	1	1	0	16
1	0	0	0	1	1	0	1	1	1	17
1	0	0	1	0	1	1	0	0	0	18
1	0	0	1	1	1	1	0	0	1	19

When the output carry is equal to 0, nothing is added to the uncorrected digit. When it is equal to 1, binary 0110 is added to the uncorrected digit. The output carry generated from the second sum may be ignored, since it supplies information already available in another line.

The fact that a four-bit adder circuit is available commercially in IC form in one MSI package gives the circuit of Fig. 12-1 a practical advantage over other forms of BCD adder implementations. The carry propagation delay in such a circuit is quite long since the carry must propagate through four FA circuits. To achieve a shorter delay, it is necessary to either use MSI packages that include some form of carry speed-up technique (see Sec. 9-3) or design the circuit with a different configuration.

12-3 CONTROL LOGIC DESIGN

A digital system may be specified by a list of register-transfer relations. The list consists of all elementary operations for the registers together with the control functions that specify the conditions for their execution. Control functions are Boolean functions whose variables are the system variables ANDed with a variable generated by a timing sequence. In a simple control, the timing sequence may follow a binary count, which can be implemented with a binary counter and a decoder. In a more complicated control, the timing sequence follows a different count for different values of system variables. A circuit that generates an arbitrary timing sequence is called a *control register*. A control network would normally consist of a control register that generates the timing sequence and combinational circuits that implement the various control functions for the other registers in the system.

A control register is a sequential circuit and may be described by a state diagram or by register-transfer relations. An example of a state diagram for a control register is shown in Fig. 12-2. The state of the register at any clock-pulse period is either p_0, p_1, p_2, or p_3, These states are used as

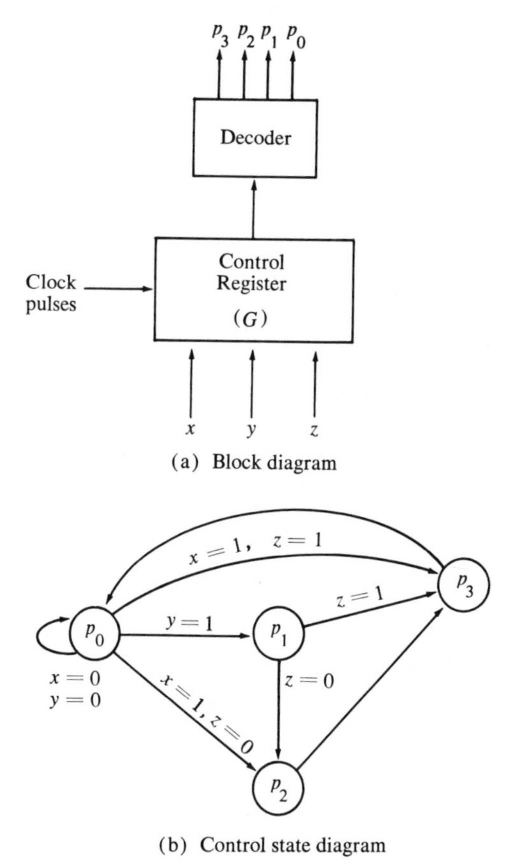

Figure 12-2 An example of a control register

timing variables in control functions. (They correspond to variables t_0 - t_3 in the control functions of Table 11-10.) Let us assume that p_0 is an initial state, x and y are input variables, and z is an internal variable. We further assume that inputs x and y are mutually exclusive and that both cannot be equal to 1 simultaneously, otherwise the state diagram is ambiguous. The sequence of states that the register undergoes depends on the values of the variables. When variables x and z are both 1, the control register goes from the initial state p_0 to state p_3 and back to the initial state. On the other hand, when $x = 1$ and $z = 0$, the control register follows the sequence of states p_2, p_3, and back to p_0. When $y = 1$ and $z = 1$, the sequence is p_1, p_2, p_3, and back to p_0. There are four different state sequences for this control register; the one generated depends on the variables x, y, and z. The four sequences of the control register represent the timing signals of a particular control network. For any clock-pulse period, one and only one p_i is logic-1 and, when p_i is included in a control function, it determines the elementary operations to be executed during that clock pulse.

Since a control register is a sequential circuit, it can be designed by the method outlined in Ch. 7. However, the design of a control register can be simplified when one realizes that a decoder must be provided and, since the outputs of the decoder are available anyway, it is convenient to use them as present state variables instead of flip-flop outputs. The recommended procedure is to derive the Boolean functions for the decoder and then obtain the flip-flop input functions in terms of decoder outputs and system inputs.

The formal design procedure with excitation tables and maps may not be effective when a control register has a large number of states and input variables. Also, a state diagram for such a sequential circuit may be too complicated and involved. The design of a control register can be simplified if register-transfer relations are used instead of a state diagram. For example, the control state diagram of Fig. 12-2 may be represented with symbolic relations. Denote the control register by G and assign binary states $p_0 = 00$, $p_1 = 01$, $p_2 = 10$, and $p_3 = 11$. The transition between states is specified with control functions and elementary operations as follows:

yp_0:	$(G) + 1 \Rightarrow G$
$xz'p_0$:	$10 \Rightarrow G$
xzp_0:	$11 \Rightarrow G$
$z'p_1$:	$(G) + 1 \Rightarrow G$
zp_1:	$11 \Rightarrow G$
p_2:	$(G) + 1 \Rightarrow G$
p_3:	$00 \Rightarrow G$ (or $(G) + 1 \Rightarrow G$ when two flip-flops are used)

We can collect control functions that specify identical elementary operations to obtain:

$$yp_0 + z'p_1 + p_2 + p_3: \quad (G) + 1 \Rightarrow G$$
$$xz'p_0: \quad 10 \Rightarrow G$$
$$xzp_0 + zp_1: \quad 11 \Rightarrow G$$

From this list of register-transfer relations, we see that the design of a control register involves an increment elementary operation, transfer of constants, and the generation of control functions in terms of system variables and decoder outputs.

Note that the straight binary assignment may not be the one that leads to the simpler combinational circuit for the register. However, the binary assignment facilitates the description with register-transfer relations. Also because this is the most natural assignment, it would probably simplify the check-out and maintenance of the circuit. In practice, the design and maintenance costs of using a natural binary assignment are less when compared to the savings in hardware that could be achieved from some other state assignment.

An alternative approach to control register implementation is to remove the decoder altogether and use a number of flip-flops for the register equal to the number of states with each flip-flop representing a state. Only one flip-flop is set at any particular time; all others are cleared. A single bit is made to propagate from one flip-flop to the other under control of decision logic. Each flip-flop in such an array represents a state and is activated only when the control bit is transferred to it. For example, the four states in Fig. 12-2 can be represented by four flip-flops F_0, F_1, F_2, F_3 with the following state assignment:

state	F_0	F_1	F_2	F_3
p_0	1	0	0	0
p_1	0	1	0	0
p_2	0	0	1	0
p_3	0	0	0	1

This type of assignment leads to control register configurations that are easier to standardize, a very important factor when digital systems are constructed with ICs.

When the above state assignment is used, a control register can be simply specified by means of register-transfer relations. In fact, the conditions for setting and clearing of control flip-flops are the same as the functions marked along the directed lines in the state diagram and the state from which they originate. Thus for the example of Fig. 12-2 we have:

$x'y'F_0$: $1 \Rightarrow F_0$
yF_0: $1 \Rightarrow F_1$, $0 \Rightarrow F_0$
$xz'F_0$: $1 \Rightarrow F_2$, $0 \Rightarrow F_0$
$z'F_1$: $1 \Rightarrow F_2$, $0 \Rightarrow F_1$
xzF_0: $1 \Rightarrow F_3$, $0 \Rightarrow F_0$
zF_1: $1 \Rightarrow F_3$, $0 \Rightarrow F_1$
F_2: $1 \Rightarrow F_3$, $0 \Rightarrow F_2$
F_3: $1 \Rightarrow F_0$, $0 \Rightarrow F_3$

The control register can be implemented with D flip-flops in a simple manner. Each of the above control functions sets one and only one flip-flop. Since the control function is not applied to any other flip-flop, the D inputs of all other flip-flops will be 0 and they will be cleared. Collecting control functions that set the same flip-flop, we obtain the following D flip-flop input functions:

$$DF_0 = x'y'F_0 + F_3$$
$$DF_1 = yF_0$$
$$DF_2 = xz'F_0 + z'F_1$$
$$DF_3 = xzF_0 + zF_1 + F_2$$

We must make sure that initially F_0 is set and all other flip-flops are cleared. If we start with all flip-flops cleared, there is no way to return to one of the defined states.

The design of a digital system that requires a control sequence starts by assuming the availability of a timing sequence. We designate each sequence by a state, say p_i for $i = 0, 1, 2 \ldots$, and proceed to form a state diagram for the state transitions between sequences. If the system is too complicated for a state diagram, we list the control register transfer relations by means of control functions and elementary operations as explained above.

Concurrent with the development of control sequences, we develop a list of control functions and elementary operations for the other registers in the system. These register-transfer relations can be derived directly from the word specification of the problem. However, it is sometimes convenient to use an intermediate representation to describe the needed sequence of operations for the system. Two other representations are helpful for the design of systems that need a control circuit. One is a timing diagram; the other is a flow chart.

A timing diagram clarifies the timing sequences and relations among the various signals in the system. In a clocked sequential circuit, the clock pulses synchronize the operations and cause signal transitions. In an asynchronous system, a given signal transition causes a change in value of another signal. A timing diagram is a pictorial representation of required changes and transitions of all system variables.

Flow charts are extensively used by programmers to specify the sequence of computer operations and decision paths needed to solve a particular problem. A flow chart is a diagram that shows the procedure for finding a solution to a problem with a given piece of equipment. For programmers, the equipment is a computer and its instructions. To the logic designer, the equipment consists of registers, flip-flops, and input-output variables. Flow charts are a convenient representation of algorithms and procedures and facilitate the design of digital systems.

A flow chart is a block diagram, where each block specifies a step in the process as the information flows from one part of the system to another. The directed lines between blocks designate the path to be taken from one step to the other. The two major types of blocks are (a) function blocks that show the operations to be performed and (b) decision blocks that have two or more alternate paths dependent on certain conditions.

The usefulness of timing diagrams and flow charts in the design of digital systems is demonstrated in the examples of the next three sections. These examples also illustrate the procedure for obtaining state diagrams for control registers.

12-4 ASYNCHRONOUS TRANSFER

Although most digital computers and systems operate synchronously, the transfer of information between two units, each having its own independent clock-pulse source, must be done asynchronously. An asychronous signal arriving from an external unit may be in a transition state during a clock pulse; therefore, special precautions must be taken for its correct detection. The design example introduced in this section demonstrates a method of asynchronous data transfer between units. It also illustrates the derivation of a control logic network.

The system to be designed is a code converter that performs the following three functions:

1. Accepts decimal data in BCD from an input device.

2. Converts the decimal data to the 2, 4, 2, 1 code (Table 1-2).

3. Transfers the converted data to an output device.

The three operations are repeated as long as the input device has data available.

A block diagram of the system is shown in Fig. 12-3. Data words are transferred from the input device to a register, where they are converted and then transferred to the output device. The control network specifies the

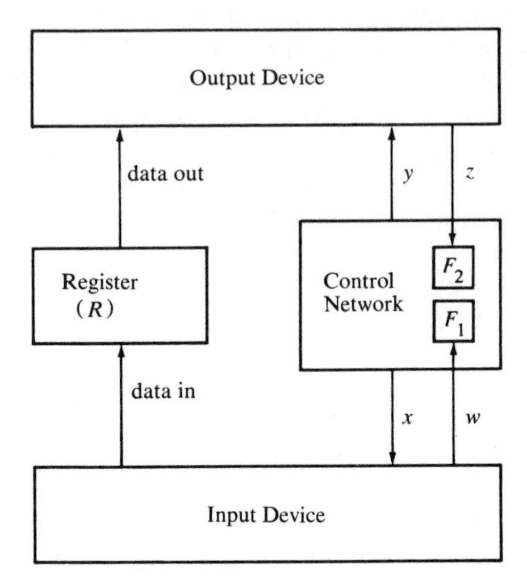

Figure 12-3 Block diagram of code converter

required sequence of operations and synchronizes the input signals with flip-flops F_1 and F_2 as explained below.

When designing a system that communicates with other devices, it is necessary that exact specifications be formulated for the interface between the system and the external devices. Since the code converter circuit does not know when the input device has data available for transfer, it is necessary that a control signal furnish this information. Moreover, since the input device would probably want to be informed when data has been accepted by the system, we must provide a control signal to inform the input device that data has been accepted. The same pair of control signals are also needed for the communication between the code converter and the output device. The four control signals shown in Fig. 12-3 provide this information in the following manner:

when $w = 1$ the system is informed that a word is available in the input device,

when $x = 1$ the system informs the input device that a word has been accepted,

when $y = 1$ the system informs the output device that a word is available for transfer,

when $z = 1$ the output device informs the system that a word has been accepted.

This sequence of control information is common between systems connected asynchronously.

An unreliable operation occurs if a control signal from an external device is used for inputs to more than one flip-flop. This is because the asynchronous signal may be changing from 0 to 1 or from 1 to 0 exactly at the time when a clock pulse occurs. The transmission will be in error if some flip-flop inputs detect the signal as a logic-1 while others detect it as logic-0. To prevent this uncertainty from happening, it is customary to provide a synchronizing flip-flop for each asynchronous external control signal. Then, if the flip-flop is not set during a given clock pulse (because of signal transition), it will set on the next pulse. Only after the synchronizing flip-flop is set does the system recognize the occurrence of the external signal. The two flip-flops used for this purpose are F_1 and F_2.

So far in the discussion we have stated in words the requirements for the code converter circuit. The combinational circuit for the register that performs the conversion can be obtained by using the design procedure for sequential circuits outlined in Ch. 7. On the other hand, to design the control network, we need some convenient representation, other than the word description, to describe the needed sequence of operations for the system. We shall now obtain a timing diagram and a flow chart and show how they can help to derive the register transfer relations and the control network.

Timing Diagram

The timing diagram for the code converter is shown in Fig. 12-4. The clock pulses are drawn on top of the diagram and all system signal transitions occur at the trailing edge of a pulse. The asynchronous signal w from the input device may come at any time. Flip-flop F_1 is set with a clock pulse if $w = 1$, and is cleared with the next clock pulse. When $F = 1$, data is transferred to the register from the input device and output x becomes logic-1 for one clock-pulse period. The decimal number in the register is converted to the 2,4,2,1 code with the next clock pulse and output y becomes logic-1; informing the output device of the availability of a word. The asynchronous signal z from the output device may come at any time. Flip-flop F_2 is set with a clock pulse if $z = 1$ and cleared with the next clock pulse. When $F_2 = 1$, the register is cleared and output y becomes logic-0. At this point the system is ready to repeat the cycle.

The control flip-flops for the system consist of F_1, F_2, and some other flip-flops that supply the required sequence. Let us designate the states of those other control flip-flops by p_0, p_1, p_2 etc. Each clock-pulse period is then distinguished by a given state p_i and the state of flip-flops F_1 and F_2. An identification sequence for each clock-pulse period can be assigned in terms of control states and flip-flops as shown in the timing diagram.

The timing diagram gives the information needed for the design of the code converter circuit. Although this representation is very convenient in

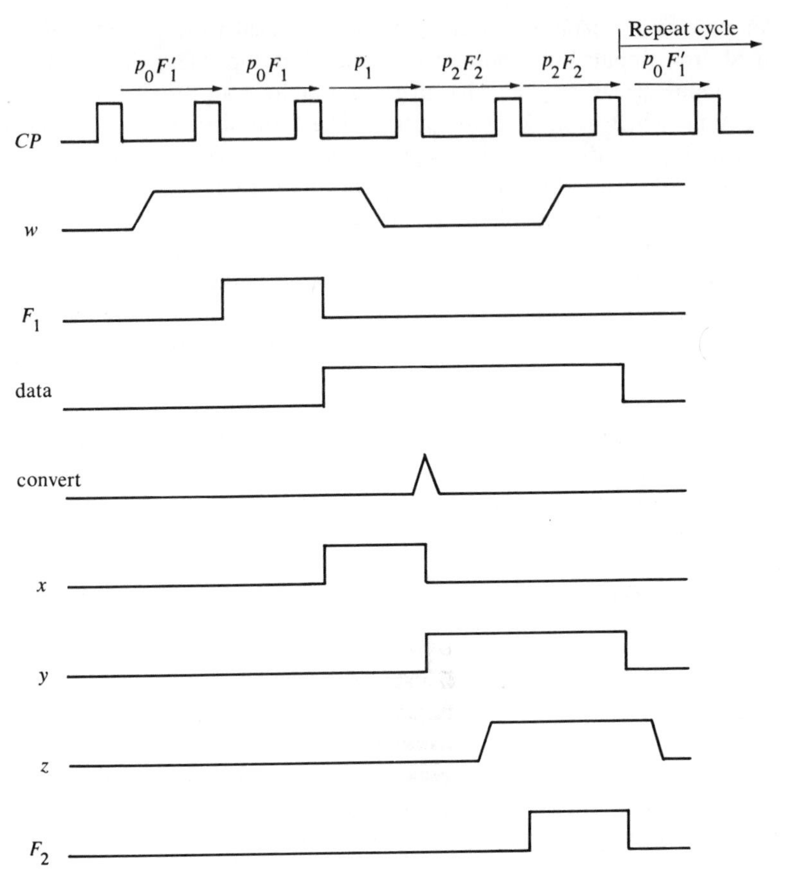

Figure 12-4 Timing diagram for code converter

the initial phase of the design, it is somewhat ambiguous and imprecise. For this reason we shall delay the derivation of the Boolean functions until a better representation is formulated.

Flow Chart

The flow chart for the code converter is shown in Fig. 12-5. It consists of function blocks, with the operation within each block specified by symbolic relations. The flow chart can be derived directly from the word statement of the problem or from the timing diagram.

To start the equipment operating properly when power is turned on, we must have some means of bringing the circuit to an initial state. This may be done by clearing all flip-flops with a manual switch. We shall assume that each flip-flop has a direct clear line from a manual switch and that these lines are independent of other flip-flop inputs. (Many commercial flip-flops have such a "direct clear" input.)

The circuit stays in the initial state until the input signal w becomes a 1. This causes flip-flop F_1 to set, which in turn causes a transfer of an input word into register R. The next step is to convert the decimal code and then send an output signal from y. When input z becomes a 1, flip-flop F_2 is set which in turn clears the register and the circuit goes back to its initial state. The information available in the flow chart is just another way to represent the required sequence specified in the timing diagram. However, this representation is somewhat more convenient for deriving the Boolean functions.

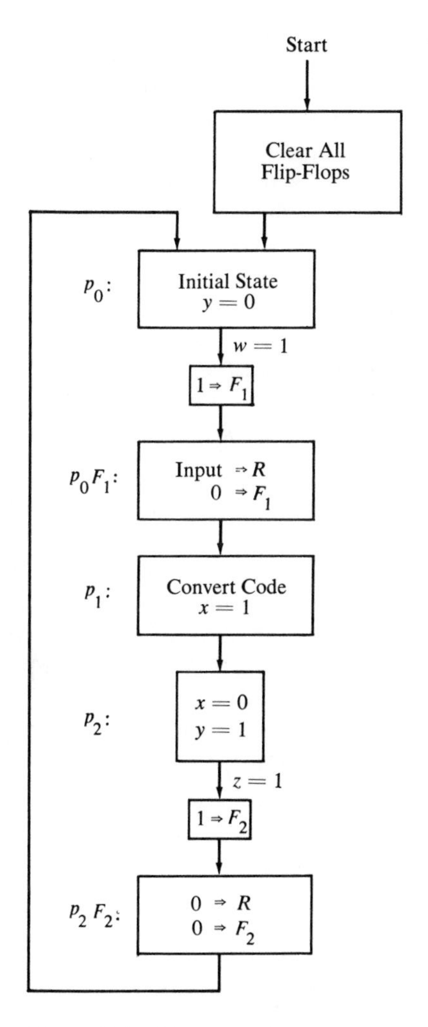

Figure 12-5 Flow chart for code converter

To determine the control sequence for the circuit, we assume control states p_i, $i = 1, 2, 3, \ldots$, and mark on the other side of each function box the control state at which the operation is to be activated. The Boolean functions can now be derived from the flow chart. However, we can achieve a clearer representation with a control register state diagram and a list of control functions and elementary operations as shown below.

Register-Transfer Relations

The control functions and elementary operations of a digital circuit can be derived directly from the word specification of the system. However, a timing diagram and/or a flow chart may sometimes facilitate the derivation. Figure 12-6 shows the state diagram for the code converter sequence. We start with initial state p_0. Input w should send the controller to the next state, but since F_1 must be included as part of the control, we can take the condition p_0F_1 for the next sequence. When F_1 equals 1, the control goes to state p_1 at which time the conversion takes place. At state p_2, we wait for the output device to respond through input z and then go to a state specified by p_2F_2. When F_2 is a 1, the control goes back to state p_0.

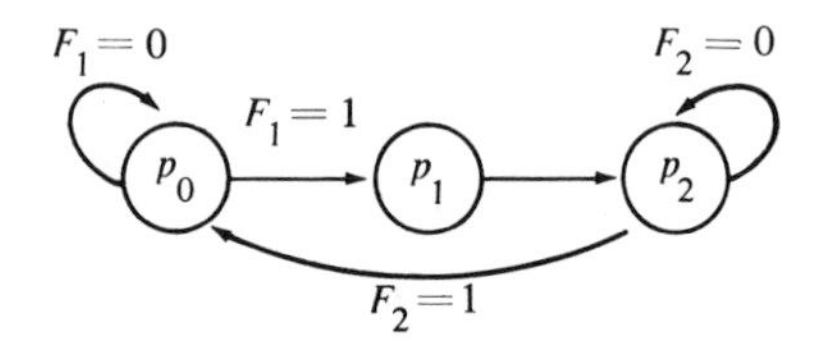

Figure 12-6 State diagram for control of code converter

Concurrent with the development of the state diagram for the control register, we derive a list of control functions and elementary operations for the system. These can be derived directly from the word specification or from the flow chart. The derivation will not be repeated again because all the information is already available in the flow chart. The control functions and elementary operations for the system are:

$p_0F'_1w$:	$1 \Rightarrow F_1$,	$0 = y$
p_0F_1:	input $\Rightarrow$ R,	$0 \Rightarrow F_1$
p_1:	convert,	$1 = x$
p_2:	$0 = x$	$1 = y$
$p_2F'_2z$:	$1 \Rightarrow F_2$	
p_2F_2:	$0 \Rightarrow$ R,	$0 \Rightarrow F_2$

Derivation of Boolean Functions

The elementary operations and control functions listed above, together with the control state diagram of Fig. 12-6, supply all the information needed for deriving the Boolean functions for the system. The information from the state diagram is used to derive the excitation table for the control register. As shown in Table 12-2, the two flip-flops for the control register are designated by F_3 and F_4, while the outputs of F_1 and F_2 constitute the input variables. When an input variable does not determine the next state, it is marked with an X and is taken as a don't-care condition. The

Table 12-2 Excitation Table for Control Register of Code Converter

	Present State		Inputs		Next State		Flip-Flops Inputs			
	F_4	F_3	F_2	F_1	F_4	F_3	JF_4	KF_4	JF_3	KF_3
p_0	0	0	X	0	0	0	0	X	0	X
p_0	0	0	X	1	0	1	0	X	1	X
p_1	0	1	X	X	1	0	1	X	X	1
p_2	1	0	0	X	1	0	X	0	0	X
p_2	1	0	1	X	0	0	X	1	0	X

input excitation for the JK flip-flops is obtained from the state transition. The simplified input functions for the control register are listed below.

The three states $p_0 = 00$, $p_1 = 01$, and $p_2 = 10$ are obtained by decoding the outputs of flip-flops F_3 and F_4, noting that state 11 is not used. The requirements for activating flip-flops F_1 and F_2 and outputs x and y are specified in the list of elementary operations. We can now collect all the Boolean functions that describe the operation of the control circuit and list them together:

$$\begin{aligned}
JF_1 &= p_0 w & KF_1 &= p_0 \\
JF_2 &= p_2 z & KF_2 &= p_2 \\
JF_3 &= F_4' F_1 & KF_3 &= 1 \\
JF_4 &= F_3 & KF_4 &= F_2 \\
p_0 &= F_4' F_3' & x &= p_1 \\
p_1 &= F_3 & y &= p_2 \\
p_2 &= F_4
\end{aligned}$$

Thus the control circuit is just another sequential circuit with four flip-flops and a simple decoder.

It is now necessary to design the combinational circuit part of the data register. We need four flip-flops for each decimal digit since the input data are decimal numbers represented in BCD. Each group of four flip-flops will contain the same combinational circuit; it is sufficient to show the design of a group of four flip-flops that form a decade. All other decades will have identical circuits. Let us designate the four flip-flops of a typical decade by R_8, R_4, R_2, R_1, where the subscript number designates the weight of the corresponding bit of the code. Similarly, we designate the input terminals of one decade by I_8, I_4, I_2, I_1.

The elementary operations to be performed on the register were listed above and are repeated here using the defined symbols:

p_0F_1:	$I \Rightarrow R$	transfer in
p_1:	$(R_{BCD}) \Rightarrow (R_{2,\ 4,\ 2,\ 1})$	convert code
p_2F_2:	$0 \Rightarrow R$	clear register

The transfer into and clearing of the register are two elementary operations that can be implemented by inspection. The flip-flop input functions that achieve the code conversion can be obtained by means of an excitation table using the procedure outlined in Sec. 7-4. The excitation table for the code conversion is shown in Table 12-3. The present states are the decimal digit in BCD; the next states are the corresponding digits in the 2, 4, 2, 1 code. Note that flip-flop R_1 does not change state and therefore, no input is required for this flip-flop during the conversion.

The simplified flip-flop input function can be derived from the information available in Table 12-3. These functions must be ANDed with control

Table 12-3 Excitation Table for Code Converter

Present State $R_8\ R_4\ R_2\ R_1$	*Next State* $R_8\ R_4\ R_2\ R_1$	JR_8	KR_8	JR_4	KR_4	JR_2	KR_2	JR_1	KR_1
0 0 0 0	0 0 0 0	0	X	0	X	0	X	0	X
0 0 0 1	0 0 0 1	0	X	0	X	0	X	X	0
0 0 1 0	0 0 1 0	0	X	0	X	X	0	0	X
0 0 1 1	0 0 1 1	0	X	0	X	X	0	X	0
0 1 0 0	0 1 0 0	0	X	X	0	0	X	0	X
0 1 0 1	1 0 1 1	1	X	X	1	1	X	X	0
0 1 1 0	1 1 0 0	1	X	X	0	X	1	0	X
0 1 1 1	1 1 0 1	1	X	X	0	X	1	X	0
1 0 0 0	1 1 1 0	X	0	1	X	1	X	0	X
1 0 0 1	1 1 1 1	X	0	1	X	1	X	X	0

Flip-Flop Inputs: JR_8 through KR_1

state p_1. Including the transfer and clear elementary operations for the register, we obtain the following input functions:

$$JR_8 = (R_4R_2 + R_4R_1)p_1 + I_8p_0F_1 \qquad KR_8 = p_2F_2$$
$$JR_4 = R_8p_1 + I_4p_0F_1 \qquad KR_4 = R_2'R_1p_1 + p_2F_2$$
$$JR_2 = (R_8 + R_4R_1)p_1 + I_2p_0F_1 \qquad KR_2 = R_4p_1 + p_2F_2$$
$$JR_1 = I_1p_0F_1 \qquad KR_1 = p_2F_2$$

This concludes the design of the code converter circuit. Note that the most difficult part of the design is the derivation of the timing sequence and control functions. Once the elementary operations and their sequence are specified, the rest of the design is mechanized by the procedures presented in Chs. 7 and 9.

12-5 DESIGN ALGORITHMS

An algorithm is a procedure for obtaining a solution to a problem. A design algorithm is a procedure for implementing the problem with a given piece of equipment. The development of a design algorithm cannot start until the designer is firmly certain of two things. First, the problem at hand must be thoroughly understood. Second, an initial configuration of equipment must be assumed for implementing the procedure. Starting from the problem statement and equipment availability, a solution is then found and an algorithm formed. The algorithm is stated by a finite number of well-defined procedural steps.

The derivation of a design algorithm is a creative endeavor that requires reasoning and the ability to think logically. Many common digital system design algorithms can be found in books and in the professional literature, so it is usually wise to search the literature for existing algorithms before starting to develop new ones.

We shall now demonstrate the development of a design algorithm by going through an example. We shall start from the statement of the problem and proceed through the design until the Boolean functions are obtained.

Statement of the Problem

In Sec. 9-9 we stated two algorithms for the addition of binary numbers when negative numbers are represented in sign-complement. The problem now is to implement with hardware the addition and subtraction of two fixed-point binary numbers represented in sign-magnitude. The definition of fixed-point, sign-magnitude representation of binary numbers can be found in Sec. 8-5.

The addition of two numbers stored in finite length registers may result in a sum that exceeds the storage capacity of a register by one digit. The extra digit is said to cause an overflow. The circuit must provide a flip-flop for storing a possible overflow bit.

Equipment Configuration

The two numbers to be added or subtracted are stored in registers A and B. Each register contains n flip-flops for storing the magnitude of an n-bit number. The signs of A and B are to be stored in flip-flops A_s and B_s, respectively. The flip-flop that stores the overflow bit is labeled E. The registers and flip-flops are shown in Fig. 12-7.

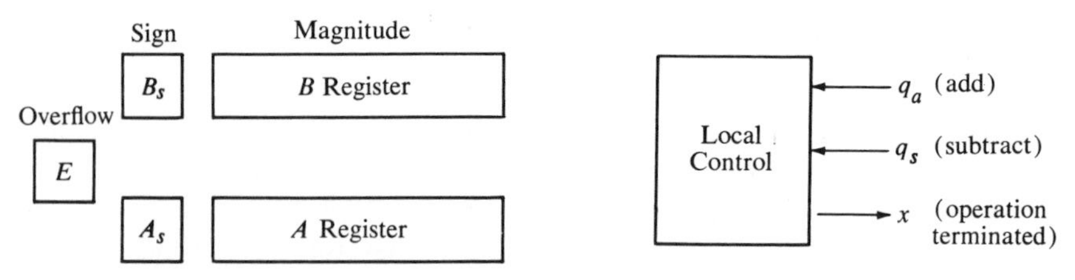

Figure 12-7 Block diagram of sign-magnitude addition and subtraction

We shall assume that the two numbers and their signs have been transferred to the respective registers and that the result of the operation is to be transferred to register A. Two input signals give the command to add (q_a) or to subtract (q_s), and an output signal x specifies the end of operation. The circuit local control communicates with the system main control through the input and output variables. The local control accepts one input signal from main control and proceeds to provide the required sequence of commands for the register operations. Upon completion of the operation, the local control informs the main control with output x that the sum or difference is in register A.

Derivation of Algorithm

The representation of numbers by sign-magnitude is familiar because it is used for paper and pencil arithmetic calculations. The procedure for adding or subtracting two signed binary numbers with paper and pencil is simple and straightforward. A review of this procedure will be helpful for deriving the design algorithm.

We designate the *magnitude* of the two numbers by A and B. When the numbers are added or subtracted algebraically, we find that there are eight different conditions to consider, depending on the sign of the numbers and the operation performed. The eight conditions may be expressed in a compact form as follows:

$$(\pm A) \pm (\pm B)$$

if the arithmetic operation specified is subtraction, we change the sign of B and add. This is evident from the relations:

$$(\pm A) - (+B) = (\pm A) + (-B)$$
$$(\pm A) - (-B) = (\pm A) + (+B)$$

This reduces the number of possible conditions to four, namely:

$$(\pm A) + (\pm B)$$

When the signs of A and B are the same, we add the two magnitudes and the sign of the result is the same as the common sign. When the signs of A and B are not the same, we subtract the smaller number from the larger and the sign of the result is equal to the sign of the larger number. This is evident from the following relations:

		if $A \geqslant B$		if $A \leqslant B$
$(+A) + (+B)$	$= +(A + B)$			
$(+A + (-B)$	$=$	$+(A - B)$	$=$	$-(B - A)$
$(-A) + (+B)$	$=$	$-(A - B)$	$=$	$+(B - A)$
$(-A) + (-B)$	$= -(A + B)$			

This concludes the algorithm for adding or subtracting two sign-magnitude numbers with paper and pencil. But how is it to be implemented with the equipment available in Fig. 12-7? For this we need a *design algorithm*.

To derive a design algorithm, it is necessary to have a broad knowledge of digital systems and their processing capabilities. Since various alternatives for performing a given task are usually available, it is wise to evaluate their relative complexity, cost, and speed of operation before making a choice. A conscientious designer does not use *any* design algorithm that comes his way. He investigates the various alternatives under the given constraints and then arrives at the best solution available.

For the example considered here, we find that we must choose among a few alternatives. For example, consider the subtraction operation. We can implement it with full-subtractor circuits (Sec. 4-4), or by the complement and add method discussed in Sec. 1-5. Moreover, the relative magnitude of the two numbers can be found with the comparator circuit of Sec. 4-6, by the method of Prob. 9-24, or by the comparison method of Sec. 9-9, by checking for a borrow after subtraction, or by some other method. The best and most convenient method must be chosen.

We shall not elaborate here on the merits and disadvantages of the various methods available. It would take us too far afield and involve a long and tedious discussion. We shall choose one method for implementing the arithmetic operations and proceed to show the other steps necessary to complete the design.

The best method for subtracting two binary numbers when adder circuits are already available is by the complement method introduced in Sec. 1-5. By this method, there is no need for comparing the relative magnitudes of the numbers since the end carry gives the required information. The algorithm is proven in Sec. 1-5 and is repeated here for convenience.

The subtraction of two positive binary numbers $A - B$ may be done as follows:

(1) Add the minuend A to the 2's complement of the subtrahend B.

(2) Inspect the result for an end carry.

 (a) if an end carry occurs, discard it;

 (b) if an end carry does not occur, take the 2's complement of the number obtained in step (1) and place a negative sign in front.

In other words, when performing $A - B$ by the above algorithm, a positive result is detected from the presence of an end carry. A negative result does not generate an end carry and gives the answer in 2's complement, so recomplementation is needed to change the number to the sign-magnitude representation.

We are now ready to state the design algorithm. We shall use some of the rules from the previous algorithm and perform subtraction by complementing the subtrahend.

(1) If the arithmetic operation specified is to subtract; change the sign of the subtrahend B.

(2) If the signs of A and B are equal, add the two numbers. Let the sign of the result be the same as the sign of A, and set the overflow flip-flop with the end carry.

(3) If the signs of A and B are not equal, take the 2's complement of the negative number and add it to the positive number. Check the end carry of this addition.

 (a) If the end carry is 1, make the sign of the result positive and discard carry.

 (b) If the end carry is 0, take the 2's complement of the number and make the sign of the result negative.

Note that we do not have a design algorithm unless we can implement every operation stated with the given hardware. We know that we can detect the operation, the sign of a number, and the end carry with outputs of flip-flops or logic gates. We also know that adding two numbers, taking the 2's complement, changing the sign, and setting the overflow flip-flop can be accomplished with register operations. So the above algorithm is a design algorithm.

Flow Chart Representation

A flow chart is a convenient way to specify the sequence of procedural steps and decision paths for an algorithm. A flow chart for a design algorithm would normally use the variable names of registers and inputs defined in the initial equipment configuration. A flow chart translates an algorithm from its word statement to an information flow diagram that enumerates the sequence of register operations together with the conditions necessary for their execution.

The flow chart of Fig. 12-8 shows how we can implement sign-magnitude addition and subtraction with the equipment of Fig. 12-7. An operation is initiated by either input q_s or input q_a. Input q_s initiates a subtraction operation, so the sign of B is complemented. Input q_a initiates an add operation and the sign of B is left unchanged. The next step is to compare the two signs. The decision block marked with A_s:B_s symbolizes this decision. If the signs are equal, we take the path marked by the symbol =. Otherwise we take the path marked by the symbol ≠. For equal signs, contents of A are added to the contents of B and the sum is transferred to A. The value of the end carry in this case is an overflow, so the E flip-flop is set if $C_{n+1} = 1$. The circuit then goes to its initial state and output x becomes a 1.

If the signs are not equal, we check for a negative sign. The sign flip-flop that is equal to 1 represents a negative sign, so the corresponding magnitude is changed to its 2's complement. Note the symbol used for 2's complement. It combines the symbol for 1's complement and the symbol for incrementing by 1. The next step is to add the two numbers and check the end carry C_{n+1}. It is convenient to transfer the value of the end carry temporarily into the E flip-flop. Then if $E = 1$, we make the result positive and clear E since this is not an overflow. If $E = 0$, we complement again and make the result negative. The circuit then goes to its initial state and output x becomes logic-1.

This concludes the derivation of the design algorithm. We shall now proceed to design the circuit and derive the Boolean functions for the control circuit and the registers.

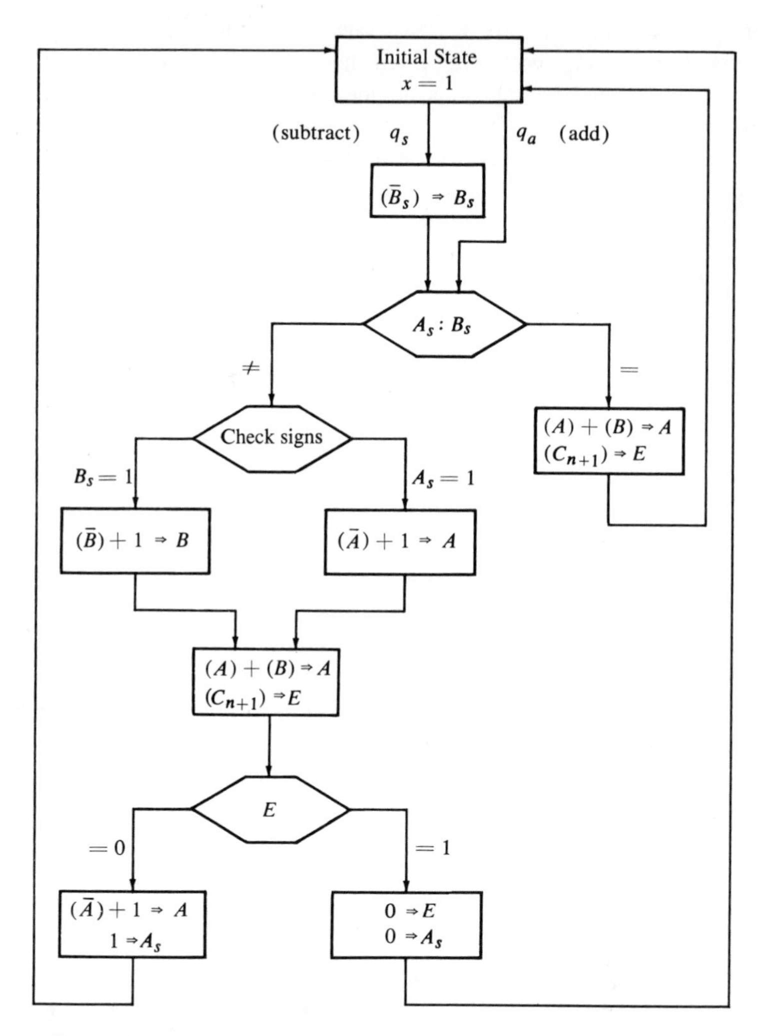

Figure 12-8 Flow chart for sign-magnitude addition and subtraction

Register-Transfer Relations

The flow chart of Fig. 12-8 gives all the information needed to design the circuit. We assume the availability of a control register with states p_i, $i = 1, 2, 3, \ldots,$ and develop the state diagram shown in Fig. 12-9. The elementary operations for the system and the control functions that cause their execution are as follows:

p_0:	$1 = x$	initial state
p_1:	$(\bar{B}_s) \Rightarrow B_s$	complement sign of B
$A_s p_2$:	$(\bar{A}) + 1 \Rightarrow A$	2's complement of A

$B_s p_2$:	$(\overline{B}) + 1 \Rightarrow B$	2's complement of B
p_3:	$(A) + (B) \Rightarrow A,\ (C_{n+1}) \Rightarrow E$	add and set overflow
Ep_4:	$0 \Rightarrow E,\ 0 \Rightarrow A_s$	end carry; sign is positive
$E'p_4$:	$(\overline{A}) + 1 \Rightarrow A,\ 1 \Rightarrow A_s$	no end carry; complement again and make the sign negative

In the state diagram, the variable s is used to designate the relation between the signs. This signal must be generated in the control circuit by forming the equivalence of A_s and B_s. Note that we have assumed that the execution of the 2's complement can be done with one elementary operation; i.e., with one clock pulse. In previous chapters this was done with two elementary operations: complement and increment. The combinational circuit that implements the 2's complement with one elementary operation is derived subsequently.

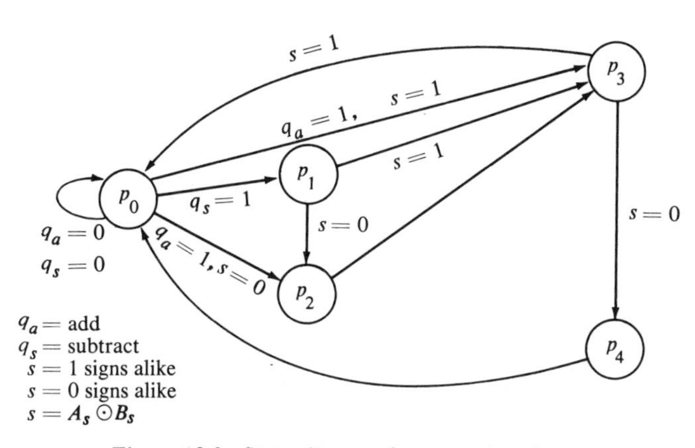

Figure 12-9 State diagram for control register

The derivation of the state diagram and symbolic relations from the flow chart requires a certain amount of logical reasoning and trial and error. An effort is made to minimize the number of states in the control register by using other flip-flop variables in the control functions. Thus two different elementary operations are listed for states p_2 and p_4; the one executed is determined from the state of flip-flop A_s, B_s, or E.

Design of Control Logic

The control register can be designed from the information in the state diagram of Fig. 12-9. Since the control register needs five states, we use three flip-flops, F_1, F_2, and F_3. We use T flip-flops for the control and other registers.

The state table for the control register is shown in Table 12-4. The inputs to the control register are q_a, q_s, and s. When an input does not

Table 12-4 Excitation Table for Control Register

	Present State			Inputs			Next State			Flip-Flop Inputs		
	F_3	F_2	F_1	q_a	q_s	s	F_3	F_2	F_1	TF_3	TF_2	TF_1
p_0	0	0	0	0	0	X	0	0	0	0	0	0
p_0	0	0	0	X	1	X	0	0	1	0	0	1
p_0	0	0	0	1	X	0	0	1	0	0	1	0
p_0	0	0	0	1	X	1	0	1	1	0	1	1
p_1	0	0	1	X	X	0	0	1	0	0	1	1
p_1	0	0	1	X	X	1	0	1	1	0	1	0
p_2	0	1	0	X	X	X	0	1	1	0	0	1
p_3	0	1	1	X	X	0	1	0	0	1	1	1
p_3	0	1	1	X	X	1	0	0	0	0	1	1
p_4	1	0	0	X	X	X	0	0	0	1	0	0

determine the next state, it is marked by an X. Note that a separate row is used for each transition. Thus, state p_0 can go to four different next states, depending on the presence or absence of certain inputs. The T flip-flop input excitations are equal to 1 for a change of state and 0 for no change of state.

The simplified input functions for the flip-flops in the control register can be derived using six-variable maps. In general, the formal combinational design procedure with maps may not be effective when the problem involves many states and many input variables. The design of a control register can be simplified when the decoder outputs are used to designate the present states. The procedure is to find the Boolean functions for the decoder and then derive the flip-flop input functions using these outputs. In this particular example, the decoder can be simplified because of the three unused states in the control register.

The list of Boolean functions below specifies the combinational circuit for the control. Variable s compares the two signs, and output x is a 1 when the circuit is in the initial state p_0. A straight binary assignment is chosen for the states. The decoder outputs p_0-p_4 identify the states of the control register. The flip-flop input functions are derived from the excitation table by inspection.

$$s = A_sB_s + A'_sB'_s \qquad p_0 = F'_3F'_2F'_1$$
$$x = p_0 \qquad p_1 = F'_2F_1$$
$$TF_3 = s'p_3 + p_4 \qquad p_2 = F_2F'_1$$
$$TF_2 = q_ap_0 + p_1 + p_3 \qquad p_3 = F_2F_1$$
$$TF_1 = (q_s + q_as)p_0 + s'p_1 + p_2 + p_3 \qquad p_4 = F_3$$

Flip-Flop Input Functions

The procedure for deriving the flip-flop input functions for registers A and B and flip-flops A_s, B_s, and E from the list of elementary operations is outlined in Ch. 9. All the elementary operations were encountered before except that for the 2's complement. The combinational circuit that changes a number stored in a register to its 2's complement during one clock pulse can be found by any one of the three methods listed in Sec. 1-5. One method states that we start from the least significant bit and check the bit values. All bits up to and including the first 1 are left unchanged; all higher significant bits after the first 1 are complemented. We now use this procedure to obtain the combinational circuit. (See also Prob. 12-20).

Let p be the control function for the 2's complement elementary operation to be executed in register A. We use variable y_i for each stage i to designate the occurrence of a 1 in some previous stage $j < i$. Flip-flop A_i is complemented if $y_i = 1$ and not complemented if $y_i = 0$. If we start with $y_i = 0$, the first flip-flop A_1 is never complemented as required (since the

Table 12-5 Sign-Magnitude Adder-Subtractor Flip-Flop Input Functions

Register Transfer Relations	*Flip-Flop Input Functions*
A_s flip-flop	
$Ep_4: 0 \Rightarrow A_s$ $E'p_4: 1 \Rightarrow A_s$	$TA_s = (E'A'_s + EA_s)p_4$
B_s flip-flop	
$p_1: (\bar{B}_s) \Rightarrow B_s$	$TB_s = p_1$
E flip-flop	
$p_3: (C_{n+1}) \Rightarrow E$ $Ep_4: 0 \Rightarrow E$	$TE = (EC'_{n+1} + E'C_{n+1})p_3 + Ep_4$
B register	
$B_s p_2: (\bar{B}) + 1 \Rightarrow B$	$TB_i = z_i$ for $i = 1, 2, \ldots, n$ $z_{i+1} = B_s p_2 (z_i + B_i)$ $z_1 = = 0$
A register	
$A_s p_2: (\bar{A}) + 1 \Rightarrow A$ $p_3: (A) + (B) \Rightarrow A$	$TA_i = (C_i B'_i + C'_i B_i)p_3 + y_i$ $C_{i+1} = A_i B_i + A_i C_i + B_i C_i$ $y_{i+1} = A_s p_2 (y_i + A_i)$ for $i = 1, 2, \ldots, n$ $C_1 = 0$ $y_1 = 0$

first bit is either 0 or the first 1). The value of y_i for all higher-order stages is determined from the relation

$$y_{i+1} = p(y_i + A_i) \qquad i = 1, 2, 3, \ldots, n$$

That is, $y_{i+1} = 1$ if control function $p = 1$ and either $A_i = 1$ (giving the first 1) or $y_i = 1$ (because the first 1 has already occurred).

The execution of the 2's complement elementary operation with T flip-flops is then:

$$\begin{aligned} TA_i &= y_i & i &= 1, 2, \ldots, n \\ y_{i+1} &= p(y_i + A_i) \\ y_1 &= 0 \end{aligned}$$

The derivation of the register input functions for the other elementary operations is straightforward. The final results are listed in Table 12-5.

12-6 BINARY MULTIPLICATION

Algorithms for arithmetic operations are very important for the design of an arithmetic unit of a digital computer. Many of these algorithms can be found in computer design books (references 1-5) or in professional journals.* Some algorithms for fixed-point binary addition and subtraction were introduced in Chs. 1 and 9 and in Sec. 12-5. In this section, we derive a binary multiplication algorithm and obtain the symbolic representation for its implementation. The purpose is to demonstrate the logical reasoning involved in the development of one other design algorithm. For a catalog of available arithmetic algorithms, including division, square root, floating-point operations and decimal arithmetic, the reader should consult the references.

Multiplication of two fixed-point binary numbers in sign-magnitude representation is done with paper and pencil by successive additions and shifting. This process is best illustrated with a numerical example. Let us multiply the two binary numbers 10111 and 10011.

```
 23          10111   multiplicand
 19    X     10011   multiplier
---          -----
             10111
            10111
           00000   +
          00000
         10111
         ---------
437      110110101   product
```

*One of the better journals is the *IEEE Transactions on Computers*.

The process consists of looking at successive bits of the multiplier, least significant bit first. If the multiplier bit is a 1, the multiplicand is copied down; otherwise, zeros are copied down. The numbers copied down in successive lines are shifted one position to the left from the previous number. Finally, the numbers are added; their sum forms the product.

The sign of the product is determined from the signs of the multiplicand and multiplier. If they are alike, the sign of the product is plus. If they are unlike, the sign of the product is minus.

When the above process is implemented in a digital machine, it is convenient to change the process slightly. First, instead of providing digital circuits to store and add simultaneously as many binary numbers as there are 1's in the multiplier, it is convenient to provide circuits for the summation of only two binary numbers and successively accumulate the partial products in a register. Second, instead of shifting the multiplicand to the left, the partial product is shifted to the right, which results in leaving the partial product and the multiplicand in the required relative positions. Third, when the corresponding bit of the multiplier is a 0, there is no need to add all zeros to the partial product since it will not alter its value. The previous numerical example is repeated here in order to clarify the proposed multiplication process.

multiplicand:	10111
multiplier:	10011
1st multiplier bit = 1, copy multiplicand.	10111
Shift right to obtain 1st partial product.	010111
2nd multiplier bit = 1, copy multiplicand.	10111
add multiplicand to previous partial product.	1000101
shift right to obtain 2nd partial product.	1000101
3rd multiplier bit = 0, shift right to obtain 3rd partial product.	01000101
4th multiplier bit = 0, shift right to obtain 4th partial product.	001000101
5th multiplier bit = 1, copy multiplicand.	10111
add multiplicand to previous partial product.	110110101
shift right to obtain 5th partial product = final product.	0110110101

Figure 12-10 shows a block diagram for implementing the binary multiplier. There are three registers, B, M, and A, that hold the multiplicand, multiplier, and partial product, respectively. In addition, there is a C counter that is initially set to hold a binary number equal to the number of bits in the multiplier. The counter is reduced by one after forming each new partial product. When the contents of the counter reach zero, the partial product becomes the required product and the process stops.

The addition is done between the A and B registers, with the partial product transferred to the A register. Both the A and M registers can be shifted to the right, with the least significant bit of A shifted into the most

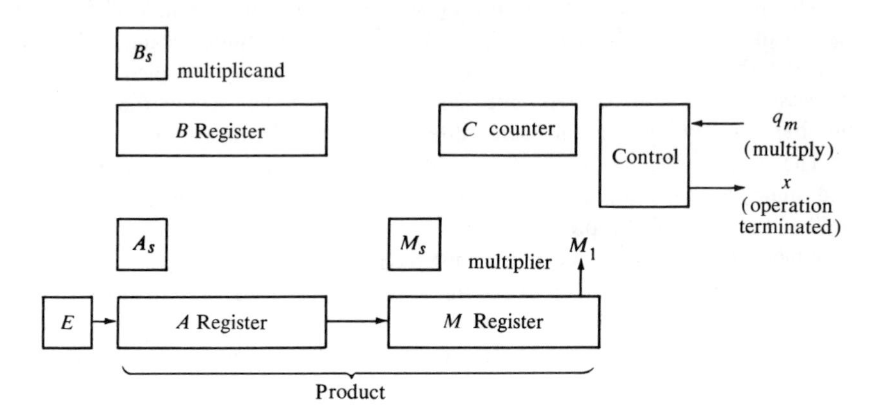

Figure 12-10 Block diagram for binary multiplier

significant bit position of M. After each shift, one bit of the partial product is shifted into M, pushing the multiplier bits one position to the right. In this manner, the least significant flip-flop in the M register, designated by M_1, will hold the corresponding bit of the multiplier which must be inspected next.

The signs of the multiplicand and multiplier are in flip-flops B_s and M_s, respectively. The sign of the product is in A_s. The overflow flip-flop E is needed for a temporary overflow of the partial product. Input q_m initiates the multiplication process; output x designates the completion of the operation.

A design algorithm for the multiplication of two binary numbers can now be stated as follows:

(1) Check the signs of the multiplicand and multiplier. If they are alike, make the sign of the product plus. If they are unlike, make the sign of the product minus.

(2) Start with an initial partial product of zero. Check successive bits of the multiplier, beginning with the least significant bit:

  (a) If the corresponding multiplier bit is a 1, add the multiplicand to the partial product.
  (b) If the corresponding multiplier bit is a 0, leave the partial product as is.

(3) Shift the new partial product once to the right to obtain the next partial product. Proceed to check the next multiplier bit and repeat 2(a) or 2(b).

(4) The nth partial product (n being the number of bits of the multiplier) is the required product.

Figure 12-11 is a flow chart for implementing the algorithm with the hardware shown in Fig. 12-10. Initially, the registers and counter are set to their corresponding values. The signs are compared, and A_s is set to correspond to the sign of the product. The low-order bit of M, M_1, is tested. If it is a 1, the multiplicand in the B register is added to the present partial product in the A register. The A and M registers, when taken as one long register, are symbolized by AM. This combined register is shifted once to the right to form the new partial product. Simultaneously, the C counter is reduced by 1. This process is repeated until the content of the C counter equals zero. Note that the partial product formed in A is shifted into M one bit at a time and eventually replaces the multiplier. The final product is then stored in both A and M, with A holding the most significant bits and M holding the least significant bits.

From the flow chart, we can derive the control state diagram and the elementary operations needed for implementing the multiplication. The state

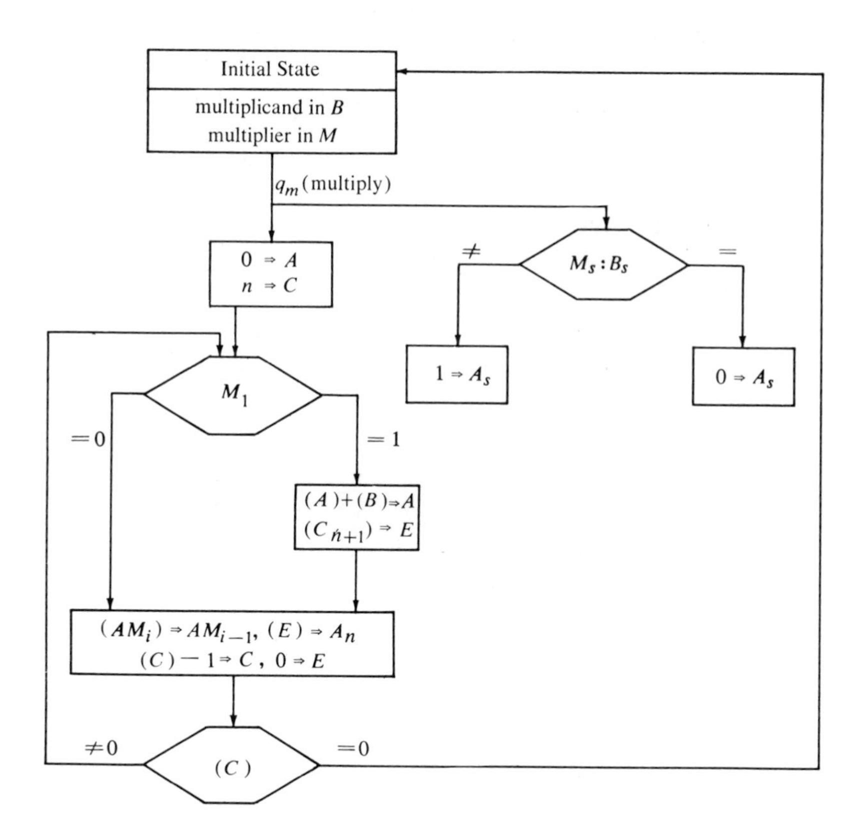

Figure 12-11 Flow chart for binary multiplier

diagram for the control sequence is shown in Fig. 12-12. The control functions and elementary operations are as follows:

p_0:	$1 = x$	initial state
p_1:	$0 \Rightarrow A,\ n \Rightarrow C$	clear A, set counter to n
$(M_s \odot B_s)p_1$:	$0 \Rightarrow A_s$	signs alike, product is positive
$(M_s \oplus B_s)p_1$:	$1 \Rightarrow A_s$	signs unlike, product is negative
p_2:	$(A) + (B) \Rightarrow A,\ (C_{n+1}) \Rightarrow E$	form partial product
p_3:	$(AM_i) \Rightarrow AM_{i-1},\ (E) \Rightarrow A_n,$ $(C) - 1 \Rightarrow C,\ 0 \Rightarrow E$	shift right both A and M registers, decrement counter

During initial state p_0, the local control generates an output $x = 1$ to inform the system control of its readiness to execute a multiplication operation. When command signal q_m is received, the control goes to state p_1. At this state, register A is cleared, the counter receives a number n equal to the number of bits in the multiplier, and the sign of the product is determined. From here on, the control stays in state p_2 or p_3 until the content of the C counter reaches zero. When the multiplier bit M_1 is a 1, the control goes to state p_2 and an ADD elementary operation is executed. Registers A and M are shifted once to the right during control state p_3, and the counter is decreased by 1. When the counter reaches zero, the control goes back to the initial state p_0 and output x becomes a 1 again.

The circuit for the control register can be obtained from the state diagram. The input functions for the registers can be obtained from the list of elementary operations. This is a straightforward process and will be left for an exercise.

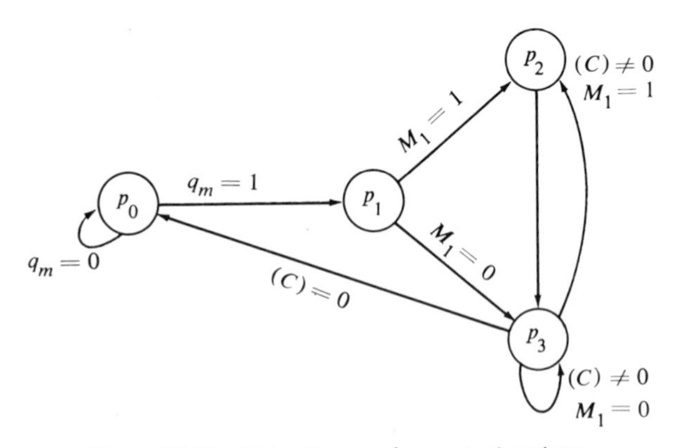

Figure 12-12 State diagram for control register

PROBLEMS

12-1. It is necessary to design an adder for decimal digits represented in the excess-3 code. Show that the correction after adding two digits with four full-adders is as follows:

(a) the output carry is equal to the uncorrected carry;
(b) if output carry = 1; add 0011;
(c) if output carry = 0; add 1101 and ignore the carry from this addition.

Show that the excess-3 adder can be constructed with seven full-adders and two inverters.

12-2. Design a decimal adder for digits represented in the 2, 4, 2, 1 code.

12-3. Design a six-bit binary to BCD converter. Use adders to add the weighted values of the binary digits. For example, binary 111111 is obtained by performing the BCD addition of the numbers 32 + 16 + 8 + 4 + 2 + 1.

12-4. Draw a block diagram of a circuit that adds four decimal digits in BCD. How many flip-flops are needed? How many BCD adders are needed?

12-5. The state diagram below represents the timing sequences of a control register.

(a) How many different state sequences are there and what values of variables generate them?
(b) Design the control register with three T flip-flops and a decoder.
(c) Design the control register with five D flip-flops and without a decoder.

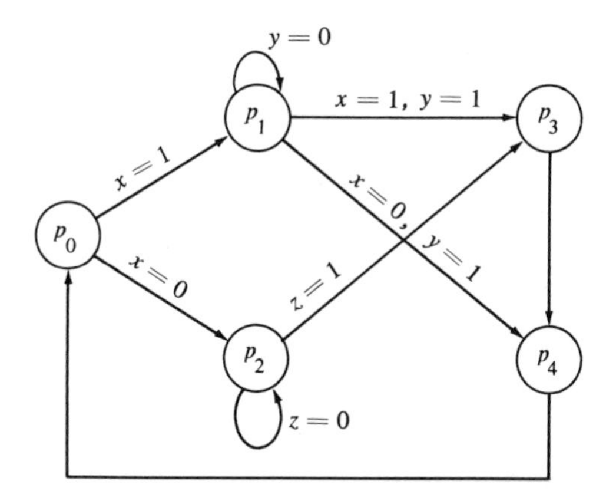

12-6. The following elementary operations specify a four-state control register G. The binary state assignment is $p_0 = 00$, $p_1 = 01$, $p_2 = 10$, $p_3 = 11$.

xp_0:	$(G) + 1 \Rightarrow G$
yp_0:	$10 \Rightarrow G$
zp_0:	$11 \Rightarrow G$
$p_1 + p_2 + p_3$:	$(G) + 1 \Rightarrow G$

(a) Draw the state diagram for the control register.
(b) Design the control register with JK flip-flops.

12-7. A system has two registers A and B and four control flip-flops F_1, F_2, F_3, F_4. Register A receives data words from input terminals I. Data words are transferred from register A to register B and from register B to an output device. Input variables x and y are logic-1 during one clock pulse period and communicate synchronously with the input and output devices, respectively. The system operation is described by means of register-transfer relations as follows:

xF_1:	$1 \Rightarrow F_2$	
F_1F_2:	$(I) \Rightarrow A$,	$1 \Rightarrow F_3$
F_2:	$0 \Rightarrow F_1$	
F_1':	$0 \Rightarrow F_2$	
F_3F_4':	$(A) \Rightarrow B$,	$0 \Rightarrow A$, $0 \Rightarrow F_3$, $1 \Rightarrow F_4$, $1 \Rightarrow F_1$
yF_4:	$0 \Rightarrow B$,	$0 \Rightarrow F_4$

(a) Draw a timing diagram for the system.
(b) Explain the function of flip-flops F_1, F_2, F_3, F_4 and inputs x, y.
(c) Obtain the input functions for the control flip-flops.

12-8. Design a digital system that performs the following sequence of operations:

(1) Accepts two binary numbers in sign-magnitude representation. Each magnitude is three bits long and is transferred into register A or B, respectively. The signs are transferred into flip-flops A_s and B_s, respectively.
(2) (a) If $(\pm A) > (\pm B)$, multiply contents of A by 2 and transfer the product out.
(b) If $(\pm A) < (\pm B)$, divide contents of B by 2 and transfer the quotient out.
(c) If $(\pm A) = (\pm B)$. transfer the number out unchanged.

12-9. Show that the code converter circuit of Sec. 12-4 can be designed with combinational circuits only; i.e., without any flip-flops.

12-10. Change the code converter of Sec. 12-4 to operate as follows: A two-position switch determines an input or output operation. When the switch is in the *input* position, it clears a flip-flop and the operation is identical to the one described in the text. When the switch is in the *output* position, it sets a flip-flop and all operations reverse; i.e., data words are transferred from the output device to the input device. The conversion is from the 2,4,2,1 code to BCD; and the function of the control line reverses. A line is provided for the external devices to inform them of the required operation.

12-11. A digital system consists of two processor registers A and B and a control register F. Each of the processor registers has seven JK flip-flops. The control register has three T flip-flops and a decoder. The system has seven input data lines designated by I_1–I_7 and an input control line s that starts the operation. The control state diagram is shown below. The elementary operations for the processor register are as follows:

p_0: initial state
p_1: $(I) \Rightarrow B$
p_2: $(A) + (B) \Rightarrow A$
p_3: $(\bar{A}) \Rightarrow A$
p_4: $(A) + 1 \Rightarrow A$
p_5: $(\bar{B}) \Rightarrow B$
p_6: $(B) + 1 \Rightarrow B$
p_7: $(A) + (B) \Rightarrow A$

Derive the Boolean functions for the system.

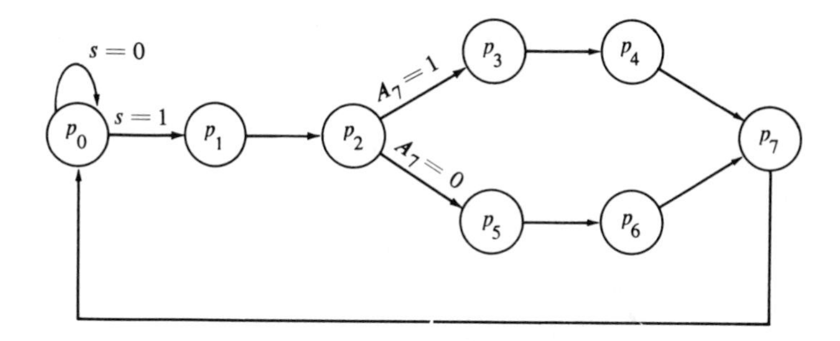

12-12. Derive an algorithm for detecting an overflow during the addition of two numbers in sign-2's complement representation. Use the carry *out* of the sign-bit position and the carry out of the most significant bit position of the binary number that goes *into* the sign-bit position. Show that an overflow occurs if these two carries are different (for another algorithm see Prob. 9-19).

12-13. Derive the flip-flop input functions of Table 12-5.

12-14. State an algorithm for adding and subtracting decimal numbers in sign-magnitude BCD representation.

12-15. Design a circuit that converts a decimal number stored in a register in BCD to its 10's complement.

12-16. Design the circuit specified in Sec. 12-5 using two elementary operations for the 2's complement; i.e., $(\bar{A}) \Rightarrow A$ and $(A) + 1 \Rightarrow A$.

12-17. The addition of a binary number stored in register A with the 2's complement of another number in register B can be done as follows: take the 1's complement of B; make the input carry C_1 into the lowest significant position equal to 1; and add the contents of A

and B (this will add 1 to the sum). Design the circuit specified in Sec. 12-5 using this method for obtaining 2's complements.

12-18. Design the circuit specified in Sec. 12-5 using adders (Sec. 4-3) and subtractors (Sec. 4-4). The required additions should be done with adders and the required subtraction with subtractors. Combine the adder and subtractor circuits to form one circuit that either adds or subtracts depending upon a given control signal.

12-19. Prove the algorithm stated in Sec. 9-9 for the addition of two binary numbers in sign-2's complement representation.

12-20. Design a four-bit register with T flip-flops to convert a binary number to its 2's complement when input $p = 1$ (this is identical to Prob. 7-9). Using the result obtained in this problem, derive the procedure adopted in Sec. 12-5 for the 2's complement elementary operation.

12-21. Derive the Boolean functions for the binary multiplier of Sec. 12-6.

12-22. Show the contents of registers A, B, and M and of counter C (Fig. 12-10) after each clock pulse during the process of multiplying the two binary numbers 10111 and 10011.

12-23. For the binary multiplication described in Sec. 12-6, explain the advantages and disadvantages of the following:

(a) forming the partial sum
(b) shifting right instead of left
(c) destroying the multiplier
(d) having a product twice as long as the multiplier or multiplicand
(e) using a counter that counts down instead of up

12-24. (a) Show that the multiplication of two n-digit numbers in base r gives a product of no more than $2n$ digits in length.
(b) Show that the above statement implies that no overflow occurs at the completion of the multiplication process described in Sec. 12-6.
(c) What similar statement can be made concerning the length of the partial products?

12-25. Derive a design algorithm and obtain the control state diagram and register transfer relations for the multiplication of two binary numbers with negative numbers represented in sign-2's complement.

12-26. Derive a design algorithm and obtain the control state diagram and register-transfer relations for each of the following arithmetic operations (consult the references listed for this chapter):

(a) Fixed-point binary division. Use any representation for negative numbers.
(b) Square-root of a fixed-point positive binary number.
(c) Addition and subtraction of two binary floating-point numbers.
(d) Multiplication of two fixed-point decimal numbers.

12-27. (a) Design a binary-to-BCD converter; i.e., a circuit that converts binary numbers to decimal numbers in BCD.

(b) Design a BCD-to-binary converter; i.e., a circuit that converts decimal numbers in BCD to binary.

REFERENCES

1. Richards, R. K. *Arithmetic Operations in Digital Computers.* Princeton, N.J.: D. Van Nostrand Co., 1955.
2. Flores, I. *The Logic of Computer Arithmetic.* Englewood Cliffs, N. J.: Prentice-Hall, Inc., 1963.
3. Gschwind, H. W. *Design of Digital Computers.* New York: Springer-Verlag, 1967.
4. Chu, Y. *Digital Computer Design Fundamentals.* New York: McGraw-Hill Book Co., 1962.
5. Braun, E. L. *Digital Computer Design.* New York: Academic Press, 1963.

APPENDIX

ANSWERS TO SELECTED PROBLEMS

Chapter 1

1-1. 0, 1, 2, 10, 11, 12, 20, 21, 22, 100, 101, 102, 110, 111, 112, 120, 121, 122, 200, 201.

1-2. (a) 1313, 102210
(b) 223.0, 11314.52
(c) 1304, 336313
(d) 331, 13706

1-3. $(100021.1111...)_3$; $(3322.2)_4$; $(505.333...)_7$; $(372.4)_8$; $(FA.8)_{16}$.

1-4. 1100.0001; 10011100010000; 1010100001.00111; 11111001110.

1-5. 2.53125; 46.3125; 117.75; 109.875.

1-6.

decimal	*binary*	*octal*	*hexadecimal*
225.225	11100001.001110011	341.16314	E1.399
215.75	11010111.110	327.6	D7.C
403.9843	110010011.111111	623.77	193.FC
10949.8125	10101011000101.1101	25305.64	2AC5.D

1-7. (a) 73.375
(b) 151
(c) 78.5
(d) 580
(e) 0.62037
(f) 35
(g) 8.333
(h) 260

1-8. 1's complement: 0101010; 1000111; 1111110; 01111; 11111.
2's complement: 0101011; 1001000; 1111111; 10000; 00000.

1-9. 9's complement: 86420; 90099; 09909; 89999; 99999.
10's complement: 86421; 90100; 09910; 90000; 00000.

1-10. $(175)_{11}$

1-14. (a) six possible tables
(b) four possible tables

1-15. (a) 1000 0110 0010 0000
(b) 1011 1001 0101 0011
(c) 1110 1100 0010 0000
(d) 1000011 0101100

1-17. 0000, 0001, 0010, 0011, 0100, 0101, 0110, 0111, 1011, 1100, 1101, 1110.

1-18. 00001, 01110, 01101, 01011, 01000, 10110, 10101, 10011, 10000, 11111.

1-20. 000, 001, 010, 101, 110, 111, representing 0, 1, 2, 3, 4, 5, respectively.

1-21. two bits for suit, four bits for number, J = 1011, Q = 1100, K = 1101.

1-23. (a) 0000 0000 0000 0001 0010 0111
(b) 0000 0000 0000 0010 1001 0101
(c) 1110 0111 1110 1000 1111 0101

1-24. (a) 597 in BCD
(b) 264 in excess-3
(c) not valid for 2421 code of Table 1-2
(d) FG in alphanumeric

1-25. 00100000001 + 10000011010 = 10100011011.

1-26. $L = (A + B) \cdot C$.

Chapter 2

2-1. closure, associative, commutative, distributive; identity for + is 2; identity for · is 0; no inverses.

2-2. All postulates are satisfied except for postulate 5; there is no complement.

2-5. (a) x
(b) x
(c) y
(d) $xz + yz$
(e) 0
(f) $y(x + w)$

2-6. (a) $A'B' + B(A + C)$
(b) $BC + AC'$
(c) $A + CD$
(d) $A + B'CD$

2-7. (a) 1
(b) $B'D' + A(D' + BC')$
(c) 1
(d) $ABC + A'D(BC' + B'C)$

2-8. (a) 1
(b) 1

2-11. (b) $F = (x' + y')' + (x + y)' + (y + z')'$ has only OR and NOT operators.
(c) $F = [(xy)' \cdot (x'y')' \cdot (y'z)']'$ has only AND and NOT operators.

2-12. (a) $T_1 = A'(B' + C')$
(b) $T_2 = A + BC = T_1'$

2-13. (a) $\Sigma\ (1, 3, 5, 7, 9, 11, 13, 15) = \Pi\ (0, 2, 4, 6, 8, 10, 12, 14)$
(b) $\Sigma\ (1, 3, 5, 9, 12, 13, 14) = \Pi\ (0, 2, 4, 6, 7, 8, 10, 11, 15)$
(c) $\Sigma\ (0, 1, 2, 8, 10, 12, 13, 14, 15) = \Pi\ (3, 4, 5, 6, 7, 9, 11)$
(d) $\Sigma\ (0, 1, 3, 7) = \Pi\ (2, 4, 5, 6)$
(e) $\Sigma\ (0, 1, 2, 3, 4, 5, 6, 7)$, no maxterms.
(f) $\Sigma\ (3, 5, 6, 7) = \Pi\ (0, 1, 2, 4)$

2-14. (a) $\Pi\ (0, 2, 4, 5, 6)$
(b) $\Pi\ (1, 3, 4, 5, 7, 8, 9, 10, 12, 15)$
(c) $\Sigma\ (1, 2, 4, 5)$
(d) $\Sigma\ (5, 7, 8, 9, 10, 11, 13, 14, 15)$

2-18. $x/y = xy'$; $y/x = x'y$. Since $xy' \neq x'y$ then $x/y \neq y/x$.

2-19. See Sec. 5-2.

2-20. $F = x \oplus y = x'y + xy'$; (dual of F) $= (x' + y)(x + y') = xy + x'y' = F'$

Chapter 3

3-1. (a) y
(b) $ABD + ABC + BCD$
(c) $BCD + A'BD'$
(d) $wx + w'x'y$

3-2. (a) $xy + x'z'$
(b) C′ + A′B
(c) a′ + bc
(d) $xy + xz + yz$

3-3. (a) $D + B'C$
(b) $BD + B'D' + A'B$ or $BD + B'D' + A'D'$
(c) $1n' + k'm'n$

(d) $B'D' + A'BD + ABC'$
(e) $xy' + x'z + wx'y$

3-4. (a) $A'B'D' + B'C'D' + AD'E$
(b) $DE + A'B'C + B'C'E'$
(c) $BDE' + B'CD' + B'D'E' + A'B'D' + CDE'$

3-5. (a) $F_1 = \Pi\ (0, 3, 5, 6); F_2 = \Pi\ (0, 1, 2, 4)$
(b) $F_1 = x'y'z + x'yz' + xy'z' + xyz; F_2 = xy + xz + yz$
(c) $F_1 = (x + y + z)(x + y' + z')(x' + y + z')(x' + y' + z);$
$F_2 = (x + y)(x + z)(y + z)$

3-6. (a) y
(b) $(C' + B)(A + B)(A + C + D)$
(c) $(w + z')(x' + z')$

3-7. (a) $z' + xy = (x + z')(y + z')$
(b) $C'D + A'B'CD' + ABCD' = (A + B' + D)(A' + B + D)(C + D)(C' + D')$
(c) $A'C' + AD' + B'D' = (A' + D')(C' + D')(A + B' + C')$
(d) $B'D' + A'CD' + A'BD = (A' + B')(B + D')(B' + C + D)$
(e) $w'z' + vw'x + v'wz = (v' + w')(w' + z)(w + x + z')(v + w + z')$

3-8. (a)

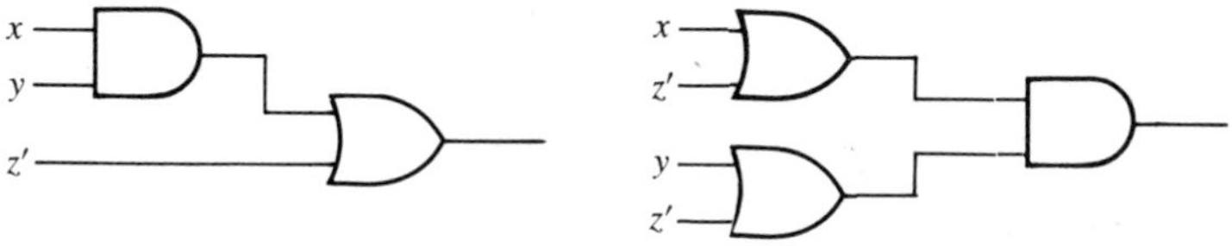

3-9. (a) $F = 1$
(b) $F = CD' + B'D' + ABC'D$

3-10. (a) $F = A'C + B'D'; A'(C + D')(B' + C)$
(b) $x'z' + w'z; (w' + z')(x' + z)$
(c) $AC + CE' + A'C'D; (A' + C)(C + D)(A + C' + D')$
or $AC + CD' + A'C'E; (A' + C)(C + E)(A + C' + E')$
(d) $A'B + B'E'; (A' + B')(B + E')$

3-12. $d = ABC'DE + AB'CDE' + ABCD'E$

3-13. $B'D'(A' + C) + BD(A' + C'); [B' + D(A' + C')]\ [B + D'(A' + C)];$
$[D' + B(A' + C')]\ [D + B'(A' + C)]$

3-14. $f \cdot g = x'yz' + w'y'z + wxy'z'$

3-15. (a) $F = A'CEF'G'$
(b) $F = ABCDEFG + A'CEF'G' + BC'D'EF$
(c) $F = A'B'C'DEF' + A'BC'D'E + CE'F + A'BD'EF$

Chapter 4

4-1. inputs: a, b, c, d
output: $F = abc + abd + bcd + acd + a'b'c' + a'c'd' + a'b'd' + b'c'd'; F = \Pi\ (3,5,6,9,10,12)$ (cannot be simplified further)

4-2. output: A = sign, BCD = magnitude; $A = wx + w'x'y' + wyz$; $B = xy$; $C = y'z' + xy'$; $D = z$

4-3. inputs: A_3, A_2, A_1
outputs: B_6 to B_1; $B_1 = A_1$; $B_2 = 0$; $B_3 = A'_1A_2$; $B_4 = A_1(A_2A'_3 + A'_2A_3)$; $B_5 = A_3(A_1 + A'_2)$; $B_6 = A_2A_3$

4-4. outputs: w, x, y, z; $w = a_0a_1b_0b_1$; $x = a_1a'_0b_1 + a_1b_1b'_0$
$y = a_1b_0b'_1 + a_0a'_1b_1 + a_0b'_0b_1 + a'_0a_1b_0$; $z = a_0b_0$

4-5. outputs: x, y, z; $x = a_1b_1 + a_1a_0b_0 + b_1b_0a_0$;
$y = a'_1a'_0b_1 + a'_1b_1b'_0 + a'_1a_0b'_1b_0 + a_1b'_1b'_0 + a_1a'_0b'_1 + a_1a_0b_1b_0$
$z = a_0b'_0 + a'_0b_0$

4-6. $S = D = (x + y + z)(x + y' + z')(x' + y + z')(x' + y' + z)$
$C = (x + y)(x + z)(y + z)$; $B = (x' + y)(x' + z)(y + z)$

4-7. inputs: A, B, C, D; $S = A \oplus B \oplus C \oplus D$; $C_1 = AB'C + AC'D + AC'D + ABD' + B'CD + BCD' + A'BD$; $C_2 = ABCD$

4-8. inputs: A, B, C, D
outputs: w, x, y, z; $w = A'B'C'$; $x = BC' + B'C$; $y = C$; $z = D'$

4-9. inputs: A, B, C, D
outputs: $F_4F_3F_2F_1$; $F_1 = D$; $F_2 = CD' + C'D$; $F_3 = (C + D)B' + BC'D'$; $F_4 = (B + C + D)A' + AB'C'D'$

4-10. inputs: $F_8F_4F_2F_1$
outputs: $\underbrace{S_8S_4S_2S_1}_{10^1}$ $\underbrace{L_8L_4L_2L_1}_{10^0}$;
$L_2 = L_8 = S_8 = 0$; $L_1 = L_4 = F_1$; $S_1 = F_2$; $S_2 = F_4$; $S_4 = F_8$

4-11. inputs: A, B, C, D
output : $F = AB + AC$

4-14. inputs: A, B, C, D
outputs: w, x, y, z; $w = AB + AC'D'$; $x = B'C + B'D + BC'D'$; $y = CD' + C'D$; $z = D$

4-15. inputs: A, B, C, D
outputs: w, x, y, z; $w = A$; $x = A'C + BCD + A'B + A'D$
$y = AC'D' + A'C'D + ACD + A'CD'$ or $y = AC'D' + B'C'D + ACD + B'CD'$ $z = D$

4-16. inputs: w, x, y, z
outputs: $\underbrace{E}_{10^1}$ $\underbrace{ABCD}_{10^0}$; $E = wx + wy$; $A = wx'y'$;
$B = w'x + xy$; $C = w'y + wxy'$; $D = z$

4-17. $(A = B) = (A_3 \odot B_3)(A_2 \odot B_2)(A_1 \odot B_1)(A_0 \odot B_0)$
$(A > B) = A_3B'_3 + (A_3 \odot B_3)A_2B'_2 + (A_3 \odot B_3)(A_2 \odot B_2)A_1B'_1 + (A_3 \odot B_3)(A_2 \odot B_2)(A_1 \odot B_1)A_0B'_0$

$(A < B)$ = same as above except that the A's are complemented instead of the B's.

4-18.

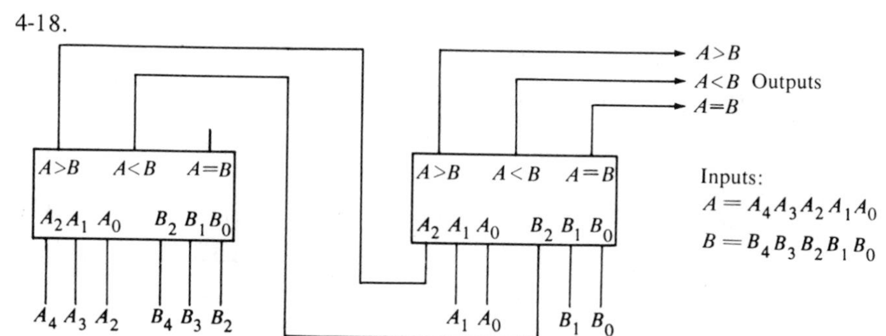

4-19 and 4-20. FA circuits.

4-21.

S_1	S_0	Y_1	Y_2	Y_3	Y_4
0	0	I_1	I_2	I_3	I_4
0	1	0	I_1	I_2	I_3
1	0	0	0	I_1	I_2
1	1	0	0	0	I_1

4-22. $Y_5 = (I_3 S_0' + I_2 S_0)\ S_1 + (I_4 S_0)\ S_1'$
$Y_6 = (I_4 S_0' + I_3 S_0)\ S_1$
$Y_7 = (I_4 S_0)\ S_1$

4-23. inputs: w, x, y, z
outputs: $D_0 = w'x'$; $D_1 = w'y'z'$; $D_2 = w'y'z$; $D_3 = xyz'$; $D_4 = xyz$; $D_5 = x'y'z'$; $D_6 = wy'z$; $D_7 = wyz'$; $D_8 = wyz$; $D_9 = wx$

4-24. inputs: A, B, C, D
outputs: $a = A + B'D' + BD + B'C$; $b = B' + CD + C'D'$;
$c = B + C' + D$; $d = B'D' + B'C + CD' + BC'D$;
$e = B'D' + CD'$; $f = A + BC' + C'D' + BD'$;
$g = A + BC' + CD' + B'C$
nine AND, seven OR, three inverters

(b)

10 11 12 13 14 15

(c) change a, f, g from A to AC', change $c = A'B + B'C' + A'D$

10 11 12 13 14 15

4-25. $A_1 = D_1 + D_3 + D_5 + D_7 + D_9$; $A_2 = D_2 + D_3 + D_6 + D_7$;
$A_4 = D_4 + D_5 + D_6 + D_7$; $A_8 = D_8 + D_9$

4-26. outputs: a, b, c, d, e, f; $a = 0$; $b = 1$; $c = H + I$; $d = D + E + F + G$; $e = B + C + F + G$; $f = A + C + E + G + I$

4-27. Use two three-bit decoders as in Fig. 4-12 and 64 AND gates with two inputs each.

4-30. $I_0 = I_1 = I_3 = I_5 = 0$; $I_2 = I_4 = 1$; $I_7 = I'$; $I_6 = I$

4-31.

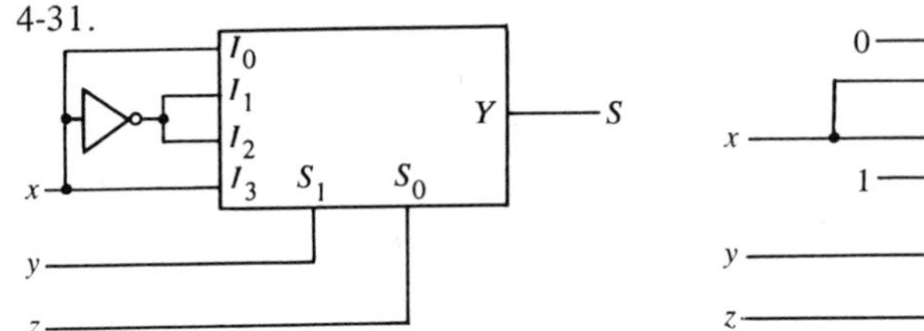

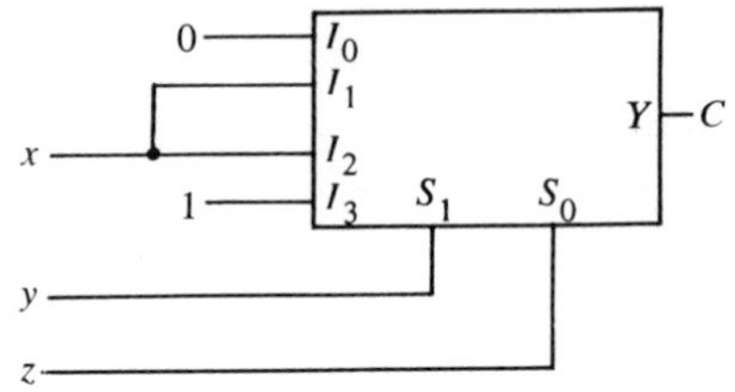

Chapter 5

5-1. Figures: 4-8, 4-10, 4-12, 4-13, 4-15, 4-16, 4-17.

5-3. (a) See Prob. 2-18.

5-4. $F = x'y'z + x'yz' + xy'z' + xyz$

5-5. Full-adder

5-6.

$x' \cdot 1 = x'$

$(x')'y = xy$

$x + y$

5-9. Six levels, five levels.

5-10. AND gate: fan-in = 6, fan-out = 10. Inverter fan-out = 10.

5-11. $F = A'CD + ABD' + BC'D + B'CD'$; $F = A(BC' + CD') + A'(B'C + BD)$

5-14.

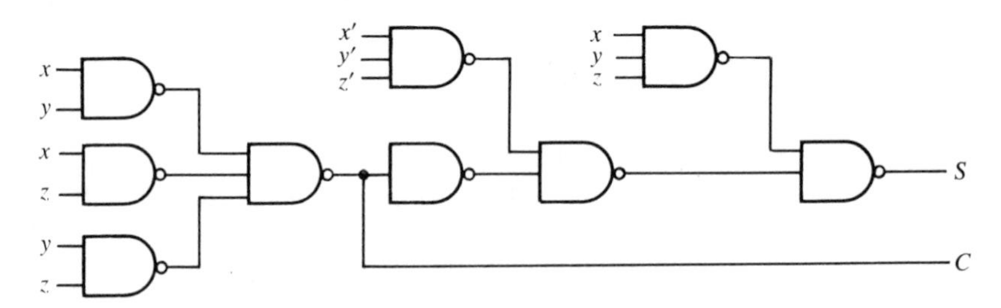

5-15. (a) $F_1 = A + D'E' + CD' = (A + D')(A + C + E')$

(b) $F_2 = A'B' + C'D' + B'C' = (B' + D')(B' + C')(A' + C')$

5-17. (a) $F = BD + D'(AB'C' + A'B'C)$

5-18. (a) $B'(A + C' + D')$
(b) $A'D + ABC'$
(c) $B'D + B'C + CD$

5-19. $F = x'y + xz$ (needs four NAND); $F = (x' + z)(x + y)$ (needs 4 NOR)

5-20. (a) $(A' + B' + C')(A + B' + C + D')(A + B + C' + D')$
(b) $(C + D)(C' + D')(A + B)(A' + B')$

5-21. $F = B'(C' + D')$

5-22. $F = ABC' + A'B + B' = A' + B' + C'$ (two NOR gates)

5-23. (a) Full-adder, F_1 is the sum, F_2 is the carry
(b) $F = A'B'C' + A'BC + AB'C + ABC'$

5-27. AND/AND → AND, AND/NAND → NAND, NOR/NAND → OR, NOR/AND → NOR, OR/OR → OR, OR/NOR → NOR, NAND/NOR → AND, NAND/OR → NAND

5-34. input variables; A, B, C, D, output variables; w, x, y, z.
$w = A$, $x = A \oplus B$, $y = x \oplus C$, $z = y \oplus D$.

5-35. $C = x \oplus y \oplus z \oplus P$, three exclusive-OR gates.

5-36.

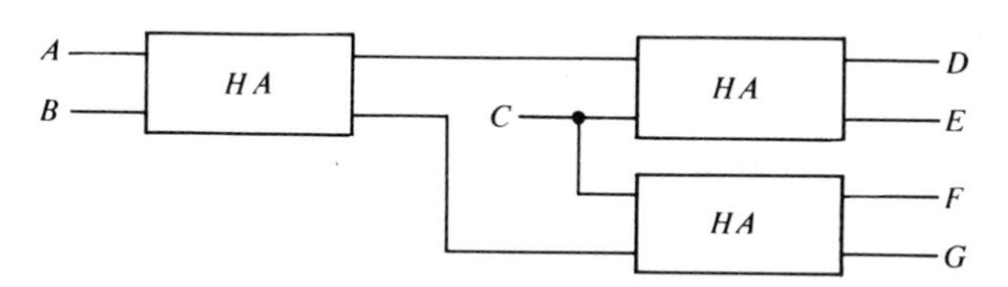

5-37. $F = (A \oplus B)(C \oplus D)$.

Chapter 6

6-5.

Q	J	K'	$Q(t + 1) = JQ' + K'Q$
0	0	0	0
0	0	1	0
0	1	0	1
0	1	1	1
1	0	0	0
1	0	1	1
1	1	0	0
1	1	1	1

6-6.

Q	SD	R	$Q(t + 1) = S + R'Q$
0	0	0	0
0	0	1	0
0	1	0	1

0	1	1	1
1	0	0	1
1	0	1	0
1	1	0	1
1	1	1	1

6-8. (a) $SA = x'A'$; $RA = xA$; $y = xA$
(c) $A(t + 1) = x'$

6-9.

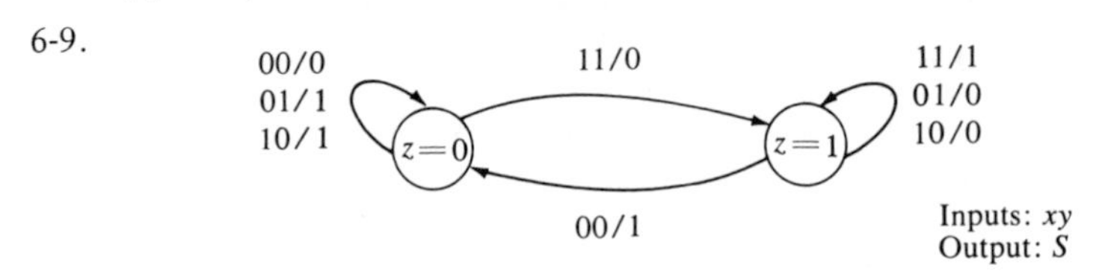

6-10. a counter with a repeated sequence: 00, 01, 10.

6-11. $x = 1$ binary sequence is: 1, 8, 4, 2, 9, 12, 6, 11, 5, 10, 13, 14, 15, 7, 3.
$x = 0$ binary sequence is: 0, 8, 12, 14, 7, 11, 13, 6, 3, 9, 4, 10, 5, 2, 1.

6-12.

P.S.	Next state				Output z			
	$xy = 00$	$xy = 01$	$xy = 10$	$xy = 11$	$xy = 00$	$xy = 01$	$xy = 10$	$xy = 11$
A B	A B	A B	A B	A B				
0 0	1 0	0 0	1 1	0 1	0	0	0	0
0 1	0 1	0 1	1 0	1 1	1	0	0	0
1 0	1 0	1 0	0 0	1 0	0	0	0	1
1 1	1 0	1 0	1 0	1 0	1	0	0	1

$A(t + 1) = xB + y'B'A' + yA + x'A$; $B(t + 1) = xA'B' + x'A'B + yA'B$

6-15. sequence: 000, 001, 010, 011, 100 and repeat; 110 → 110, 111 → 111, 101 → 001.

6-16. (c) A flip-flop is complemented when all previous flip-flops are 0.

6-17. There are ten states with each state consisting of two 1's in the five bits.

6-18. 000000, 100000, 010000, 101000, 110100, 011010, 101101

6-19. 1000, 0100, 0010, 0001

6-20. 0000, 1000, 1100, 1110, 1111, 0111, 0011, 0001, 0000, and repeat.

Chapter 7

7-1.

Present State	Next State 0	Next State 1	Output 0	Output 1
a	*f*	*b*	0	0
b	*d*	*a*	0	0
d	*g*	*a*	1	0
f	*f*	*b*	1	1
g	*g*	*d*	0	1

7-2.

State:	*a*	*f*	*b*	*c*	*e*	*d*	*g*	*h*	*g*	*g*	*h*	*a*
Input:	0	1	1	1	0	0	1	0	0	1	1	
Output:	0	1	0	0	0	1	1	1	0	1	0	

7-3.

State:	*a*	*f*	*b*	*a*	*b*	*d*	*g*	*d*	*g*	*g*	*d*	*a*
Input:	0	1	1	1	0	0	1	0	0	1	1	
Output:	0	1	0	0	0	1	1	1	0	1	0	

7-5.

J	K'	$Q(t+1)$
0	0	0
0	1	$Q(t)$
1	0	$Q'(t)$
1	1	1

$Q(t)$	$Q(t+1)$	J	K'
0	0	0	X
0	1	1	X
1	0	X	0
1	1	X	1

7-6.

SD	R	$Q(t+1)$
0	0	$Q(t)$
0	1	0
1	0	1
1	1	1

$Q(t)$	$Q(t+1)$	SD	R	
0	0	0	X	
0	1	1	X	
1	0	0	1	
1	1	X	0	either
		1	X	

7-8. (a) $TA = A + B'x$; $TB = A + BC'x + BCx' + B'C'x'$;
$TC = Ax + Cx + A'B'C'x'$

(b) $SA = A'B'x$, $RA = A$; $SB = A + C'x'$; $RB = BC'x + Cx'$;
$SC = A'B'x' + Ax$; $RC = A'x$

(c) $JA = B'x$, $KA = 1$; $JB = A + C'x'$, $KB = C'x + Cx'$;
$JC = A'B'x' + Ax$, $KC = x$; $y = A'x$

7-9. $(A = 2^3, B = 2^2, C = 2^1, D = 2^0)$; $TA = (D + C + B)x$;
$TB = (D + C)x$; $TC = Dx$; $TD = 0$

7-10. (a) $A(t+1) = AB'C'x' + A'BC'x + A'BCx + AB'C'x + AB'Cx$
$B(t+1) = A'BC'x' + A'B'Cx$
$C(t+1) = A'B'Cx' + A'BC'x' + A'BCx' + AB'C'x' + AB'Cx'$
$d(A, B, C, x) = \Sigma\,(0, 1, 12, 13, 14, 15)$ don't-care terms.

7-12. $TF_i = (F_i \oplus F_{i-1})L + (F_i \oplus F_{i+1})R$; $L = 1$ for left shift, $R = 1$ for right shift. F_i is a flip-flop in stage i; F_{i-1} and F_{i+1} are the flip-flops on its left and right, respectively.

7-13.

0000 (self-loop)

0001 → 1000 → 0100 → 0010 → 1001 → 1100 → 0110 → 1011 → 0101 → 1010 → 1101 → 1110 → 1111 → 0111 → 0011 → 0001

7-14. $JA = x, \ KA = x'; \ JB = Ax', \ KB = 1; \ JC = Bx + Ax, \ KC = Bx'$

7-17. $SA = B, \ RA = B'; \ SB = A'B' + B'x', \ RB = AB + Bx$

7-18. $DA = A'B'C + ACD + AC'D' \qquad DC = B$
$DB = A'C + CD' + A'B \qquad DD = D'$

7-19. $JA = yC + xy \qquad JB = xAC \qquad JC = x'B + yAB'$
$KA = x' + y'B' \qquad KB = A'C + x'C + yC' \qquad KC = A'B' + xB + y'B'$

7-20. $JQ_8 = Q_1Q_2Q_4 \qquad JQ_4 = Q_1Q_2 \qquad JQ_2 = Q_8'Q_1 \qquad JQ_1 = 1$
$KQ_8 = Q_1 \qquad KQ_4 = Q_1Q_2 \qquad KQ_2 = Q_1 \qquad KQ_1 = 1$

7-21 $\begin{bmatrix} 2 & 4 & 2 & 1 \\ A & B & C & D \end{bmatrix}$; $TA = BCD + A'B; \ TB = CD + A'B; \ TC = D + A'B;$
$TD = 1$

7-22. For straight binary 0000 to 1011 and variables $A \ B \ C \ D$
$TA = ACD + BCD; \ TB = A'CD; \ TC = D; \ TD = 1$
$t_3 = A'B'CD; \ t_6 = BCD'; \ t_9 = AC'D; \ t_{12} = A'B'C'D'$

7-23. (a) $JA = B, \ KA = 1; \ JB = A', \ KB = 1; \ t_0 = A'B', \ t_1 = B, \ t_2 = A$
(b) $JA = BC, \ JB = C, \ JC = A'; \ t_0 = A'B'C', \ t_1 = B'C, \ t_2 = BC'$
$KA = 1, \ KB = C, \ KC = 1; \ t_3 = BC, \ t_4 = A$
(c) $JA = BC, \ JB = C, \ JC = B' + A'; \ t_0 = A'B'C', \ t_1 = A'B'C,$
$KA = B, \ KB = A + C, \ KC = 1; \quad t_2 = A'BC', \ t_3 = BC,$
$t_4 = AB'C', \ t_5 = AC, \ t_6 = AB$

7-24. $SA = BC' \qquad SB = B'C \qquad SC = B'$
$RA = BC \qquad RB = AB \qquad RC = B$

7-25. $TA = A \oplus B; \ TB = B \oplus C; \ TC = A \odot C$

7-26. $JA = B' \qquad JB = A + C \qquad JC = A'B$
$KA = 1 \qquad KB = 1 \qquad KC = 1$

Chapter 8

8-2. (a) 7680
(b) 32
(c) no external lines

8-3. A shift-register with input x and output y.

8-4. (a) 011, 001, 001, 100, 011, 011, 001, 011
(b) 011, 001, 100, 011, 001, 100, 011

8-5. $(A_{1\text{-}8}) \Rightarrow B_{9\text{-}16}, \ (B_{1\text{-}8}) \Rightarrow A_{9\text{-}16}$

8-6. Address of operand 1 $\Rightarrow D$
$(<D>) \Rightarrow B$
$(B) \Rightarrow R_2$
Address of operand 2 $\Rightarrow D$
$(<D>) \Rightarrow B$
$(B) \Rightarrow R_1$
$(R_1) + (R_2) \Rightarrow R_3$
address of sum $\Rightarrow D$
$(R_3) \Rightarrow B$
$(B) \Rightarrow <D>$

8-7. P: $(x) \Rightarrow A$
S_R: $(I_R) \Rightarrow A_3$, $(A_i) \Rightarrow A_{i-1}$, $i = 1, 2, 3$
S_L: $(I_L) \Rightarrow A_0$, $(A_i) \Rightarrow A_{i+1}$, $i = 0, 1, 2$

8-8. (a) 16
(b) 8, 16
(c) 16
(d) $16 + 255k$, where k is the number of 1's in the word to be stored

8-9. (a) 5
(b) 18 if *BCD* is used

8-10. (c) (1) binary 75 $\Rightarrow D$
read: $(<D>) \Rightarrow B$
binary 90 $\Rightarrow D$
write: $(B) \Rightarrow <D>$

(2) binary 75 $\Rightarrow D$
read: $(<D>) \Rightarrow B$, $0 \Rightarrow <D>$
$(B) \Rightarrow <D>$
binary 90 $\Rightarrow D$
write: $0 \Rightarrow <D>$
$(B) \Rightarrow <D>$

8-16. (a) The coefficient with the smaller exponent is shifted right a number of times equal to the difference of the two exponents.
(b) The exponent of the sum is equal to the larger exponent.
(c) Shift the coefficient and adjust exponent until the most significant digit of the coefficient is non-zero.

8-17. (a) $(1 - 2^{-26}) \cdot 2^{255}$ and 2^{-256}
(b) $(2^{35} - 1) \cdot 2^{2047}$ and 2^{-2013}

Chapter 9

9-1. (a) $S_4S_3S_2S_1 = 1100$; $C_5C_4C_3C_2C_1 = 01110$
(b) Add command and one *CP*
(c) $(A) = 1100$, $S_{4-1} = 0011$, $C_{5-1} = 11000$

9-2. Same as Fig. 9-3 with J and K inputs connected together.

9-4. $SA_i = A_i'B_i'C_i + A_i'B_iC_i'$, $RA_i = A_iB_i'C_i + A_iB_iC_i'$, carry is the same as in a *FA*.

9-5. (a) Substitute *FS* for *FA* in Fig. 9-1; for (b) and (c) use Figs. 9-3. and 9-6 with output *D* being the same as *S* and borrow *K* implemented as:

(b) $K_{i+1} = A_i'B_i + A_i'K_i + B_iK_i$

(c) $K_{i+1} = A_iB_i + A_i'K_i$

9-6.

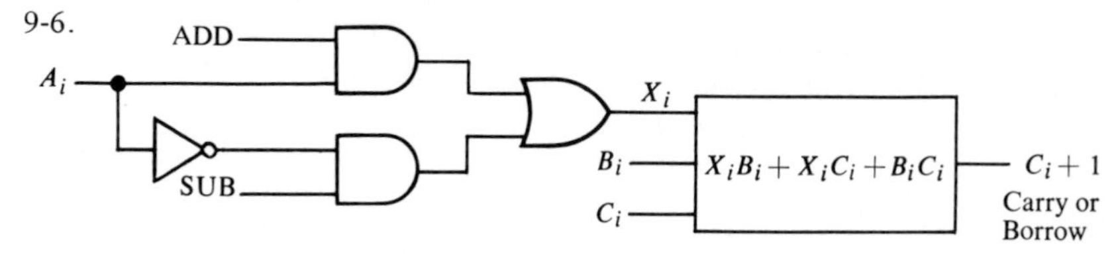

A_i
B_i
C_i
$A_i \oplus B_i \oplus C_i$
Sum or Difference

9-7. (a) $A = 0101, \quad C = 0110$

(b) $A = 0010, \quad C = 0111$

(c) $A = 1100, \quad C = 1111$

9-8. $C_i = A_6B_6 + A_5A_6B_5 + A_5B_5B_6 + A_4A_5A_6B_4 + A_4A_6B_4B_5 + A_4A_5B_4B_6 + A_4B_4B_5B_6 + C_4(A_4 + B_4)(A_5 + B_5)(A_6 + B_6)$. Twelve gates; two levels delay from C_4 to C_7.

9-9. 1600 nsec + 50 nsec for AND gate in the input of the last flip-flop.

9-11. $S_i = (A_i' + B_i' + C_i)(A_i' + B_i + C_i')(A_i + B_i' + C_i')(A_i + B_i + C_i)$

$TC_i = (A_i + C_i)(A_i' + B_i)(B_i' + C_i')$

9-12. Maximum frequency = 100,000 pulses per second; 240 μsec

9-13.

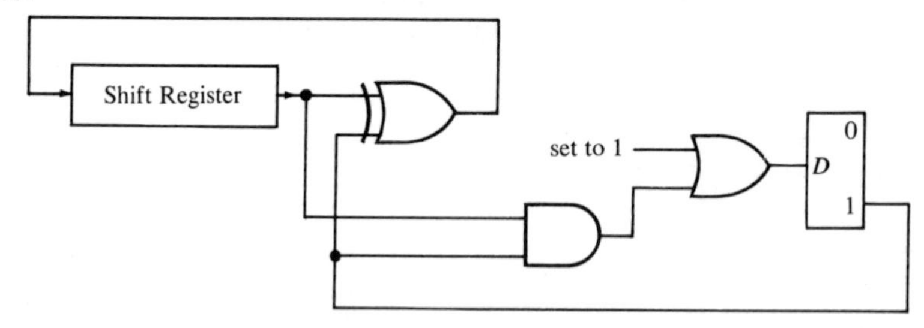

9-14. 240 nsec delay.

9-16. (a) $TA_i = (B_iC_i' + B_i'C_i)\, p_1 + A_ip_2 + p_3 + A_iB_i'p_4 + A_i'B_ip_5 + B_ip_6 + (A_{i+1} \oplus A_i)p_7 + (A_{i-1} \oplus A_i)p_8 + E_i$

(b) $SA_i = (A_i'B_iC_i + A_i'B_iC_i')p_1 + A_i'p_3 + B_ip_5 + A_i'B_ip_6 + A_{i+1}p_7 + A_{i-1}p_8 + A_i'E_i$

$RA_i = (A_iB_i'C_i + A_iB_iC_i')p_1 + p_2 + A_ip_3 + B_i'p_4 + A_iB_ip_6$
$+ A_{i+1}'p_7 + A_{i-1}'p_8 + A_iE_i$

for both (a) and (b):

$C_{i+1} = A_iB_i + A_iC_i + B_iC_i$; $E_{i+1} = E_iA_i$; $E_1 = p_9$; $C_1 = 0$

9-17. $JA_i = E_i + (B_i \oplus K_i)q_2$; $KA_i = E_i + (B_i \oplus K_i)q_2 + B_iq_3$
borrow: $K_{i+1} = A_i'B_i + A_i'K_i + B_iK_i$; $K_1 = 0$; $E_{i+1} = A_i'E_i$;
$E_1 = q_1$

9-19. (b) If the signs of augend and addend are the same and the sign of the sum is different, an overflow occurs.

9-20. (a) inputs: $ABxyz$; outputs: S, C_1, C_2;
$C_1 = ABy'z' + AB'x'z + AB'xy' + A'Byz' + A'Bxy' + A'B'xy$
$+ Bx'y'z + Ax'yz' + A'x'yz + B'xyz'$;
$C_2 = ABxy + ABxz + Bxyz + Axyz + AByz$.

9-23.

$X \Rightarrow B$	$(A) + (B) \Rightarrow A$, $(C_{n+1}) \Rightarrow L$
$0 \Rightarrow A$	L: $(A) + 1 \Rightarrow A$
$(A) + (B) \Rightarrow A$	$Z \Rightarrow B$
	$(\overline{B}) \Rightarrow B$
$Y \Rightarrow B$	$(A) + (B) \Rightarrow A$, $(C_{n+1}) \Rightarrow L$
$(\overline{B}) \Rightarrow B$	L: $(A) + 1 \Rightarrow A$

9-24. Compare sign bits: if unequal, the one with a 0 (positive) is the greater number. If both sign bits are 0, refer to Sec. 4-6 for algorithm. If both sign bits are 1, modify algorithm of Sec. 4-6 accordingly.

9-26. Use De Morgan's theorem: $AB' = (A' + B)'$
$(\overline{A}) \Rightarrow A$
$(A) \vee (B) \Rightarrow A$
$(\overline{A}) \Rightarrow A$

Chapter 10

10-1. (a) 6
(b) 12
(c) 6, 2, 1
(d) needs one fetch cycle for each two instructions.

10-2. (a) $(R1) + (R2) \Rightarrow R2$
(b) listed in Sec. 10-2
(c) add $(\langle M_1 \rangle)$ to $(R1)$ and store sum in $\langle M_2 \rangle$:
$(I[M_1]) \Rightarrow D$
$(\langle D \rangle) \Rightarrow B$
$(B) \Rightarrow \langle D \rangle$, $(R1) + (B) \Rightarrow R1$
$(I[M_2]) \Rightarrow D$, $(R1) \Rightarrow B$
$0 \Rightarrow \langle D \rangle$
$(B) \Rightarrow \langle D \rangle$
(d) $(I[M]) + (R1) \Rightarrow R1$

10-3 (1) $(P) + (A) \Rightarrow A$
$(I[M]) \Rightarrow D$, $(A) \Rightarrow B$
$(B) \Rightarrow \langle D \rangle$

(2) if $(A) > 0$: $(I[M]) \Rightarrow C$
if $(A) < 0$: $(C) + 1 \Rightarrow C$
if $(A) = 0$: nothing

(3) $(I[M]) \Rightarrow D$
$(\langle D \rangle) \Rightarrow B$
$(B) + 1 \Rightarrow B$
$(B) \Rightarrow \langle D \rangle$

(4) $(I[M]) \Rightarrow D$
$(\langle D \rangle) \Rightarrow B$
$(B) \Rightarrow A$, $(A) \Rightarrow B$
$(B) \Rightarrow \langle D \rangle$

(5) $(I[M]) \Rightarrow D$
$(\langle D \rangle) \Rightarrow B$
if $(B) = 0$: $(C) + 1 \Rightarrow C$
if $(B) \neq 0$: $(B) + 1 \Rightarrow B$
$(B) \Rightarrow \langle D \rangle$

10-5.

750	Input	0
751	Load	750
752	Increment	
753	Store	750
754	Complement	
755	Increment	
756	Add	760
757	Branch on zero	759
758	Branch unconditional	750
759	"Stop"	
760	Input	750

10-6. (a) 16, 25
(b) 9
(c) 512
(d) ± 16, 777, 215

10-10. Processor: Y Y ENQ
Terminal: Y Y ACK
Processor: SOH Y Y STX X X X. . . . . .X ETX EOT
Terminal: Y Y ACK

10-12.

Clear		0101 0000000000
Load	A	1100 0000010000
Add	B	0001 0000010001
Store	SUM	1101 0000010010
Load	DIF	1100 0000010011
Add	C	0001 0000010100
Add	SUM	0001 0000010010
Store	SUM	1101 0000010010

Assuming that operands A, B, SUM, DIF, and C are in memory registers in location 16, 17, 18, 19, and 20, respectively.

10-19. Instruction format:

Op.	d	X	M

Assume nondestructive read memory.

Condition	*Elementary Operation*	
$(I[X]) = 0$:	$(I[M]) \Rightarrow D$	no index
$(I[X]) = n$:	$(I[M]) + (Xn) \Rightarrow D$	add index number
$(I[d]) = 0$:	$(\langle D\rangle) \Rightarrow B$	direct; read operand
$(I[d]) = 1$:	$(\langle D\rangle) \Rightarrow B$	indirect; read address
	$(B[M]) \Rightarrow D$	transfer address
	$(\langle D\rangle) \Rightarrow B$	read operand

Continue to execute instruction with operand in B register.

Chapter 11

11-3. $(A) = 0$, $(\langle 103\rangle) = 5101$ (BUN 101)

11-5. (a) nothing

(b) two or more operations will be executed simultaneously with a probable error in content of accumulator.

11-7.
$$JA_i = C_i'p_1 + (B_i + C_i)p_2 \qquad KA_i = C_ip_1$$
$$JB_i = A_i'p_1 \qquad KB_i = A_ip_1 + (A_i' + C_i')p_3$$
$$JC_i = B_i'p_1 + (A_i \oplus B_i)p_4 \qquad KC_i = B_ip_1 + (A_i \oplus B_i)p_4$$

11-10. (a)
Ft_0: $(B_{1-12}) \Rightarrow D, m$
Fq_1t_2: $(A) + (B) \Rightarrow A$, carry $\Rightarrow E$
Fq_1t_3E: $(A) + 1 \Rightarrow A$, $0 \Rightarrow F$

(b) To perform the addition during t_2, it is necessary that the carry propagate in less than 200 nsec. Therefore, carry look-ahead circuits are needed.

(c) 1's complement representation generates either positive zero (all 0's) or negative zero (all 1's). Additional circuits are needed to detect a negative zero.

11-13. (a) $(A) - (\langle M\rangle) \Rightarrow A$

Fetch: $(RF = 00)$

$R'F't_0$: $(C) \Rightarrow D, m$
$R'F't_1$: $(C) + 1 \Rightarrow C$
$R'F't_2$: $(B_{13-16}) \Rightarrow I$
$R'F'q_8t_3$: $1 \Rightarrow F$

read operand: $(F = 1)$

Ft_0: $(B_{1-12}) \Rightarrow D, m$
Ft_3: $0 \Rightarrow F$
q_8Ft_3: $1 \Rightarrow R$

11-13. *execute:* $(R = 1)$

$q_8 R t_0$: $(\bar{B}) \Rightarrow B$
$q_8 R t_1$: $(B) + 1 \Rightarrow B$
$q_8 R t_3$: $(A) + (B) \Rightarrow A,\ 0 \Rightarrow R$

(b) $q_9 t_3 z_A$: $(B_{1-12}) \Rightarrow C$

(c) $(A) + (\text{<}M\text{>}) \Rightarrow \text{<}M\text{>}$

Fetch: same as in (a)

read operand and add: $(F = 1)$

$F t_0$: $(B_{1-12}) \Rightarrow D,\ m$
$q_A F t_3$: $(A) + (B) \Rightarrow A,\ 1 \Rightarrow R$
$F t_3$: $0 \Rightarrow F$

store in $\text{<}M\text{>}$: $(R = 1)$

$q_A R t_0$: m
$q_A R t_2$: $(A) \Rightarrow B$
$q_A R t_3$: $0 \Rightarrow R$

11-15. *Fetch*: same as Prob. 11-13(a)

read operand and add: $(F = 1)$

$F t_0$:	$(B_{1-12}) \Rightarrow D,\ m$	
$q_B F t_2$:	$(A) \Rightarrow B,\ (B) \Rightarrow A$	swap and store (A)
$q_B F t_3$:	$(A) + (B) \Rightarrow A,\ 1 \Rightarrow R$	sum in A register
$F t_3$:	$0 \Rightarrow F$	

swap (A) *and* $(\text{<}M\text{>})$: $(R = 1)$

$q_B R t_0$:	m	memory cycle
$q_B R t_2$:	$(A) \Rightarrow B,\ (B) \Rightarrow A$	swap to store sum and restore (A)
$q_B R t_3$:	$0 \Rightarrow R$	go to fetch cycle

Chapter 12

12-4. Each decimal digit position needs: four flip-flops in the accumulator, four flip-flops for the addend digit, and one BCD adder.

12-5. (a) Assume that either p_1 or p_2 may be an initial state. There are six possible state sequences that start from an initial state and return to an initial state.

(b) $p_0 = F_3' F_2' F_1',\ p_1 = F_2' F_1,\ p_2 = F_2 F_1',\ p_3 = F_2 F_1,\ p_4 = F_3$

$TF_1 = p_0 x + p_1 x' y + p_2 z + p_3$

$TF_2 = p_0 x' + p_1 x y + p_3$

$TF_3 = p_1 x' y + p_3 + p_4$

(c) $DF_0 = F_4,\ DF_1 = y' F_1 + x F_0,\ DF_2 = z' F_2 + x' F_0,$
$DF_3 = x y F_1 + z F_2,\ DF_4 = F_3 + x' y F_1.$

12-6. (a)

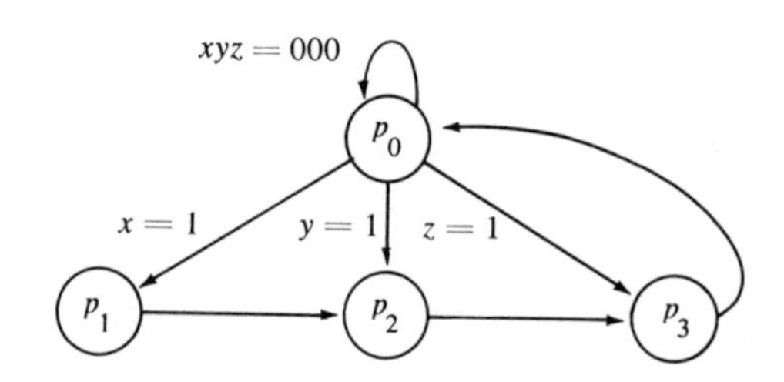

12-7. (b) $F_1 = 1$ informs input device that the system is ready to accept a word.

$x = 1$ input device informs the system that a word is available in terminals I.

F_2 is used to count two clock pulses before transfer of word.

$F_3 = 1$ when register A has a word; $F_3 = 0$ when register A is cleared.

$F_4 = 1$ when register B has a word; $F_4 = 0$ when register B is cleared.

$y = 1$ output device informs system that a word has been accepted.

12-9. Perform the conversion with a combinational circuit (Sec. 4-5) and connect the control signals directly; i.e., w to y and z to x.

12-11. (a)

$$TF_1 = p_0 s + p_0' = F_1 + F_2 + F_3 + sF_1'F_2'F_3'$$
$$TF_2 = p_1 + p_2 A_7' + p_3 + p_4 + p_5 + p_7 = F_1 + F_2'F_3 + F_2 F_3' A_7'$$
$$TF_3 = p_2 A_7' + p_3 + p_7 = F_1 F_2 + F_2 F_3' A_7'$$

12-12.

carry out of sign-bit position	carry in sign-bit position	overflow
0	0	no
0	1	yes
1	0	yes
1	1	no

12-24. Maximum number is $(r^n - 1)$; it is necessary to show that the maximum product is less than $(r^{2n} - 1)$.

$$(r^n - 1)(r^n - 1) = (r^{2n} - 2r^n + 1) < (r^{2n} - 1)$$

example: $r = 10,\ n = 2,\ 10^2 - 1 = 99$

$$99 \times 99 = 9801 = 10^4 - 2 \times 10^2 + 1 < 9999$$

INDEX

C

P

Q

R

S